Practical Ethics for Food Professionals

Press

The *IFT Press* series reflects the mission of the Institute of Food Technologists — to advance the science of food contributing to healthier people everywhere. Developed in partnership with Wiley-Blackwell, *IFT Press* books serve as leading-edge handbooks for industrial application and reference and as essential texts for academic programs. Crafted through rigorous peer review and meticulous research, *IFT Press* publications represent the latest, most significant resources available to food scientists and related agriculture professionals worldwide. Founded in 1939, the Institute of Food Technologists is a nonprofit scientific society with 18,000 individual members working in food science, food technology, and related professions in industry, academia, and government. IFT serves as a conduit for multidisciplinary science thought leadership, championing the use of sound science across the food value chain through knowledge sharing, education, and advocacy.

IFT Press Advisory Group

WILEY-BLACKWELL

A John Wiley & Sons, Ltd., Publication

Practical Ethics for Food Professionals

Ethics in Research, Education and the Workplace

Edited by

J. Peter Clark and Christopher Ritson

 | Press

A John Wiley & Sons, Ltd., Publication

1006926713

Library of Congress Cataloging-in-Publication Data has been applied for.

ISBN 978-0-4706-7343-0 (hardback)

A catalogue record for this book is available from the British Library.

Wiley also publishes its books in a variety of electronic formats. Some content that appears in print may not be available in electronic books.

Cover image: Apple © iStockphoto/JuSun; Scales © thinkstock
Cover design by Meaden Creative

Set in 10.5/12.5pt Times by Aptara® Inc., New Delhi, India
Printed and bound in Singapore by Markono Print Media Pte Ltd

1   2013

Titles in the IFT Press series

- *Sensory and Consumer Research in Food Product Design and Development* (Howard R. Moskowitz, Jacqueline H. Beckley, and Anna V.A. Resurreccion)
- *Sustainability in the Food Industry* (Cheryl J. Baldwin)
- *Thermal Processing of Foods: Control and Automation* (K.P. Sandeep)
- *Trait - Modified Oils in Foods* (Frank T. Orthoefer and Gary R. List)
- *Water Activity in Foods: Fundamentals and Applications* (Gustavo V. Barbosa - Cánovas, Anthony J. Fontana Jr., Shelly J. Schmidt, and Theodore P. Labuza)
- *Whey Processing, Functionality and Health Benefits* (Charles I. Onwulata and Peter J. Huth)

A John Wiley & Sons, Ltd., Publication

Contents

PART III EXAMPLES AND CASE STUDIES

PART IV CONCLUSION

Contributors

Timothy F. Bednarz
Majorium Business Press
2025 Main Street
Stevens Point, WI 54481-3019
USA

Herbert J. Buckenhüskes
European Federation of Food Science
and Technology (EFFoST)/DLG e.V.
(Deutsche Landwirtschafts-
Gesellschaft/German
Agricultural Society)
Frankfurt
Germany

Vinh Sum Chau
Senior Lecturer in Strategy
Kent Business School, University of Kent
Parkwood Road
Canterbury CT2 7PE
UK

Louis B. Clark
Computer Consultant,
Swampscott, MA 01907
USA

J. Peter Clark
Consultant to the Process Industries
644 Linden Avenue
Oak Park, IL 60302
USA

Mark F. Clark
World Bank (Retired)
Silver Spring, MD
USA

Charlie Clutterbuck
Research Fellow, Centre for Food Policy,
City University,
10 Mayville Road,
Brierfield
Nelson BB9 5RP
UK

Sue Davies
Chief Policy Adviser
Which? (UK)
2 Marylebone Road
London NW1 4DF
UK

Temple Grandin
Dept. of Animal Science
Colorado State University
Ft. Collins, CO 80523
USA

Jeanette Longfield
Co-ordinator
Sustain: the alliance for better
food and farming
94 White Lion Street
London N1 9PF
UK

Daryl Lund
Editor in Chief Institute of Food
Technologists' Peer-Reviewed
Journals
Professor Emeritus University of
Wisconsin-Madison
Madison, WI 53715-1149
USA

Ben Mepham
Special Professor in Applied Bioethics
Centre for Applied Bioethics
School of Biosciences
University of Nottingham
Sutton Bonington Campus
Loughborough LE12 5RD
UK

Thomas A. Nairn
Catholic Health Association of the US
4455 Woodson Road
St. Louis, MO 63134
USA

Richard Norman
Emeritus Professor of Moral Philosophy
University of Kent,
34 St Michael's Place
Canterbury CT2 7HQ
UK

Christopher Ritson
Emeritus Professor of Agricultural
    Marketing
Newcastle University
Agriculture Building
Newcastle upon Tyne NE1 7RU
UK

Edmund G. Seebauer
Department of Chemical & Biomolecular
    Engineering
University of Illinois
Urbana, IL 61801
USA

Chris Sutton
Director, Little Purple Dot CIC
6 Ivy Mews, Ivy Place
Hove BN3 1BG
UK

Preface

This volume was inspired by a symposium, 'Everyday ethics for the food scientist: Ethics in research, education and the workplace', co-sponsored by the student programming and the education divisions of the Institute of Food Technologists at the 2009 Annual Meeting. Additional topics and authors with an international perspective have been introduced to produce a comprehensive treatment that, it is hoped, will serve a wide audience.

Ethical considerations are involved in every aspect of a food professional's education and career, but the subject is rarely taught explicitly. Unfortunate examples of ethical lapses abound in news reports. Large issues of strategy have ethical components, but guiding principles are sometimes difficult to articulate.

Some business and engineering colleges have adopted an emphasis on values and ethics, but these need to be communicated to food professionals who have the opportunity to apply them in a practical sense.

The editors hope that this book will be the foundation for a seminar in colleges of agriculture, where departments of food science and technology are usually found. We think it should also be on the desk of every executive or would-be executive responsible for important decisions about marketing, resources, sustainability, the environment and people in the food industry.

There are numerous broad issues relating to food that have been discussed and debated at great length. Some of these are treated in this volume as well, but the emphasis here is on practical issues that individuals face and can affect.

The book has sections on: principles, issues, examples, and a concluding chapter.

- **Principles**
  Four chapters address principles of ethics from several points of view: the use of virtues, lessons from medical ethics, ethical principles derived from moral philosophy, and an East Asian view.
- **Issues in food industry ethics**
  Issues and applications include: ethics in publishing, humane treatment of livestock, sustainable food production and consumption, health claims, and worker exploitation.
- **Examples and case studies**
  Examples and case studies include: ethical practices in the workplace, ethical thinking and education, the fair trade movement, the Peanut Corporation of America *Salmonella* case, nanotechnology and commodity speculation.
- **Conclusion**
  Finally, a concluding chapter summarizes and synthesizes the individual chapters.

WAYS TO USE THIS BOOK

This book could serve as a primary text or supplementary resource for a one or two semester course or seminar on ethics for undergraduate students in food science or related fields. An instructor will need to generate his or her own assignments and discussion questions, but they should flow naturally from the material provided.

The editors hope as well that it will serve as an educational and inspirational resource for people at all phases of a career in the food industry. There are many other sources of information on the topics, some of which are listed as references. Students, readers and instructors must stay current in this field, as issues evolve and new challenges arise. The principles are timeless, but their application requires constant self-education and vigilance.

EDITORS' ACKNOWLEDGMENT

The editors thank David McDade, Samantha Thompson and Mark Barrett of the staff of John Wiley & Sons for their patience and skill in shepherding this project. We also thank all the author contributors for their generosity in participating.

J. Peter Clark
Christopher Ritson

I Principles

1 Fundamentals of ethics: the use of virtues

Edmund G. Seebauer

Virtue is its own reward. Cicero (1064–1063 CE),

De Finibus

Habits change into character. Ovid (43 BCE–18 CE),

Heriodes

1.1 THE IMPORTANCE OF ETHICS IN FOOD SCIENCE

There are both instrumental and intrinsic reasons why those involved in the study and production of food should cultivate ethical skills and practices. Instrumentally, good ethical behavior usually leads to good consequences for ourselves, our organizations, and the larger world. For example, Temple Grandin describes in Chapter 7 why humane treatment of livestock is good for business. It is true that unethical behavior can pay big dividends at times, especially in the short term. However, injustice and careless ethics lead mainly to suffering in the end. The case of peanut recall due to salmonella described in Chapter 14 provides a good example. Those involved in the study and production of food make decisions crucial to society at large, and therefore shoulder an enormous burden of public trust. From an intrinsic perspective, satisfaction comes from reasoning through an ethical problem, choosing a good course of action, and following through. Of course, people can do good based upon gut instinct alone. As thinking beings, however, many people find more satisfaction in understanding why they do what they do. Over 2000 years ago, Aristotle identified good ethical thought and action as the ultimate source of human happiness.

Many professional societies and corporations use formal codes of ethics. These codes have real value as reminders of the ethical standards expected in the work place, and as ways to instill those standards into new members. As public documents, codes serve as a basis for taking formal or legal disciplinary action against violators. However, codes are limited in what they can accomplish. For example, no list of guidelines can possibly cover all the complex situations that can arise. Moreover, code-based ethics sometimes leads to minimalism, which is the idea: "If it's not specifically forbidden, it

Practical Ethics for Food Professionals: Ethics in Research, Education and the Workplace, First Edition.
Edited by J. Peter Clark and Christopher Ritson.
© 2013 John Wiley & Sons, Ltd. Published 2013 by John Wiley & Sons, Ltd.

must be allowed." In addition, some situations call for on-the-spot decisions, with no time to consult a guidebook. These shortcomings point to a need for ethics that spring habitually from inside the individual, and do not depend upon some external list of rules. Strong ethical character makes it easier to rapidly and consistently handle complicated situations not listed in a code.

1.2 ANTHROPOLOGICAL FOUNDATIONS OF ETHICS

Ethical principles and methods of reasoning necessarily rest upon important presuppositions about the nature of human thinking and how it influences behavior. In other words, there are anthropological foundations to ethics. An anthropology is basically a model for the person. Even among groups of individuals who are well intentioned, differing anthropologies will lead to different principles and methods, which in turn will lead to different ethical conclusions. For example, anthropological considerations underlie much of the discussion in Chapter 4 on East Asian perspectives in food ethics, and again in Chapter 7 on the humane treatment of livestock.

The presentation of virtue ethics in this chapter uses an anthropology that conceptualizes the psyche as a unity of mind, emotions, and will. Other anthropologies exist as well, some having origins that are very ancient. Here are a few examples of anthropologies that currently find use.

- *Anthropologies based upon psychology:* A comprehensive anthropology should account for the psychology of human development. In fact, many models exist to account for the development of ethical behavior. A good summary of developmental theories can be found in Helminiak, *Spiritual Development: An Interdisciplinary Study* (1987). Choosing among them affects the attribution of moral responsibility. For example, at what point do teenagers become fully responsible for their eating habits, and how does this impact the ethical marketing of food? A complete anthropology should also account for psychological disorders like psychosis, depression, compulsion and autism. Once again, various models exist. But there is evidence (often controversial) that certain components of a person's diet can influence the severity of such disorders.
- *Anthropologies based upon natural observation:* Some anthropologies incorporate only observations that can be made in the natural world. In this view (sometimes called positivism), people represent no more than the aggregation of their constituent molecules, and disappear completely at death. Ethical behavior is then framed in terms of human pleasure, survival of the species, and the like. Such anthropologies often have a scientific appeal, but also suffer from problems with justifying why people should do good in the face of undeserved suffering and uncertain rewards.
- *Anthropologies based upon the supernatural:* Some anthropologies presuppose a realm that exists beyond the observable world. The most well-known of these anthropologies stem from long-standing religious traditions. Others include shamanism and witchcraft. Several support ethical systems that prohibit certain foods. Although such anthropologies can fill the gaps in purely natural anthropologies, these systems cannot be verified by systematic measurement. Thus it becomes difficult to choose among them and the ethical systems they imply.

Many arguments about ethics are fundamentally rooted in differing anthropologies. However, since the anthropology underlying a particular ethical position is sometimes only tacit, it is easy to wrongly assume that the opponents are ignorant, obstinate, or malicious.

Ethical reasoning must presuppose not only an anthropology but also a method. Space does not permit a detailed treatment of the methods commonly used today, but some examples include deontology, casuistry, utilitarianism, rights-based approaches, and intuitionism. For a convenient and detailed summary, see Frankena (1972). Among all these methods (other than virtue ethics), only intuitionism pays significant attention to aspects of ethics that are internal to the person.

How should we choose among all these different anthropologies and methods? Although they often conflict with each other, each brings a perspective that contains an important kernel of truth. One approach focuses on similarities in the approaches, and argues that each perspective represents just one portion of a single, deeper ultimate reality. This view is tantamount to monism – the belief that reality is a single fundamental entity. Whatever truth this idea might hold, it tends to gloss over major differences in practical moral rules. Such glossing represents a serious problem that often leads to a superficial approach to ethical living. Another approach is to assert that all ethical systems have equal validity. This view is tantamount to relativism. Relativism has had many defenders over the centuries, beginning with the Sophists of ancient Greece. More recently, this view has sprung from a belief that truth represents no more than a culturally conditioned phenomenon with no objective validity. One major danger of relativism has been known since the time of the Sophists. Thrasymachus held that the appearance of justice serves only as a veil to protect the interests of the strong (Stumpf, 1982). In other words, a world where all forms of ethics are considered equivalent devolves a world dominated by raw power.

Yet ethical diversity in the present world remains an established fact. It probably makes sense to just accept this fact, and to choose a good anthropology and methods. Having made this choice, we should try to adhere to it consistently.

What about those whose approaches differ from one's own? The analogy between furniture making and ethical living may prove helpful. Both undertakings represent a craft. Several good ways may exist to build a cabinet, but some ways are better than others and some ways miscarry completely. Likewise, there are several good approaches to crafting an ethical life. Experts in this craft admire and learn from each other's actions, in the way that skilled furniture makers can admire and learn from each other's handiwork. There should be no issue of trying to "convert" someone else from his or her fundamental perspective. Of course, even casual observation of the world shows that experts in the ethical life are few. It makes sense for those who are not yet expert (but want to be) to find a master artisan and attend closely to what he or she does.

1.3 THE VIRTUE ETHICS MODEL FOR THE PERSON

We cannot expect any single anthropology to depict every aspect of how people behave. Instead, we must employ a simplified model. For a more complete treatment of this and other subjects in this chapter, see Seebauer and Barry (2001). As with any scientific

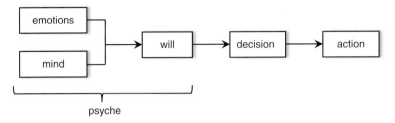

Figure 1.1 A simple model for ethical action.

model, it is useful to decide the degree of accuracy to which the subject should be represented. More accuracy usually entails more complexity in the model. In general, good models correspond to reality under most circumstances with only modest effort. Well-known examples from the natural sciences include the ideal gas law, classical Newtonian mechanics and the "lock-and-key" model for enzyme action.

The anthropology that underlies virtue ethics is quite simple and was originally developed by the ancient Greeks. The person comprises the senses and the psyche. The five senses provide raw data about the outside world. The psyche compiles these data into a coherent perception and understanding that we typically call "consciousness." As shown in Figure 1.1, the psyche has three components:

- *Mind*: The mind corresponds in some ways to a computer; with logic and memory functions. The mind classifies abstract concepts and uses them according to logical rules. In ethical decision-making, the mind integrates data from the senses with past memories to predict what could happen in the future.
- *Emotions*: The emotions are conscious, nonrational psychic responses to data from the senses and to various sorts of internally-driven neurochemistry (sickness, hormonal swings, psychoactive drugs, and the like). Many emotions induce physical responses, such as sweating or blushing.
- *Will*: The will decides among alternatives presented to it by the mind as influenced by the emotions. Using the will typically involves rational thought, and so the will might seem to be part of the mind. But emotions cause the decision-making process in humans to differ greatly from that in computers, so most ethics writers consider the will to be distinct from the mind.

How does this unity of mind, emotions and will function in an ethical decision? Many observers over the centuries have recognized that the ability to make an ethical decision often does not involve a lengthy, drawn-out mental process for each choice. Rather, many simple ethical choices occur with little thought because they have become habitual. That is, the will, mind and emotions regularly coordinate for nearly effortless ethical action.

Such habits regulate more complicated aspects of ethical decision-making as well. In Aristotle's view, a category of good habits exists for each of the three parts of the psyche. Each category represents a *virtue*. In other words, a virtue is the habitual direction of one part of the psyche toward ethical good. In simple situations, a virtue makes good ethical action nearly effortless. In more complex situations where the best choice may not be clear, a virtue makes discerning a good solution easier.

1.4 THE FOUR CLASSICAL VIRTUES

Because of the interplay between the mind, will, and emotions, good actions rely upon all the virtues. Nevertheless, classical ethical thought identifies four primary virtues. Each one is rooted in a particular component of the psyche. These four typically bear the name "cardinal virtues" or "natural virtues." The ancient Greeks observed that each of the natural virtues promotes actions lying somewhere between excess and deficiency.

- *Justice:* Justice connects with the will and has two aspects: truth and fairness. Acting in truth recognizes the world as it actually is and rejects mere appearances. Acting with fairness seeks to give that which is due to everyone involved.
- *Prudence:* Prudence connects with the mind. A prudent mind thinks about a moral problem lucidly and thoroughly. The mind must also supply itself with enough time to think and must apply itself at the appropriate level of detail. Hence, prudence involves forethought and practicality.
- *Temperance and Fortitude:* Temperance and fortitude represent opposite sides of the same coin; both connect with the emotions but from opposite perspectives. Pleasant emotions (like elation, affection, and enjoyment) draw us toward their origin, while unpleasant ones (like dread and sorrow) push us away. Temperance controls our reaction to appeal, tamping down the impulse to move recklessly toward something we like. Fortitude controls our antipathy, tamping down the impulse to move recklessly away from something we dislike.

In describing the virtues, we need to maintain a balance between their aspects of choice and habit. Exercising the virtues can be likened to breathing the air; both activities transpire largely unconsciously. Yet breathing can be brought under conscious control if we choose. For example long-distance runners consciously regulate their breathing. In the same way, people typically act habitually, but retain the capability to consciously regulate their ethical behavior. That is, although we might routinely act with prudence, temperance, fortitude, and justice, we can also elect to do so.

1.5 THE ROLE OF INTENTION

As mentioned above, virtue ethics concerns itself not only with habits but also with internal intentions. Intentions are especially important in cases wherein the virtues must be consciously brought to bear. In such situations, a person's intention plays a key role in determining whether a particular action is ethical. Indeed, ethicists over the centuries have attached great importance to intention in assessing the goodness of an action. A few simple examples will illustrate why. People typically become much angrier with someone who deliberately enters incorrect data into a spreadsheet than with someone who does so accidentally. Most legal systems account for intention by differentiating between voluntary and involuntary manslaughter, for example, and between tax evasion and accidental underpayment of taxes.

Note that "intention" can be understood in multiple ways. Many ethicists define intention as the purpose of an action. That is, intention is the answer to the question, "What is this action attempting to accomplish?" However, sometimes people do things with several goals in mind. In other words, an action can have several intentions. In such cases, it is useful to employ a different definition of intention. The goals of an action can almost always be described as specific consequences. For example, if someone has the goal of developing a larger vocabulary, the action might be the study of word lists, and a consequence might be learning the definitions of two hundred new words. Phrasing goals as specific consequences enables the systematic identification of intentions as answers to the question, "To which consequences do we give approval?" (Some ethicists use the word "consent" instead of "approval.") That is, we can identify intentions by examining attitudes toward the consequences of an action. We can categorize such attitudes in one of four ways: approval, disapproval, mixed, and indifferent. These attitudes differ from mere feelings, be they positive, negative, conflicted or neutral. For example, it is possible to feel bad when reporting a coworker for stealing, but it is also possible to simultaneously approve of the consequence of justice that follows. Since the ability to choose resides in the will, approval and disapproval are chosen intentions of the will rather than reflexive responses of the emotions.

1.6   MAINTAINING THE HABIT OF ETHICAL BEHAVIOR

Virtue focuses on the theme of habit. That is, acting voluntarily tends to internalize a behavior pattern that makes this action easier to do in the future. This process of internalization undergirds the old saying, "You become what you do." In the same way, doing something that has ethical significance has the consequence of imprinting a behavior pattern. Such imprinting occurs because acting on a decision, and to some extent even making the decision, exerts internal effects. This latter point is recognized in criminal law, which punishes conspiracy to commit murder as well as the act of murder itself. Even if a plan for murder never comes to fruition, the planning itself imprints the conspirator in a way that makes murder more likely in the future.

Human behavior does not obey fixed mathematical rules, of course, and under most circumstances people can choose to control their actions in spite of habit. As suggested above, an analogy exists between acting ethically and breathing. Ordinarily we breathe without thinking, but within limits we can chose to control our rate and depth of breathing. Even though a decision may affect others that follow, this effect occurs only in terms of likelihoods, not certainties. Yet studies from the social sciences show that a convicted criminal is more likely to commit a crime in the future than someone with no record. And a one-time user of illicit drugs is more likely to continue to use such drugs than a zero-time user.

Although acting according to the virtues in one case makes it easier to behave this way again, a certain degree of effort is always required. Muscles weaken without such effort, for example, and so does the ethical will. Thus, if we just try to "get by" in the ethical arena, without continually trying to improve, backsliding typically occurs sooner or later. Rarely can an "ethical athlete's" good conditioning remain static at a fixed level; usually an athlete's capabilities are either improving or declining – sometimes rapidly and sometimes barely perceptibly. Thus, it is important to avoid doing the bare minimum just to get by in the ethical life. For example, such minimalism was an important issue at play in the recent firing of the famed collegiate sports coach Joe Paterno at Pennsylvania State University (Seebauer and Barry, 2001).

But does acting above the bare minimum call for the practice of benevolence? Benevolent actions seek what is best for others without any anticipation of return. Defined this way, benevolence certainly deserves admiration, but is it required? This question has been debated for centuries, and some ethical systems promote benevolence as the highest good. Most, however, concede that some degree of self-interest can legitimately enter into an ethical decision. In other words, benevolence is a worthy aspiration to pursue insofar as possible, as guided by the virtue of prudence.

1.7 APPLICATIONS OF THE VIRTUES

The following sections offer specific examples of each of the four classical virtues. In the case of the virtue of justice, examples are given focused on both key dimensions: truth and fairness.

1.7.1 Application of the virtue of justice (truth): scientific publication

Most scientists pursue their work at least partly from intrinsic desires to learn, and to share discoveries for the benefit of all people. However, many scientists are also motivated by tangible rewards as well. Since the communal ownership of scientific knowledge sometimes limits the profits a scientist can earn from a discovery, the primary tangible reward for innovation often lies in public recognition. This recognition accrues to the individuals who show they were first to observe and recognize the significance of valuable new knowledge.

The reward for priority in publishing infuses a healthy energy and originality into many scientific endeavors. While recognizing this fact, scientists often remain suspicious of efforts to win professional or public fame (Vera, 2011). As an example of such modesty, Isaac Newton wrote, "If I have seen farther, it is by standing on the shoulders of giants." (in a letter to Robert Hooke, who was challenging Newton's claim to have invented the theory of colors) (Koyre, 1952). Sigmund Freud described the reward for publishing first as an "unworthy and puerile" motivation for scientific effort (Merton, 1973). Nevertheless, many people (including scientists) crave tangible expressions of approval for what they do. As a result, the drive for priority in publishing can become an end in itself rather than a means to inspiration. Is it really so important when one scientist publishes a few days or weeks ahead of someone else? The distinguished sociologist of science Robert Merton (1973) put it this way:

> The fact is that all of those firmly placed in the pantheon of science – Newton, Descartes, Leibniz, Pascal or Huygens, Lister, Faraday, Laplace or Davy – were caught up in passionate efforts to achieve priority and to have it publicly registered.

Clearly these pressures of publishing can corrode the truth aspect of the virtue of justice. Several sorts of actions offend against these ideals. For example, fraud or falsification in public work is clearly wrong. Such actions strike at the heart of scientific truth. It is of course necessary to distinguish between reports that are unintentionally incorrect and those that are deliberately counterfeited. The literature is replete with results that ultimately proved to be incorrect due to accidental errors. Falsified results are simply fabricated out of nothing. Fraudulent data are genuine, but were not collected by the stated method. The fraud usually involves removing or manipulating in a way that Charles Babbage once called "cooking" and "trimming"(Merton, 1973). While cooking or trimming data without good reason is clearly unethical, more complicated ethical problems arise when scientifically plausible reasons exist to eliminate or recalculate certain data. Sometimes measurement instruments have easily recognizable but sporadic problems. Other times, changes in procedure slink into the experiment, either by intention or by error. Some experiments are just too arduous or costly to repeat. Given the length limitations imposed by some journals, it may prove impossible to describe the analysis fully. Prudent judgment must be exercised – the literature does not benefit from avalanches of questionable data.

High-profile cases of fraud and falsification have arisen in recent years. For example, in 2011 the Office of Research Integrity at the US Department of Health & Human Services released a report detailing an elaborate scheme of fraud and falsification perpetrated by a chemistry graduate student at Columbia University in order to publish work related to C–H bond functionalization (Schultz, 2011). As a result of a lengthy and extensive investigation by the university and the Office of Research Integrity, the student's doctoral advisor had to retract roughly a half dozen published papers, and the student's PhD degree was revoked.

Plagiarism also counts as a clear offense against truth. Plagiarism involves paraphrasing or directly copying the words or results of someone else without appropriate citation. Some researchers include in their definition of plagiarism the frequent practice of intentionally failing to cite closely related work by others. The ethical question of course depends upon how strongly related the work is. It is difficult to determine the incidence of plagiarism in the literature. However, it is notable that charges of stealing scientific ideas seem to be more common than the actual theft itself (Merton, 1973)! Accusations of plagiarism have been well-known since the time of Descartes in the 1500s, who was falsely accused of pilfering ideas from Harvey, Snell, and Fermat respectively in physiology, optics, and geometry (Merton, 1973). Not all such claims are spiteful. The human imagination often takes new ideas and packages them into well-established boxes, thereby making them seem familiar.

1.7.2 Application of the virtue of justice (fairness): resource allocation

Resource allocation takes place continually in the workplace, and is regulated by a principle that ethicists call "distributive justice," although "distributive fairness" might be a more accurate expression. Ethicists have wrestled with the problems connected with resource allocation for a very long time. No general resolution has materialized. Therefore, this discussion will only highlight some of the principal lines of thought on this issue. Circumstances and intentions make considerable difference in deciding which approach to use.

This discussion uses the word "resource" to refer to an item that is measurable and is available to several or many people. This perspective excludes qualitative items like credit or prestige. Examples of resources include:

- *Money*: Organizations of all kinds need to allocate salary among employees and revenue streams among operating units. Funds may need to flow into research and development, daily operations, taxes, investments, or materials and supplies. Closely related to money are jobs and the salaries that go with them. Chapter 10 of this book examines in detail some questions revolving around casual employment and low pay in the food/catering industry. Ultimately those questions originate in resource distribution.
- *Consumables*: These include food, water, medicine, energy, and virtually any other tangible item needed for human existence. In food production, typical examples of consumables include feedstocks, electricity, and fuel.
- *Time*: On the job, scientists in the food industry decide how much time to give to laboratory experiments, group meetings, training opportunities, safety and cleanup, and socializing. This distribution involves not only the hours themselves but also the ability to offer genuine attention.
- *Space*: Space allocation for laboratories and offices presents a continual challenge in most work environments.
- *Services*: Service resources may include biological or chemical analyses, administrative support, and technical support from machinists and consultants.
- *"Negative resources"*: Negative resources spring from collective responsibility for a debt or other financial liability. Any quantifiable liability can be considered from this point of view: requirements to provide space, administrative services, and the like. Some dangers like process effluents and waste streams can also count as negative resources.

Numerous methods have been described for resource allocation. Sometimes more than one method is used in the same situation. Examples of the primary methods include (Outka, 1974):

- *Allocation by merit*: Allocation by merit views resources as rewards that should be distributed according to effort or demonstrated ability. Examples include job offers, salary raises, promotions, and protection from layoff. However, allocation by merit breaks down for resources that are necessary for living, such as food, water, and housing. Use of merit principles under unsuitable circumstances can lead to the abrogation of basic human rights, and discriminates against people or groups who are disadvantaged through no fault of their own. In destitute regions of the world, for example, it would be impermissible to deny children food because they are not as productive as adults.
- *Allocation by social worth*: Allocation by social worth directs resources toward those who appear most likely to contribute to the common good, usually in ways that do the greatest good for the largest number of people. Possible standards for social worth include age, seniority, rank, and expertise. In exceptional cases of natural disaster or

war, for example, social worth preferentially directs food, water, and medical attention to political and military leaders and to physicians to help preserve overall social order and health. Social worth sometimes correlates with merit, so the distinction between these two approaches can become indistinct. Allocation by social worth breaks down when the criteria for worth disregard basic human rights. For example, if wealth is used to measure social worth, fundamental resources such as food, clean water, energy, and education will "flow uphill," exacerbating imbalances that already exist.

- *Allocation by need*: Allocation by need views resources as basic human rights, meaning that every person has the same right to some modest level of a given resource. Examples include the needs for food, clothing, and shelter. This approach is often active after natural disasters, where the weakest and sickest receive the most consideration. Allocation by need breaks down when this standard is applied so rigorously that it removes the motivation to produce. Those who produce the most sometimes have the fewest needs, and in such cases allocation by need can greatly reduce this motivation. When the most productive members of an organization lose their incentive, the organization suffers.

- *Allocation by ability to pay*: Allocation by ability to pay views resources in terms of market forces, and in certain ways represents a cross between merit and need. Merit factors in because those who produce more can afford to pay more. Need factors in because those in greater necessity are often willing to pay a greater price. An obvious example of distribution by ability to pay is a black market. Allocation by ability to pay typically works well for nonessential items, but fails when there is not a fair distribution of wealth at the outset. An ability-to-pay approach merely perpetuates these established injustices. The problem is especially severe when ability to pay governs the distribution of essentials like food, clean water and medicine.

- *Allocation by equal or random assignment*: Allocation by equal or random assignment presupposes that no impartial way exists to distribute resources. Equal assignment can be used for items that can be divided into very small amounts, such as food, water and money. The resource is partitioned into as many identical portions as there are people. Random assignment finds use for items that cannot be divided, including homes, jobs, and doses of medicine. This method typically uses a lottery system that offers everyone an equal chance. Equal or random assignment are free from personal biases. However, these methods abandon any attempt to account for authentic differences in other factors such as merit, need, and the like. Equal or random assignment both seek to avoid the challenging ethical choices presented by these factors. Sometimes good practical reasons exist to justify this approach, but other times the practice is grounded in ethical spinelessness. Equal assignment breaks down when each portion of a resource is too small to be useful. For example, dividing a store of infant formula into small portions during a famine could make each portion so small that no one benefits. Random assignment fails when by lucky fortune certain people wind up with the lion's share of the resources while others receive almost nothing. Sometimes resources are distributed only rarely – for example, chances to interview with employers through college placement offices. A lottery scheme for distribution can occasionally give some students several times as many interviews as others.

- *Allocation by similarity*: Allocation by similarity is actually an approach to utilizing other methods of distribution (by merit, need, etc.) rather than an independent method.

Allocation by similarity only says that cases that appear to be the same should be treated in the same way. For example, if during the aftermath of a hurricane it is decided to allocate food according to need basis, similarity indicates that clean drinking water be distributed the same way. Although allocation by similarity may seem quite reasonable, there can be problems with deciding which cases are similar and which are not. Real-world situations are typically complicated and are rarely exactly alike.

No simple algorithm for allocation can assure fairness in all cases. The best approach depends upon what needs to be distributed and on the specific circumstances of each situation. Intentions also need to be taken into account. Yet one important principle should be kept in mind when selecting an allocation method: the obligation to avoid what is bad outweighs the obligation to do what is good. This principle underlies the long-held rule for the physician treating an ailing patient: "Do no harm." Tom Nairn discusses related ideas in Chapter 2 on medical ethics. Other examples include early systems of law like the Code of Hammurabi, which focused upon which evils to shun rather than which goods to pursue. This practice continues to the present; most codes of law say much more about what people should *not* do than about what they *should* do. In addition, it has been observed that people are more willing to accept risk to avoid harms than to preserve benefits (e.g. Rowe, 1979), and to exert more effort to avoid a loss than to secure a gain.

1.7.3 Application of the virtue of prudence: cooperating in the unethical behavior of others

Ethical questions pose special difficulties when we are pressured to cooperate with an injustice pursued by someone else. For example, you may see a coworker doing something wrong in a situation where the ethical stakes are very high. You may see (or be pressured to facilitate) blatant offenses against safety, environmental standards, work-place equality, and the like. Sometimes these wrongdoings take place with the tacit or explicit endorsement of management. To stop them would require you to appeal to very senior levels of management or to oversight agencies outside your organization. Such an appeal is commonly termed "whistleblowing."

Whistleblowing presents a difficult ethical choice. Often the options reduce to "should I say something to stop these wrongs and risk retribution, or should I remain quiet and stay out of trouble?" Several studies of whistleblowing show that whistleblowers typically face hostility within their organization and often leave their jobs, voluntarily or not. In academic research, about 12% of whistleblowers who face opposition ultimately lose their jobs – according to a 1995 survey conducted by the US Department of Health and Human Services. However, the most common response to whistleblowing is pressure to drop the charges and/or countercharges (see Glazer, 1997). Lasswell and Harmes (1995) present a compelling example case in which a whistleblower was harassed and fired by his employer. Eventually the whistleblower sued and was awarded $13.7 million.

Thus, lawsuits are often involved. But remaining silent in the face of wrongdoing does not stop the misconduct and gnaws at the conscience.

Few general rules exist to promote or discourage whistleblowing, and most situations must be examined carefully on a case-by-case basis according to the virtue of prudence. For example, the perceptions of others must be taken into consideration. There can be situations where someone can perform an action that by itself would not be wrong, but where others might misunderstand what is happening and conclude that the action is wrong. Most ethicists agree that because of the potential severity of the retaliation, whistleblowing is praiseworthy but is not required under all circumstances. However, one standard does apply: the duty to blow the whistle increases as the gravity of the wrongdoing increases (Seebauer, 2001, 2004).

Such problems of cooperating in wrongdoing arise frequently, leading to the development of well-defined concepts for classifying degrees of cooperation. This taxonomy introduces no new principles into virtue ethics, but provides a useful means for looking at such problems. Cooperation with injustice can be classified into three categories:

- *Mediate material cooperation*: Here we disapprove of the injustice we see another person doing, and our own actions would customarily be considered good or neutral. Also, our actions should provide nothing essential for the injustice to occur, and should be only remotely connected with the situation. Classical moral writers generally consider this kind of cooperation to be acceptable given adequate reasons.
- *Immediate material cooperation*: Here we disapprove of the injustice we see another person doing, and our own actions would customarily be considered good or neutral. However, now our actions provide something essential for the injustice to occur. Such cooperation is generally considered acceptable only for serious reasons. Crucial factors include the degree of your role, the degree of harm caused by not cooperating, and the likelihood of giving a bad example to others.
- *Formal cooperation:* Here we approve of the injustice and also provide something essential for it to occur. Furthermore, under ordinary circumstances our actions would be considered morally bad. Such cooperation is generally considered to be unacceptable.

1.7.4 Application of the virtue of temperance: risk

The virtue of temperance is required when the potential for profit (especially in the short term) is strong, but the possibility for harm is significant. In the food industry, for example, harms can arise from low-level chemical and biological contamination, or from severe food allergies that are rising in the general population. The harms may be inherent in the products themselves, or may arise through the production process. Determination of the acceptability of risk involves the virtues of prudence and justice. Persistence in undertaking such determinations may involve fortitude. But abiding by an adverse determination requires temperance against the allure of short-term gain. For an account of the broader risk that technology poses to the way people themselves, see Lewis (1946), and for a more recent account of the risks of technology, see Childress (1981).

Risk in the food industry is often considered within the context of "safety." "Safety" is an abstract term with physical, psychological, and economic aspects. "Risk" is similarly abstract, and in common speech refers to virtually any threat to safety. Well-grounded ethical analysis requires a common and precise understanding of what these terms mean. This common understanding may be difficult to achieve. It has been well documented, for example, that technologists understand "risk" differently from the general public. The common technical understanding of risk is the probability that some given harm will occur. For example, risk might refer to the probability that a particular level of food additive causes cancer in a large population. By contrast, laypeople tend to incorporate ethical importance into the concept of risk. Thus, the risk that harm will occur includes both likelihood (loosely defined) and ethical importance. Recently some ethicists have started to view risk as the "mathematical" product of importance and probability; see for example, Lowrance (1980). Furthermore, laypeople often do not distinguish clearly the risk itself from the acceptability of risk. Risks tend to be interpreted as less acceptable if they are difficult to understand, unfairly distributed, in close proximity (Martin and Schinzinger, 1986), or not within direct control.

Martin and Schinzinger follow Lowrance (1980) in plotting a harmbenefit function, which has thresholds near the neutral point of small harms or benefits. The lack of effort below the threshold on the harm side originates from the human propensity to ignore small harms in order to avoid anxiety overload. Martin and Schinzinger propose a (smaller) threshold on the gain side that represents a combination of inertia and "generosity" that inhibits people from instantly seeking selfish gain. These thresholds vary with the circumstances of individuals. Unfortunately, in some cases the difference in perspective between technologists and laypeople is too big to span in a timely way (Slovic *et al.*, 1980), causing real problems for technical projects.

Risk has become a front-burner social issue only during the last few decades, and no generally accepted approach has emerged. But certain guidelines for handling risk operate in almost all cases. First, increasing risks may be accepted only when the possible benefits increase proportionately. In other words, one cannot take large risks to secure small benefits. Second, there must be diligent attempts to obtain informed consent when consumers are involved. Fully informed consent is sometimes very difficult to obtain. If the risk is significant, the requirement to obtain fully informed consent increases accordingly.

1.7.5 Application of the virtue of fortitude: ethical responsibility

The virtues of justice and prudence often give clear guidance about what the proper course of action should be. But sometimes, fully assuming the ethical responsibility to take that path is arduous or rouses opposition. Pursuing the right path with consistency often requires the virtue of fortitude. Several other chapters in this book describe such cases. For example, Chapter 8 discusses environmentally sustainable food production, which often requires investments in extra research or manufacturing processing., Such well-intentioned investments do not always pay off. In one case, Frito Lay created a new type of compostable food packaging material as a way to reduce waste in landfills. But consumers did not like the packaging because it was perceived as too noisy when being removed. Considerable fortitude was required for Frito Lay to persist in trying to gain consumer acceptance, although the material ultimately had to be abandoned. Food manufacturers have devoted efforts to develop ingredients and processes to reduce the amount of salt and fat in foods. As consumers have shown they like the taste of salty foods with high fat, fortitude is required to make the foods more healthful. Several beverage manufacturers are trying to put more healthful vegetables into consumer drinks – again in the face of uncertain rewards.

It is worth noting, however, that there are factors that can limit the ethical responsibility an individual or an organization has in a given situation. In classical moral thought, ethics concerns the goodness of voluntary human conduct that impacts the self or other living beings. The word "voluntary" is very significant, because it implies there is adequate control over what is being done. Assuming there has been no deliberate attempt to remain ignorant, powerless, or indifferent, an individual or organization has complete ethical responsibility for what is done with adequate knowledge, freedom, and approval. This criterion for responsibility points to three potentially limiting factors.

- *Lack of knowledge*: Lack of knowledge that limits ethical responsibility takes two forms: legitimate ignorance of key aspects of a situation, or ignorance that an action is unethical.
- *Lack of freedom*: Threats of physical or psychological violence by others can remove freedom, as can the influence of alcohol or drugs or overwhelming passion (fear, guilt, or grief).
- *Lack of approval*: Many ethicists also use the term "deficient consent." A common example is when there is insufficient time for reflection before acting. If circumstances force a snap judgment, so that there is not enough time to think through ethical consequences, there is no ability to approve or disapprove of them.

1.8 VIRTUE ETHICS IN A BROADER CONTEXT

Many models for ethical action are based on the workings of the mind and will, largely ignoring the emotions. These "rationalist" models take varied forms. Some focus on adults that are fully developed as moral creatures. These models are called "static," and are often used by philosophers, and include deontology, utilitarianism, rights-based theories, and casuistry. Other models speak directly to questions of development. Most focus on moral development in children, but some also address the changes that take place throughout adult life. Many of these models treat areas of human psychic growth that include more than just the ethical, although the ethical is certainly included. For example, Eric Erickson lays out eight "psycho-social" stages (Erikson, 1963; Helminiak, 1987). Each stage involves the resolution of a type of psychological tension, like trust versus mistrust, intimacy versus isolation, and social responsibility versus stagnation. Among all these models, Lawrence Kohlberg's (Kohlberg, 1968, 1969, 1977) concerns itself most directly with ethical growth. Kohlberg's original theory includes six stages, ending at the point where a person acts according to universal rational principles.

Rationalist approaches have significant benefits for day-to-day ethical decision-making and for larger more complicated questions. These benefits include: comprehensiveness and completeness, as well as ease of use in discussion, debate and legislation. The specific reasons why people disagree become readily apparent, which is very useful in a pluralistic society where disagreements about ethics abound. These disagreements often require compromises, which can be made more easily when everyone understands exactly where the compromises must lie.

As significant as these benefits are, a purely rationalist approach also has several drawbacks. For example, ethicists going back at least to the Middle Ages have pointed out that rational analysis has limitations when dealing with truly complex cases. Thomas Aquinas (c. 1225–1274), the leading philosopher of the Middle Ages in the West, asserted that people have a "natural judgment" concerning certain ethical goals such as life, truth, and fairness. Aquinas's "natural judgment" refers to an ability to distinguish good from bad by means other than pure reason – that is, by intuition (Aquinas, 1981). Problems with concretely describing many of the abstract, ambiguous features of complicated situations plague all rationalist approaches to ethics. Many rationalist theories come across as dry, abstract, wrongly framed. For example, Carol Gilligan has roundly criticized Lawrence Kohlberg's theory on these grounds (Gilligan, 1982), arguing that women often "underperform" men in sociological studies of ethics because the tests are written only in terms of abstract duties and obligations.

In addition, rationalist approaches usually assume that people will naturally do the ethical thing if they know what it is. Unfortunately, this idea does not accord with much human experience. Rationalist approaches typically struggle to answer the question of why it is important to act ethically. For example, Kohlberg's original theory ends when an individual thinks in terms of universal rational principles, but does not explain why life be ordered according to such principles. In a later attempt to resolve this problem, Kohlberg added an extra "Stage 7" that is explicitly spiritual (Kohlberg and Power, 1981). Rationalist theories provide little defense against rationalization – wherein irrational jealousy, envy, bias, and malice creep in unnoticed to skew moral analysis. Food scientists, who normally work in an environment that prizes rational thought, can certainly be vulnerable to this kind of thinking.

The disadvantages to rationalist ethics have inspired several attempts to fill in the gaps. Although this chapter focuses specifically on ethics in food science, we note in passing that some writers have attempted to set the whole endeavor of science and engineering in a more complete (and less rationalistic) context. Some of these writers are scientists themselves. For examples of this extensive literature, see Polanyi (1974), Jaki (1993), and Smith (1984). The psychologists James Rest and coworkers (1986) have created a "four-component model" for moral decision-making that is easy to understand. The model proposes four steps to ethical decision-making: sensing the presence of ethical issues, reasoning through them, making a decision, and following through on the decision. Thus, the four-component model shares important affinities with classical virtue theory. Prudence underlies sensitivity and reasoning, justice underlies judging, and temperance/fortitude underlies doing. Rest's model also allows for "affect" (in our terminology, emotions) to influence decision-making in addition to the mind.

Virtue ethics in public discourse has been resurgent after lying dormant for several decades (Wallace, 1978; McIntyre, 1980; Meilaender, 1984; Hittinger, 1987; Cessario, 1991). As indicated earlier, virtue ethics employs the idea of habit to place specific acts within the context of a broader orientation of life. Present-day virtue theorists propose that this approach represents an advance because the goal of the ethical life focuses on development of habits that form "character." This character offers the power to do what is good more easily.

ACKNOWLEDGMENT

This material is based upon work supported by the National Science Foundation under Grant No. DMR 100-5720. Any opinions, findings, and conclusions or recommendations expressed in this material are those of the author and do not necessarily reflect the views of the National Science Foundation.

REFERENCES

Aquinas Thomas, *Summa Theologica* (Westminster, Md., Christian Classics, 1981) Question 47, Article 15.

Cessario R, *The Moral Virtues and Theological Ethics* (South Bend, Ind., University of Notre Dame Press, 1991).

Childress JF, "The art of technology assessment," *Priorities in Biomedical Ethics* (Philadelphia, Westminster Press, 1981) 98–118.

Erickson E, "Eight ages of man," in *Childhood and Society*, 2nd ed. (New York, W.W. Norton, 1963).

Frankena W, *Ethics*, 2nd ed. (Englewood Cliffs, N.J., Prentice Hall, 1973).

Gilligan C, *In a Different Voice: Psychological Theory and Women's Development* (Cambridge, Mass., Harvard University Press, 1982).

Glazer S, "Dealing with threats to whistleblowers," *CQ Researcher* **7** (1997) 6.

Helminiak DA, *Spiritual Development: An Interdisciplinary Study* (Chicago, Loyola University Press, 1987), ch. 3.

Hittinger R, *A Critique of the New Natural Law Theory* (South Bend, Ind., University of Notre Dame Press, 1987).

Jaki SL, *Is There a Universe?* (New York, Wethersfield Institute, 1993).

Kohlberg L and Power C, "moral development, religious thinking, and the question of a seventh stage," *The Philosophy of Moral Development: Moral Stages and the Idea of Justice, Essays on Moral Development*, Vol. **1** (San Francisco, Harper & Row, 1981) 311–372.

Kohlberg L, "Stages and sequence: the cognitive-developmental approach to socialization," in *Handbook of Socialization Theory and Research*, D. A. Goslin, ed. (Chicago, Rand McNally, 1969).

Kohlberg L, "The Child as Moral Philosopher," *Psychology Today*, September 1968, 25–30.

Kohlberg L, "The Implications of Moral Stages for Adult Education," *Religious Education* **72** (1977) 183–201.

Koyre A, "An Unpublished Letter of Robert Hooke to Isaac Newton," *Isis* **43** (December 1952) 312–337.

Lasswell M and Harmes J, "Justice deferred," *People*, June 12, 1995, pp. 101–102, E. G. Seebauer, "When Do You Blow the Whistle?" *Chemical Engineering*, 108 (April, 2001) 123–126.

Lewis CS, *The Abolition of Man* (London, William Collins Sons, 1946).

Lowrance WR, "The nature of risk," in *Societal Risk Assessment: How Safe Is Safe Enough?*, Richard C. Schwing and Walter A. Albers, Jr., eds. (New York, Plenum, 1980) 6.

MacIntyre A, *After Virtue: A Study in Moral Theory* (South Bend, Ind., University of Notre Dame Press, 1980).

Martin MW and Schinzinger R, *Ethics in Engineering*, 3rd ed. (New York, McGraw-Hill, 1996) 132 ff.

Meilaender G, *The Theory and Practice of Virtue* (South Bend, Ind., University of Notre Dame Press, 1984).

Merton RK, "Behavior patterns of scientists," *The Sociology of Science* (Chicago, University of Chicago Press, 1973) 325–342.

Merton RK, "Priorities in scientific discovery," *The Sociology of Science* (Chicago, University of Chicago Press, 1973) 286–324.

Outka G, "Social justice and equal access to health care," *Journal of Religious Ethics* **2** (Spring 1974) 11–32.

Polanyi M, *Personal Knowledge* (Chicago, University of Chicago Press, 1974).

Rest J, Bebeau M, and Volker J, "An overview of the psychology of morality," in *Moral Development: Advances in Research and Theory*, James Rest, ed. (Westport, Conn., Greenwood Publishing, 1986) 1–27.

Rowe WD, "What is an acceptable risk and how can it be determined?" in *Energy Risk Management*, G. T. Goodman and W. D. Rowe, eds. (New York, Academic Press, 1979) 327–344.

Schultz WG, "A puzzle named Bengu Sezen," *Chemical & Engineering News* **89** (August 8, 2011) 40–43.

Seebauer EG, "Whistleblowing: is it always obligatory?" *Chemical Engineering Progress*, **100** (2004) 23–27.

Seebauer EG and Barry RL, *Fundamentals of Ethics for Scientists and Engineers* (New York, Oxford University Press, 2001).

Slovic P, Fischhoff B, and Lichtenstein S, "Risky assumptions," *Psychology Today* **14** (June 1980) 44–48.

Smith W, *Cosmos and Transcendence* (Peru, Ill., Sherwood Sugden, 1984).

Stumpf SE, *Socrates to Sartre: A History of Philosophy*, 3rd ed. (New York, McGraw-Hill, 1982) 30–32.

Vera M, "Paterno ousted with President by Penn State," *The New York Times*, November 10, 2011, p. A1.

Wallace JD, *Virtues and Vices* (Ithaca, NY, Cornell University Press, 1978).

2 Lessons from medical ethics

Thomas A. Nairn

2.1 INTRODUCTION

It might be appropriate to begin this chapter with a qualification – I am a medical ethicist, with no expertise in food production. Therefore this chapter is principally one that analyzes the brief history and current state of biomedical ethics in the hope that this can be of use to those with the expertise who hope to develop similar analyses dealing with ethical issues in food production. The chapter is divided into two sections. The first section will provide a brief history of contemporary medical ethics by identifying four contemporary methods of bioethical decision making – a principles-based ethical analysis, casuistry, a method which emphasizes the medical facts that affect ethical decisions, and preventive ethics. Although these four do not exhaust the methods available to medical ethics,[1] they were chosen both for their historical importance and for their potential relevance to the ethics of food production. Each method will be placed within a context that shows the contributions of that particular method to the larger conversations in bioethics. (In this chapter, *medical ethics*, *bioethics*, and *biomedical ethics* will be used as synonyms.). I will use a case regarding the ethics of end of life care to compare and contrast the four methods in order to ascertain what the lessons are that can be learned from that particular method of medical ethics. In the second section I will try to demonstrate how the lessons learned from these methods may be applied to the ethics of food production. This section will be more evocative than analytical, suggesting how the questions and methods discussed in the first section might have relevance for the ethics of food production.

2.2 FOUR METHODS IN MEDICAL ETHICS

For much of its history, medical ethics was a field within the discipline of theology rather than philosophy, and consequently it framed the ethical and medical issues within

---

[1]See, for example, Jeremy Sugarman and Daniel Sulmasy, eds., *Methods in Medical Ethics, Second Edition* (Washington, DC: Georgetown University Press, 2010). The authors identify 13 distinct methods with variations within each method.

Practical Ethics for Food Professionals: Ethics in Research, Education and the Workplace, First Edition. Edited by J. Peter Clark and Christopher Ritson.
© 2013 John Wiley & Sons, Ltd. Published 2013 by John Wiley & Sons, Ltd.

specifically religious contexts. Theologians addressed the questions regarding sickness and health from the point of view of the Bible's Ten Commandments or by interpreting the great religious thinkers of the past, such as Ibn Sīnā (known in the West as Avicenna) for Islam, Moses Maimonides for Judaism, or Thomas Aquinas or John Calvin for Christianity.

By the middle of the 20th century, religious thinkers in North America began more and more to weigh in regarding the ethical aspects of current practices of medicine. Theologians and religious philosophers, such as Gerald Kelly (1957) or later Richard McCormick (1984) from the Catholic point of view, Paul Ramsey (1970) or Joseph Fletcher (1954) from a Protestant perspective, Immanuel Jakobovits (1959) or Fred Rosner (1977) from a Jewish viewpoint, or Fazlur Rahman (1987) from an Islamic perspective moved beyond a strictly religious context and engaged in a larger dialog between their religious traditions, contemporary philosophy, and developing medical practice.

The movement of bioethics from a discipline within a predominantly religious context to a more secular discipline took place in the 1960s and especially in the 1970s with the creation in the United States of the National Commission for the Protection of Human Subjects of Biomedical and Behavioral Research. Congress created this Commission as a result of a series of revelations concerning the unethical treatment of subjects in government-funded medical research, the most infamous of which was the Tuskegee syphilis study, a 40-year longitudinal study conducted by the US Public Health Service (Jones, 1981). The US Public Health Service conducted this study between 1932 and 1972 in Macon County, Alabama to document the long-term effects of syphilis. The subjects of the study, all African-American men, were left untreated for syphilis long after a cure for the disease was discovered. Senator Edward Kennedy sponsored the legislation creating the Commission, and President Richard Nixon signed the National Research Act into law on July 12, 1974. (For a more detailed history see Jonsen (1998)). The law mandated that:

> the Commission shall (i) conduct a comprehensive investigation and study to identify the basic ethical principles which should underlie the conduct of biomedical and behavioral research involving human subjects, [and] (ii) develop guidelines which should be followed in such research to assure that it is conducted in accord with such principles." Public Law 93-348, section 202[a][1][A], quoted in DuBose et al., 1994.

The Commission did not begin the process of ethical reflection on experimentation with human subjects, for at their disposal was a variety of international documents such as the Nuremberg Code (1947) and the Helsinki Declaration (1964) that had been produced years and even decades earlier (Jonsen, 2005). What was novel for the development of the discipline of bioethics, however, was the manner in which the Commission went about its task of identifying "the basic ethical principles which should underlie the conduct of biomedical and behavioral research." What the Commission accomplished had significant consequences for the dominant form that bioethics would take in the United States and beyond.

2.2.1 A principles-based ethics

The members of the Commission understood that their task went beyond merely a restatement of the conclusions of previous documents; rather they analyzed the *"ethical foundations* for human research" (Jonsen, 2005). The members of the Commission eventually agreed on the three principles articulated in the Belmont Report – respect for persons, beneficence and justice (National Commission, 1978). This report marks the birth – or at least an important turning point – of bioethics as an independent discipline in the United States and provided bioethics with a particular ethical methodology, one which is equated with medical ethics in many schools and medical centers in the United States.

The Belmont Report ushered into American medicine an ethical seriousness regarding practice and research. It also utilized a particular form of ethical reflection – an ethics of principles. Looking at previous professional codes of ethics, the Report concluded:

> Such rules often are inadequate to cover complex situations; at times they come into conflict, and they are frequently difficult to interpret or apply. Broader ethical principles will provide a basis on which specific rules may be formulated, criticized and interpreted (National Commission, 1978, p. 1).

It further claimed that the three general principles it articulated – respect for persons, beneficence, and justice – were precise enough yet comprehensive enough to "provide a framework that will guide the resolution of ethical problems arising from research involving human subjects" (National Commission, 1978, p. 2).

The Report went on to explain how each of the principles related to the Commission's task. Respect for persons "incorporates at least two ethical convictions: first, that individuals should be treated as autonomous agents, and second, that persons with diminished autonomy are entitled to protection" (National Commission, 1978, p. 4–5). Beneficence comprises two complementary imperatives, "(1) do not harm and (2) maximize possible benefits and minimize possible harms" (National Commission, 1978, p. 6). Justice is equated to "fairness in distribution" and is described by means of five widely-accepted though somewhat divergent formulas of fair distribution:

> (1) to each person an equal share, (2) to each person according to individual need, (3) to each person according to individual effort, (4) to each person according to societal contribution, and (5) to each person according to merit (National Commission, 1978, p. 9).

The Commission then linked each of the principles to particular behaviors deemed important in the area of biomedical research. The Report explained:

> Just as the principle of respect for persons finds expression in the requirements for consent, and the principle of beneficence in risk/benefit assessment, the principle of justice gives rise to moral requirements that there be fair procedures and outcomes in the selection of research subjects (National Commission, 1978, p. 18).

Thus the Commission related the principles to the specific purpose mandated by Congress, that of the protection of human research subjects. Each of the principles relates to a specific concern previously demonstrated – the need for informed consent (autonomy), for addressing the burden and benefits of experimentation (beneficence), and for the appropriate selection of research subjects (justice).

Tom L. Beauchamp was a staff member to the Commission and the major author of the *Belmont Report*. At the time he wrote the Report he was also collaborating with James Childress in writing a text that was to become the primary text book in American bioethics, *Principles of Biomedical Ethics* (1979).[2] In the text, the authors enumerated four ethical principles, closely mirroring those of the *Belmont Report*. These principles have become the basis of contemporary bioethics, familiar to bioethicists and medical personnel alike. Two of the principles are virtually the same as those expressed in the Report – the respect for autonomy (which the Report called "respect for persons") and justice. The other two principles separate into the two elements what the Report had referred to as the complementary imperatives of beneficence – the negative principle of nonmaleficence (Do no harm) and the positive principle of beneficence.

Early editions of the text simply gave a summary of contemporary ethical theory and then explained the four principles, along with an analysis of how each principle relates to particular issues in bioethics. For example, within the broader treatment of the principle of autonomy, the authors discussed such bioethical issues as informed consent, patient competence, and refusal of medical treatment (Beauchamp and Childress, 1983). What the text did not do (similar to the *Belmont Report* before it) was to justify or give the philosophical foundations for the four principles and their interrelation. In response to criticism, however, beginning with the fourth edition, the authors did articulate a justification both for a principle-based morality in general and for the four ethical principles themselves. They justified their principles by means of what they called "common morality." In their understanding, common morality theory is comprised of several elements:

> Two or more nonabsolute (prima facie) principles form the general level of normative statement. Second, common-morality ethics relies heavily on ordinary shared moral beliefs for its content, rather than relying on pure reason, natural law, a special moral sense, and the like. The principles embedded in these shared moral beliefs are also usually accepted by rival ethical theories. Although not the most general principles in many normative theories the [four] principles are nonetheless accepted in most types of ethical theory. The four principles . . . should be understood as principles of this description (Beauchamp and Childress, 1994).

For an account of common morality, see Donagan (1977) and Gert (2004).

In the understanding of the authors, such a common-morality theory is therefore not based on absolute moral principles but rather upon the weighing and balancing of

---

[2]There have been seven editions of the book, from 1979 through 2012. Tom Beauchamp joined the staff of the National Commission in 1976 and was given the task of drafting the Belmont Report. He explains that by that time he and James Childress had already drafted substantial portions of the book. He adds, "The two manuscripts were drafted simultaneously, often side by side, the one inevitably influencing the other." James F. Childress, Eric M. Meslin, and Harold T. Shapiro, eds., *Belmont Revisited: Ethical Principles for Research with Human Subjects* (Washington DC: Georgetown University Press, 2005), p. 12.

what can be at times competing principles. Responding to criticisms regarding how such differences in balancing can be adjudicated, the authors suggest that there are eight criteria of adequacy by which to judge any moral theory – clarity, coherence, comprehensiveness, simplicity, explanatory power, justificatory power, output power, and practicability (Beauchamp and Childress, 2012 pp. 352–354). Although most of these criteria are relatively straightforward, some may need further explanation. The criterion of simplicity, for example, suggests that an ethical theory should have (1) no more norms than are necessary and (2) only that set of norms that can be used without confusion. By "justificatory power" the authors mean that an appropriate moral theory ought to provide the basis for reasoned judgments and also enable one to criticize defective ethical judgments. They describe the "output power" of a moral theory as that which enables the theory to "produce judgments that were not in the original database of considered moral judgments on which the theory was constructed" (Beauchamp and Childress, 2012, p. 354). Thus a moral theory should not simply repeat moral judgments already believed sound prior to the development of the theory.

In the 30 years since the publication of the first edition of this text, there has been much debate regarding the four principles and their interrelation. Many believe that in practice the principle of autonomy has become the overriding principle in the practice of medicine, at least in the United States. Others have criticized the notion of "common morality" on which the method has been based. Nevertheless, this remains perhaps the dominant method of medical ethics today.

To investigate further how such a principles-based method has been employed in bioethics as well as its strengths and limits, it may be helpful to picture the following scenario:

> Jim receives a phone call that his mother has just been rushed to the local hospital. He rushes there as quickly as he can, and when he arrives he is greeted with the news that she has suffered a severe stroke. At this point, the doctors do not know how long the brain has been deprived of oxygen or how much damage has been done. Jim's mother is in a coma. The doctors tell him that things do not look good but that there is still a lot that they don't know yet. Right now they are trying to stabilize her, but they want Jim's input regarding how aggressive they should be in treating the aftermath of the stroke. Years ago, his mother named him as her agent when she filled out a durable power of attorney for health care, but they have not had any detailed conversation. To make matters worse, Jim remembers that one of the few times they even addressed the matter was a conversation he had with her just a couple weeks ago, after she had visited a close friend of hers in the intensive care unit of the same hospital – the same intensive care unit where she is now. At that time she made him promise her that she would not end up like her friend, spending her last days surrounded by tubes and monitors in a sterile and seemingly unfriendly environment.

The case represents a fairly common occurrence in most medical centers today, given the complexity of health care and the reluctance of people to discuss in detail their concerns regarding life sustaining treatments. Can the four principles be of help to Jim in making his decision?

The very fact that Jim's mother had legally named him as agent in her durable power of attorney already indicates the importance of the principle of autonomy in this country. The principle sets limits on what the physicians can do without the expressed authorization of Jim's mother or of Jim as her agent. The principle also addresses the question regarding who has competence to make decisions and the information that they need (and therefore what the physicians need to disclose) in order for them to make informed decisions. The principle of autonomy could also regulate other issues such as any possible coercion on the part of the physicians or under what circumstances it would be possible for Jim to refuse treatment for his mother

The related principles of nonmaleficence and beneficence raise the question of whether making use of a particular treatment or of refusing treatment is harmful or helpful to the patient. These related principles can encompass a spectrum of issues. At one end of the spectrum is medical neglect, and at the other end may be futile treatment. Traditionally, consideration of beneficence entails a balancing benefits and burdens for the patient. With few exceptions, medical ethics has understood the proscription not to harm as having greater moral weight than the positive demand of doing good for the patient. As we return to Jim's situation, we may see that the duty not to harm may supersede a positive moral claim for a particular treatment.

It might be difficult to envision how the principle of justice can be of help to Jim. The principle of justice is invoked when people believe that discrimination has occurred, for example when they point to racial or ethnic disparities in medical care or when they document differences in treatment because of age or gender. The principle of justice is foundational for those who advocate for the right to health care.

Although these principles have influenced both the ethical training of physicians and US public policy, it might be more difficult to see how they directly help Jim in making his decisions regarding treatment for his mother. These principles have affected the broad contours of what most believe is essential to medical ethics, but they do not seem specific enough to provide clear guidance to Jim and his family. We need to look to other methods for help.

2.2.2 Casuistry

Not all members of the National Commission agreed with the principles-based approach of Beauchamp and the *Belmont Report*. Two members of the original Commission, Stephen Toulmin and Albert Jonsen, have continued to challenge the centrality of a principle-based method in bioethical decision making. In what has become a classic article on the subject, Toulmin has raised questions regarding the nature, scope, and force of a morality of principles and explained his distrust of such a method, suggesting that the contemporary preoccupation with principles may stem from a distrust of individual discretion in ethical decision making (Toulmin, 1981, p. 34). Although he accepts that ethical principles have a limited and conditional role in resolving ethical problems (Toulmin, 1981, p. 33), he maintains that, because of the existence of genuine moral complexity, one best addresses moral issues not deductively by applying more and more highly developed ethical rules and principles, but rather inductively, on a case by case basis (Toulmin, 1981, pp. 31–39). Noting what he considers the limitations of Beauchamp's common-morality theory, he maintains that "moral wisdom is exercised ... by those who understand that, in the long run, no principle can avoid running up against another equally absolute principle, and by those who have the experience and discrimination needed to balance conflicting considerations in the most humane way" (Toulmin, 1981, p. 34). Since a principles approach to ethics can oversimplify the ethical issues, Toulmin concludes that many forms of principles-based ethics can too easily become tyrannical and disproportionate (Toulmin, 1981, p. 38).

In an effort to articulate an alternative, Toulmin and Jonsen have retrieved a seventeenth-century method of ethics called casuistry, a method emphasizing practical reasoning and the use of moral analogy rather than moral principles and their application.[3] Although for many the term "casuistry" conjures up all sorts of pejorative connotations,[4] these authors explain that in its most basic form casuistry is simply the task of choosing appropriate moral paradigms and ordering the relation between these paradigms and more ambiguous or complicated cases by developing appropriate taxonomies. The casuist looks to paradigms because "in unambiguous ('paradigmatic') cases we can recognize an action as, say, an act of cruelty or loyalty, as directly as we can recognize that a figure is triangular or square" (Jonsen and Toulmin, 1988, p. 66). One then moves from these clear paradigms to circumstances of greater complexity and likely more conflict. In explaining this method, they quote the 17th century ethicist, Gabriel Daniel[5], who commented that "the resolution of a difficult case – for example, whether one can kill another in anticipation of deadly danger – is a consequence drawn by analogy with the earlier decision on self-defense under attack, the truth of which no one doubts" (Jonsen and Toulmin, 1988, p. 252). In many ways, what Toulmin and Jonsen propose for bioethics is not very different from developments in medical research or the use of precedents in law.

They describe how this method was practiced by the casuists of the 16th and 17th centuries:

> The paradigm cases enjoyed both 'intrinsic and extrinsic certitude' . . . . Then, in succession, cases were proposed that moved away from the paradigm by introducing various combinations of circumstances and motives that made the offense in question less apparent . . . . This gradual movement from clear and simple cases to the more complex and obscure ones was standard procedure for the casuist; indeed, it might be said to be the essence of the casuistic mode of thinking (Jonsen and Toulmin, 1988, p 252).

They maintain that this method has "little resemblance to those forms of moral reasoning that seek to 'deduce' a particular conclusion from a moral premise that serves as a universal premise" (Jonsen and Toulmin, 1988, p. 256). One may speak of the application of principles, but only "so long as their applicability to new cases is clear, meaningful and unambiguous" (Jonsen and Toulmin, 1988, p. 322).

In this understanding of morality, a principle such as autonomy or beneficence can be meaningful, and in straightforward situations it may be helpful. It cannot, however, govern all applications, especially as one moves away from the rather clear-cut paradigms. Questions arise regarding the very meaning of the principle in its applicability to particular circumstances. The further one moves from the paradigm, the more questions there

---

[3]Casuistry is a method of moral reasoning that compares a particular situation to a paradigm case or paradigm cases and, by developing a taxonomy of similarities and differences between the given case and the paradigm, reaches an ethical conclusion.

[4]The first definition for casuistry in *The American Heritage Dictionary of the American Language* is "specious or excessively subtle reasoning intended to rationalize or mislead."

[5]The authors are quoting Gabriel Daniel, *Entretiens de Cléanthet Eudoxe* (1694), p. 358.

are that may arise. Supporters of casuistry claim that this is not a flaw in this method of reasoning but rather is an aspect of practical morality itself. It was Aristotle who maintained that morality ought not be considered a science but rather practical wisdom. He reminded his readers that "matters concerned with conduct and questions of what is good for us have no fixity, any more than matters of health. The general account being of this nature, the account of particular cases are [sic] yet more lacking in exactness." The movement from principle to practice, or from paradigm to application, is therefore never mechanical or formulaic. Nevertheless it continues to maintain distinctive features: the ethicist begins the process with clear and relatively unambiguous cases in which the determination of the rightness or wrongness of an action is rather straightforward and then moves from these clear cases to those of greater complexity and therefore of possibly greater moral conflict. Jonsen and Toulmin identify seven levels to this process:

1. Similar type cases ("paradigms") serve as final objects of reference in moral arguments, creating initial "presumptions that carry conclusive weight, absent "exceptional" circumstances."
2. In particular cases the first task is to decide which paradigms are directly relevant to the issues each raises.
3. Substantive difficulties arise, first, if the paradigms fit current cases only ambiguously, so the presumptions they create are open to serious challenge.
4. Such difficulties arise also if two or more paradigms apply in conflicting ways, which must be mediated.
5. The social and cultural history of moral practice reveals a progressive clarification of the "exceptions" admitted as rebutting the initial moral presumptions.
6. The same social and cultural history shows a progressive elucidation of the recognized type cases themselves.
7. Finally, cases may arise in which the factual basis of the paradigm is radically changed (Jonsen and Toulmin, 1988, pp. 306–307).

How do these levels of casuistry work in practice? Let us return to Jim's situation and suppose that several days have passed with his mother showing hardly any progress. The physicians have suggested that they begin tube feeding and want permission to insert a PEG (percutaneous endoscopic gastrostomy) tube into the stomach. Even though Jim is his mother's agent, he consults with his siblings. An argument breaks out, with his brother demanding that they must insert the tube because they cannot starve their mother and his sister insisting that his mother had told them that she does not want tubes!

The analysis of casuistry demonstrates that what has occurred in this situation is that each of Jim's siblings has chosen a different and conflicting paradigm by which to make his or her initial ethical judgment. His sister relates his mother's wishes to what she has said regarding medical interventions. She has equated the feeding tube with other tubes that are part of medical interventions and has concluded that, as a medical intervention, this is not obligatory. In fact, it would be counter to her mother's wishes. Jim's brother, on the other hand, has taken the clear case of feeding (in the ordinary sense of that term as nourishing someone by means of food and drink) as paradigm. Because each has – probably unreflectively – chosen a different paradigm, it is likely that they will continue to speak past each other, with each unable to appreciate the other's argument.

This has also been the case among bioethicists. Over the past three decades there has been a major debate concerning the appropriate paradigm to use in assessing the morality of withholding or withdrawing the medical means of administering nutrition and hydration (Lynn, 1986; Hamel and Walter, 2007). Valparaiso University's Gilbert Meilaender, for example, has maintained that withdrawal of medically assisted nutrition and hydration is not ethical. He explains, "when we stop feeding ... we are not withdrawing from the battle against any illness or disease; we are withholding the nourishment that sustains all life (Meilaender, 1986)." Other bioethicists disagree, noting the possible medical risks and burdens of such treatment (Sulmasy, 2007). The casuistic method explains the source of the disagreement (substantive difficulties). It also suggests that the eventual resolution will depend upon the social and cultural evolution of moral practice. However, even though it explains why the problem exists, it does not offer immediate help to Jim or his siblings. We move to the third method.

2.2.3 The "ordinary language of moral thinking"

In 1969, Daniel Callahan, a philosopher, and Willard Gaylin, a psychiatrist, together founded one of the nation's most prominent think tanks dealing with bioethics, the Institute of Society, Ethics and the Life Sciences, now known as the Hastings Center. The Center was established to pursue interdisciplinary research and education in bioethics, to engage a broad audience in the field and to collaborate with policy makers to identify and analyze the ethical dimensions of their work.[6] Soon after its founding, the Center began to publish the journal, *Hastings Center Studies*, later the *Hastings Center Report*. In one of the first issues of the journal, Callahan wrote about his vision of bioethics as incorporating the *ordinary* language of moral thinking rather than a technical, academic jargon, and thus be of practical help to those confronted with actual bioethical questions. Within this context, he enunciated three tasks for bioethics: (1) explaining which medical problems raise ethical issues, (2) providing some sort of systematic method for thinking through these ethical issues, and (3) helping physicians and members of related disciplines to make correct ethical decisions (Callahan, 1997, pp. 66–73). In contradistinction to the two approaches already discussed, Callahan insisted that the bioethicist must be at service to the discipline of medicine rather than the other way around. Continuing his emphasis on "ordinary language," he stressed that "the source and importance" of bioethics "lie not in the academy but in private and public human life, where what people think, feel and do make all the difference there is" (Callahan, 1997, p. 90).

The first task that Callahan articulates seems rather straightforward. Not every medical issue will necessarily become an ethical issue. In explaining this task further, however, Callahan suggests that there is a danger for the ethicist of reading ethical problems into what are really medical ones, especially if one begins with ethical theory rather than with the medical facts. He indicates that bioethicists can be guilty of actually changing the nature of a medical issue to fit their theory, rather than allowing ethics to be of service to the medical community and its ethical needs. Within this context, he also raises the possibility that the manner in which an ethicist understands science and medicine

---

[6]The Hastings Center mission statement can be found at http://www.thehastingscenter.org/ (accessed 10 December 2012).

might simply be wrong, and he recommends that the bioethicist therefore be humble in providing specific ethical solutions to medical problems.

In looking to the second task, Callahan maintains that a systematic ethical method is necessary for medicine, but he distances himself from both of the methods previously discussed. Although he demands that bioethical analysis be rigorous, he qualifies his understanding of what such rigor entails. Rather than emphasis on logical consistency, coherence and comprehensiveness, he claims that "thinking straight" about bioethical questions demands attention to three areas of ethical activity – thinking, feeling and behaving. Most ethical theories concentrate almost exclusively on the first and the third and often disregard moral import of feeling when raising ethical questions. Callahan, on the contrary, insists that "a passion for the good is not inappropriate for ethicists" (Callahan, 1997, p. 90).

The first two tasks lead to the third, that of helping physicians and other clinicians make good, practical, ethical decisions. Again Callahan emphasizes that the bioethicist must help the medical community, patients, and their families make the real ethical decisions that face them. He warns against the use of overly philosophical concepts and of "very broad and general thinking that is of limited use" (Callahan, 1997, p. 91). Rather, if bioethicists are truly to be of service to the medical community, they will need more than simply a theoretical knowledge of ethical theory. They also need exposure to the kinds of practical issues that physicians and families must make, the psychological and cultural factors involved, and the limitations as well as possibilities that are inherent in every theoretical analysis. He emphasizes that perhaps the most important test of bioethics is "the extent to which it is called upon by scientists and physicians" (Callahan, 1997, p. 92).

How might Callahan's method help Jim? On the one hand, Callahan's emphasis on feeling raises some important issues. In a book that Callahan wrote on death and dying, he spoke of the sorts of questions that surround care for the seriously ill and dying, questions such as losing control, of pain and suffering, becoming dependent upon others, and of mourning the loss of one's idealized self (Callahan, 2000, pp. 127–148).

He also raises questions regarding physicians and technology. Part of the explanation for this phenomenon lies in the success of science and medical technology, which has heightened the expectation that all that is needed to eradicate disease is more knowledge and better technology. This belief can lead people to see sickness and death as factors they can, and ought to, control. Daniel Callahan explains:

> The use of technology is ordinarily the way, in modern medicine, that action is carried out: to give a pill, to cut out a cancerous tumor, or to use a machine to support respiration. With an ethos of technological monism, all meaningful actions ... are technological, whether technological acts or technological omissions. What nature does, its underlying natural causes and pathologies, becomes irrelevant. No death is "natural" any longer – the word becomes meaningless – no natural successful choices in an cause necessarily determinative, no pathology fatal unless failure to deploy a technology makes it so (Callahan, 2000, p. 68).

Technological monism, this belief that all meaningful actions are technological, can in turn lead to what Callahan calls technological brinkmanship, "pushing aggressive treatment as far as it can go in the hope that it can be stopped at just the right moment if it turns out to be futile" (Callahan, 2000, p. 192)

Unfortunately, medicine lacks the precision necessary for such brinkmanship to succeed. Physicians pursue aggressive treatment for their patients beyond reasonable hope for success, because once they have begun a course of action, they do not know when or how to stop. Patients dread an impersonal death, surrounded by tubes, wires, and machines, but also are reluctant to refuse such treatments for fear of becoming "hopeless cases." They may be afraid that others will no longer respond to their medical and emotional needs and abandon them. Medical technology, with its promise of prolonged health and human flourishing, thus can become a threat to such flourishing. Jim and his siblings must be helped to deal with the real limits of medical technology as well as its promise.

2.2.4 The ethics of prevention and preventive ethics

The final method to be examined is that of preventive ethics. The term entered the bioethical literature in 1993 (Forrow *et al.*, 1993) and can be defined as a proactive approach to ethical issues that develops policies and practices designed to prevent ethical problems from occurring rather than responding to them after they occur. With different nuances, the broad contours of this method have been utilized both in public health ethics and by the hospitals of the US Department of Veteran Affairs (VA). Within public health ethics, the ethics of prevention has been examined within the larger rubric of justice. Norman Daniels of Harvard University, for example, has chided the medical field for its fascination with biomedical technology geared toward the high-tech cure of disease and what he considers its consequent failure to deal with the social determinants of population health, including the reduction of health care risks that lead to disease (Daniels, 2008). He maintains that the primary social obligation is to protect health and not merely develop cures for sickness (Daniels, 2012). Similarly, the World Health Organization in its 2002 report on reducing health care risks described what it considered an appropriate decision-making process that includes monitoring possible health risks, identifying risk factors, and communicating prevention strategies (WHO, 2003).

From a different perspective, the VA has incorporated preventive ethics in its integrated ethics model, which has become a highly imitated model today. It developed the model as a response to what it perceived as a weakness of many of the other models of bioethics (including those mentioned above) that tended to deal with prohibited behavior rather than inspire beneficial ethical practices. It criticized these models for reacting to ethical problems rather than developing an ethical culture within the institution. The VA established its National Center for Ethics in Health Care to develop a "national, standardized, comprehensive, systematic, integrated approach to ethics in health care" (Fox *et al.*, 2007, p. 3).

The integrated ethics model endeavors to move beyond the presenting ethical issue to explore the systems and processes that exist in an organization that give rise to the ethical issue in the first place. The VA explains:

> Preventive ethics aims to produce measurable improvements in an organization's ethics practices by implementing systems-level changes that reduce disparities between current practices and ideal practices (Fox *et al.*, 2007, p. 7).

The model employs six steps to address ethical issues proactively and to improve the ethical culture of a health care institution. First, it encourages all employees to identify potential ethical issues that the institution ought to address, that is, "ongoing situations that involve organizational systems and processes that give rise to uncertainty or conflicts about values" (Fox *et al.*, 2007, p. 28). In a manner similar to that of Callahan, it stresses that every problem that the institution encounters is not necessarily an issue that can be addressed by ethics. Once these issues have been named and prioritized, the second step involves a study of the issue by all involved. Most medical and health care processes are complex and there may be a variety of possible solutions. Those involved would be asked to gather specific data about best practices and compare these data to the organization's current practices. The next step entails brainstorm strategies to narrow the gap between the current practice and the ideal that has been identified. The fourth step is to develop and execute a specific plan to carry out the desired strategy. The fifth step includes evaluation of how well the plan was executed to ascertain whether the desired results were obtained. The step would also raise the question regarding whether execution of the plan also had negative effects. Finally, once it has been determined that the plan was successful in narrowing the ethics gap, the changes would be systematically integrated into the standard operating procedures of the organization and implemented as widely as practical. The cycle would then begin again.

Returning to Jim's situation for a last time, one sees that both understandings of preventive ethics have an indirect bearing on the issues that Jim and his siblings face, even though they would not immediately affect the decisions that the family has to make. It is at least possible that had Jim's mother and her physician been more attentive to the ethics of prevention, they might have collaborated to find a way – for example, through diet and medication – to prevent the stroke in the first place. In the larger public health arena, there might have been a discussion regarding hypertension as a public health risk and a concerted effort to do something about it on a population level, since within this model, health care ethics is not simply about personal decisions or individual choices.

The other understanding of preventive ethics may have even more direct relevance, although Jim may never know it. Let us assume that Jim and his siblings have agreed on a course of action regarding his mother. Let us now suppose that the physician feels that the decision is premature. There is a difference of opinion, perhaps even an argument, and the family requests an ethics consultation. The ethics committee comes to realize that there have been a large number of similar consultations requested in similar situations in this particular department of the medical center. The committee might raise the question regarding why the same issue continues to surface in this particular ward of the medical center. It might begin the process mentioned above in an attempt to deal with such issues proactively and to develop further the ethical culture of the medical center.

2.2.5 Convergence or divergence?

Even though the methods of biomedical ethics discussed above were often developed in opposition to one another and continue to be part of an ongoing debate regarding appropriate methodologies within the discipline, they actually may all contribute to a fuller and more robust understanding of the breadth of the field of bioethics. Each of the methods comes from a particular perspective and raises questions that are important from that perspective. As one attends to these various perspectives and appreciates why a particular perspective gives rise to the ethical questions it does, one gains access to a range of necessary ethical issues that need to be addressed. Thus the methods discussed when taken together may in fact be more helpful to the discipline of biomedical ethics than any individual method used in isolation.

2.3 POSSIBLE LESSONS FROM MEDICAL ETHICS FOR THE ETHICS OF FOOD PRODUCTION

From this brief analysis of four methods in bioethics and the questions they raise for contemporary medicine, it might not be obvious that there are *any* lessons that one can learn for the ethics of food production. Nevertheless, it is often when one allows oneself to reflect upon perspectives different from one's own that a person is able to find answers to the questions at hand. We will therefore revisit the four methods to discover what, if anything, they might have to say to those involved in the ethics of food production. As mentioned in the Introduction, this section will be more suggestive of possible ethical questions and issues than a fully developed analysis. The application of methods in bioethics to food production is developed further in the next chapter, and most of the questions posed in this section are confronted in Part II of this book.

2.3.1 Differences and similarities

At first glance, there seem to be many more differences than similarities between the fields of medicine and food production, and therefore it would seem that the bioethical reflection on the field of medicine that I have developed has little to say to this industry. Food production is a complicated affair, involving many industries with a variety of levels of responsibility, including the growing of food and the raising of livestock, transportation, the processing of food, and selling it to the consumer. Each of these levels of production does not seem to be as closely related to the public good as medicine obviously is. It seems more appropriately associated with the free market, driven by supply and demand. Thus, if there is an ethical imperative associated with the goal of food production, it is to develop those technologies and processes to feed a growing population as efficiently as possible. Since much of food production is part of a competitive enterprise, it would also seem that food production ought to be as attentive to consumer preference as possible and also to be responsive to the needs and desires of shareholders.

The past few decades, however, have shown that there are important relationships between food production and issues central to biomedical ethics. There have been direct intersections between food production and issues of health and sickness, as, for example, concerns regarding the increase of *Escherichia coli* infections related to food and food production attest. In the last two decades the medical community has raised questions regarding the relation between the use of human growth hormone in livestock and possible links to breast cancer. Similarly, some have suggested that the non-therapeutic use of antibiotics in livestock has aided the development of disease-resistant strains of harmful bacteria. Even more recently, the media has declared that obesity has become an epidemic in the United States, as researchers have shown the link between obesity and increased rates of heart disease, strokes, diabetes and cancer. Although these statements are themselves controversial and demand further analysis, perhaps they demonstrate that the methods of bioethics mentioned above might have something to say regarding ethical questions related to food production.

2.3.2 Casuistry

The first set of questions relevant to the ethics of food production would probably best be presented within the context of casuistry. The adherents to this method of biomedical ethics might review the goals of food production mentioned earlier in this section and re-ask the question: To whom is the food industry responsible and for what is it responsible? The casuistic method might challenge the paradigm that seems dominant – that the food industry is responsible to the consumer and the shareholder to produce and deliver food to an increasingly large population as efficiently as possible, offering the most consumer choice and a return on investment to the shareholder. It would acknowledge that, according to this paradigm, the choices that have been made by the industry are understandable. But it would raise the question whether other paradigms are also appropriate. It might suggest that other paradigms would emphasize corporate responsibility to the common good. By using the latter paradigm, the ethical responsibilities of those involved in food production broaden. In enumerating these possible additional ethical responsibilities, one would have recourse to the other biomedical ethical methods discussed above.

2.3.3 A principle-based ethics

The four principles articulated by Beauchamp and Childress raise questions that might be relevant to the ethics of food production. For example, one might invoke the principle of *respect for autonomy* in either of two contradictory ways. One possible interpretation would be for the industry simply to respect consumer choice, that is, to give the customers what they want. Although this attitude does demonstrate a respect for autonomy, it does so only on a superficial level. When we looked at the use of this principle in medical ethics, we saw that it affects issues like freedom of choice, competence, and informed decision making. At the very least, it would seem to imply a responsibility to consumers that keeps them free from coercion (advertising and marketing), and gives them enough information to ensure informed choices.

The principle of *nonmaleficence* suggests that the food production industry has a responsibility to do no harm. Within this context, one of the areas that would need to be investigated further is the justification of the non-therapeutic use of antibiotics and hormones in livestock. The World Health Organization, for example, has developed principles regarding the overuse and misuse of antimicrobials in food animals, which it believes contributes to the emergence of resistant forms of disease-causing bacteria (WHO, 2000). The principle of *beneficence* would move beyond the question of harm and ask regarding positive contributions that the food industry should make in creating more nutritious products, for example.

The principle of *justice* raises the question of access to healthy food. It might raise questions regarding the locations of markets and determine whether disparities among populations are just or not. It might also ask whether the sorts of foods available to the poor or to racial and ethnic minorities are as wholesome as those offered to wealthier consumers.

2.3.4 The use of everyday language

At this point in our enumerating the variety of ethical questions that might be asked of those in the food production industry, it might be good to raise again Callahan's statement that not every medical issue is necessarily an ethical problem. As those trained to view the industry from the point of view of ethical analysis, we may in fact see ethical problems where they do not exist. Callahan's observation that medical ethics be humble also has a place in the ethics of food production.

Those dealing with the ethics of food production may also want to consider Callahan's analysis of the tasks of ethics, especially the third task of assisting people who need to make practical choices. What is the responsibility of those in food production to the ordinary practical choices that people make? Does the food industry have a responsibility to encourage people to make healthy choices? Those involved with the ethics of food production may need to look at the availability and cost of healthy foods, especially in poorer neighborhoods. They might also look to areas such as education and marketing, especially to children.

2.3.5 Prevention and preventive ethics

In many ways, the ethics of prevention has already been addressed by many of the questions raised in the previous sections. If, as Daniels claims, there is a primary social obligation to protect health, then the links between food safety and health should be analyzed even more closely by the food industry. Referring again to the principle of beneficence, it might not be enough simply to deal with sickness. Rather there needs to be a concerted strategy of keeping populations healthy. Food production becomes part of this larger strategy for creating healthier populations.

Attending to the VA's understanding of preventive ethics might raise the most difficult question for the ethics of food production: Does the food industry have the desire to do what is necessary to develop a culture of ethics within the industry? Such an ethics demands what in health care has come to be known as continuous quality improvement. Movement toward such a culture of ethics, dedicated to the improvement of the health and well being of the population can easily come in conflict with the development of those technologies and processes needed to feed a growing population as efficiently and cost-effectively as possible.

2.4 CONCLUSION

This brief essay on the methods of biomedical ethics and their possible relevance to the ethics of food production has now come full circle. We return to the debate between Tom Beauchamp and Stephen Toulmin, having noted both the usefulness of current ethical methodologies but also their limitations. As Toulmin has suggested, the social and cultural history of ethical reflection has in fact revealed a greater clarification of moral principles and practices to guide the industry, be it medicine or food production, in arriving at ethical solutions to issues presented. Yet, the task of ethics is always unfinished. As many ethical issues are clarified, other issues arise that may call into question solutions previously considered ethical. And so the methods of ethics continue to develop.

REFERENCES

Aristotle, *Nichomachean Ethics* II, 2, 1104a.

Beauchamp TL, and Childress JF, *Principles of Biomedical Ethics* (New York: Oxford University Press, 1979).

Beauchamp TL, and Childress JF, *Principles of Biomedical Ethics*, 2nd edition (New York: Oxford University Press, 1983), 66–102.

Beauchamp TL, and Childress JF, *Principles of Biomedical Ethics*, 4th edition (New York: Oxford University Press, 1994), 100.

Beauchamp TL, and Childress JF, *Principles of Biomedical Ethics*, 7th edition (New York: Oxford University Press, 2012), 352–354.

Callahan D, "Bioethics as a discipline," in Jecker NS, Jonsen AR, and Pearlman RA, eds., *Bioethics: An Introduction to the History, Methods, and Practice* (Sudbury, Massachusetts: Jones and Bartlett Publishers, 1997), 88.

Callahan D, *The Troubled Dream of Life* (Washington, DC: Georgetown University Press, 2000), 127–148.

Daniels N, "Treatment and prevention: what do we owe each other?" in Faust HS, and Menzel PT, eds., *Treatment Vs. Prevention: What's the Right Balance?* (New York: Oxford University Press, 2012).

Daniels N, *Just Health: Meeting Health Needs Fairly* (New York: Cambridge University Press, 2008), 102.

Donagan A, *The Theory of Morality* (Chicago: The University of Chicago Press, 1977).

DuBose ER, Hamel R, and O'Connell LJ, eds., *A Matter of Principles: Ferment in US Bioethics* (Valley Forge, PA: Trinity Press International, 1994), xiv.

Fletcher J, *Morals and Medicine* (Princeton, NJ: Princeton University Press, 1954).

Forrow L, Arnold RM, and Parker LS, "Preventive ethics: expanding the horizons of clinical ethics," *Journal of Clinical Ethics* 4, 4 (1993): 287–294.

Fox E, Bottrell M, Foglia MB, and Stoeckle R, *Preventive Ethics: Addressing Ethics Quality Gaps on a Systems Level* (Washington DC: National Center for Ethics in Health Care, 2007).

Gert B, *Common Morality: Deciding What To Do* (New York: Oxford Univeristy Press, 2004).

Hamel R, and Walter JJ, eds., *Artificial Nutrition and Hydration and the Permanently Unconscious Patient: The Catholic Debate* (Washington, DC: Georgetown University Press, 2007).

Jakobovits I, *Jewish Medical Ethics* (New York: Bloch Publishing Company, 1959).

Jones JH, *Bad Blood: The Tuskegee Syphilis Experiment* (New York: The Free Press, 1981).

Jonsen AR, *The Birth of Bioethics* (New York: Oxford University Press, 1998), 90–99.

Jonsen AR, "On the origins and future of the *Belmont Report*," in Childress JF, Meslin EM, and Shapiro HT, eds., *Belmont Revisited: Ethical :Principles for Research with Human Subjects* (Washington DC: Georgetown University Press, 2005), 3.

Jonsen AR, and Toulmin S, *The Abuse of Casuistry: A History of Moral Reasoning* (Berkeley: University of California Press, 1988).

Kelly G, *Medical Moral Problems* (St. Louis: The Catholic Hospital Association, 1957).

Lynn J, ed., *By No Extraordinary Means: The Choice to Forgo Life-Sustaining Food and Water* (Bloomington: Indiana University Press, 1986).

McCormick R, *Health and Medicine in the Catholic Tradition* (New York: The Crossroad Publishing Co., 1984).

Meilaender G, "Caring for the permanently unconscious patient" in Lynn J, ed, *By No Extraordinary Means*, (Bloomington: Indiana University Press, 1986), 196.

National Commission for the Protection of Human Subjects of Biomedical and Behavioral Research, *The Belmont Report: Ethical Principles and Guidelines for the Protection of Human Subjects of Research* (Washington, DC: US Government Printing Office, 1978).

Public Law 93-348, section 202[a][1][A].

Rahman F, *Health and Medicine in the Islamic Tradition* (New York: The Crossroad Publishing Co., 1987).

Ramsey P, *The Patient as Person* (New Haven: Yale University Press, 1970).

Rosner F, *Medicine in the Bible and the Talmud* (New York: Yeshiva University Press, 1977).

Sulmasy D, "End of life care revisited" in Hamel RP and Walter JJ, eds, *Artificial Nutrition and Hydration and the Permanently Unconscious Patient* (Washington, DC: Georgetown University Press, 2007), 187–199.

Toulmin S, "Tyranny of principles," *The Hastings Center Report* 11, 6 (December 1981).

World Health Organization, "WHO Global Principles for the Containment of Antimicrobial Resistance in Animals Intended for Food" (2000). Available at http://whqlibdoc.who.int/hq/2000/WHO_CDS_CSR_APH_2000.4.pdf (accessed 15 December 2012.

World Health Report 2002: Reducing Risks, Promoting Healthy Life (New York: The World Health Organization, 2003).

3 Ethical principles and the ethical matrix

Ben Mepham

3.1 INTRODUCTION

A common perception of ethics is that it deals with vague, rather subjective opinions, over which there is rarely any consensus – unlike scientific conclusions or technical assessments, which are based on hard, objective data. Consequently, it is often argued, ethical judgments carry little authority because most people are not convinced that they are soundly based. One reason for this perception may lie in confusion over the different ways in which the word *authority* is used.

People may be considered to possess authority for either of two, often contrasting, reasons. Thus, police and customs officers, and members of armed forces often exert authority because they perform roles in which they are required to ensure that individuals behave in accordance with the law or other enforceable directives. Ignoring or attempting to flout the authority they wield can result in severe penalties. The same applies in some undemocratic states in which, for example, the people in power impose on all members of society a strict observance of their particular interpretation of certain scriptural texts.

On the other hand, some experts, such as medical epidemiologists, who may have devoted considerable effort and skill over a long period in seeking to identify the causes of a serious disease condition, also possess undoubted authority. But in this case, while there is no compulsion to comply with their advice on, say, smoking cigarettes, you might be foolish to ignore it. Clearly, such experts "speak with authority" as a result of experience gleaned from reasoned, systematic research.

If authority has any place in respect of ethical considerations, it unquestionably conforms to the latter characterization, because arriving at and enacting ethical decisions must surely depend on reasoned conviction rather than externally imposed compulsion. But accepting that interpretation, suggests two important questions: "Are there objective criteria on which to base rational ethical deliberation?" and, if so, "Is it possible to acquire relevant expertise in reaching sound ethical decisions?" Or, to contextualize the issue, and put the latter question more bluntly: "Can the discipline of food ethics serve a useful role in promoting more ethical practices in the various sectors of the food industry?" Perhaps unsurprisingly, my answer to that question is affirmative; but it is

Practical Ethics for Food Professionals: Ethics in Research, Education and the Workplace, First Edition.
Edited by J. Peter Clark and Christopher Ritson.
© 2013 John Wiley & Sons, Ltd. Published 2013 by John Wiley & Sons, Ltd.

nevertheless an answer to which it is necessary to attach several caveats, the significance of which will become apparent in the course of this chapter.

3.2 THE COMMON MORALITY

The fundamental nature of ethics (*metaethics*) is a matter of continuing debate among philosophers, which is beyond the scope of this discussion. Even so, a definition that is likely to command support from virtually everyone is that ethics is concerned with moral behavior that is directed to achieving "the right and/or the good." Some people see ethics principally as a matter of our duty to live according to certain rules of virtuous behavior, others as the attempt to promote happiness and relieve suffering; and while these two motives may often suggest similar courses of action, in some cases the emphasis assigned to one or the other may result in quite different decisions.

This is perhaps a good point at which to draw a distinction between morals and ethics, two words often coupled in pronouncements that are designed to emphasize the gravity of such concerns. Although not absolutely definitive, it is generally accurate to use *morals* to describe a society's general attitudes to standards of acceptable behavior, whereas *ethics* refers to the disciplined, systematic enquiry into the nature of morality. According to this distinction, a third term, the *common morality*, has been introduced to identify the moral code shared by members of society in the form of unreflective common-sense and tradition. While some academic philosophers might disparage any acknowledgment of the role of the common morality in ethical deliberation, it is arguably a good starting point in seeking to arrive at ethically sound judgments – a presumption that is adopted here and it is one of the four methods of bioethical decision making discussed in Chapter 2.

3.3 ETHICAL THEORIES AND ETHICAL PRINCIPLES

For over two millennia philosophers have been debating the nature of ethics. This can be seen as the attempt to discern the theoretical underpinnings of one of the two big questions facing humanity, namely, "How should we act?" (the other momentous question is "What is all this?") – the answers to which are generally sought by forms of scientific enquiry.

Ethical behavior is clearly the result of considered judgments, arrived at voluntarily, with the aim of achieving rightful and/or good outcomes. Although a few people argue that we should act solely in our own interests (so-called *ethical egoism*), the vast majority of us believe that selfish desires (*psychological egoism*) should be subject to constraints in the interests of others, attitudes that might be characterized as ethical *obligations* that are, as Darwin put it, "summed up by that short, but imperious word *ought*" (Darwin, 1883).

Greatly simplifying the situation in the interests of brevity, two major theoretical strategies have emerged with respect to these obligations, namely those asserting the primary role of our duties to others (*deontological* theory) and those based on the motive of achieving desirable outcomes (*consequentialism*). While both of these strategies may be considered altruistic, it is also likely that most people would themselves prefer to live in a society not dominated by selfish individualism.

3.3.1 Deontological theory

This is most famously associated with Immanuel Kant, the 18th century German philosopher. Rather than making any assumption about the nature of the "good," Kant sought to establish, by a process of reasoning, principles that would be applicable irrespective of people's desires or social relationships. The distinctive features of Kant's theory (Kant, 1932) might be summarized as:

- Morality consists of the duty to perform rightful actions, such as always telling the truth and refraining from inflicting harm (described as *categorical imperatives*), without seeking to predict possible outcomes.
- Each person has a duty to respect the inherent dignity (or *autonomy*) of others, and to treat them as *ends in themselves*, and not as means to one's own ends (instrumentally).
- The rights possessed equally by each person (*universal rights*) imply each person's correlative *duty* to observe others' rights.

The categorical nature of such principles may be said to amount to a form of *moral law*, but in contradistinction to externally imposed laws, or those ascribed to scriptural sources, they are arrived at solely by a process of reasoning, and may be said to correspond to the ancient edict "Do as you would be done by," which although often attributed to Christian sources is actually represented in most ethical traditions (e.g. including that of Confucius in the 5th century BCE).

But straightforward as this theory is in principle, it encounters some serious difficulties in practice. For example, the principle of honesty would suggest that you (A) should never deceive another person. But what if that person (B) appears to have bad intents, such that B seems likely to seriously harm a third person (C) if you tell B of C's whereabouts? The duty not to cause harm is thus in conflict with your duty to "tell the truth" and most people would consider in this case that you should tell a (white) lie to B as to C's whereabouts. Deontological principles are thus sometimes in conflict, and might depend on a superhuman strength of will, or callous obstinacy, to fully observe.

3.3.2 Utilitarian theory

Utilitarian theory contrasts with deontological theory in relying on what Kant called *hypothetical imperatives*, in which the word "ought" applies to acting in ways that are aimed at achieving a satisfactory *result*. It is a form of consequentialist theory in which the object is to maximize utility, which is classically interpreted as the surfeit of happiness over unhappiness. The theory is usually attributed to Jeremy Bentham, the 18th century English legal and social reformer, who characterized human life as subject to two factors, pleasure and pain (Bentham, 1948). Thus, according to Bentham, ethical (i.e. morally "good") actions seek to achieve *the greatest good for the greatest number*. Because the intended end result affects the precise actions one performs, utilitarianism relies on a form of predictive cost–benefit analysis. But because happiness is a slippery concept, which e.g. might entail short-term thrills (like inebriation) or serious injury (e.g. from masochistic acts) utilitarianism has been refined in various ways e.g. by J.S. Mill, in the 19th century, to distinguish between different qualities of pleasure (Mill, 1910).

Utilitarian thinking is undoubtedly a common form of reasoning, certainly at the personal level. For example, seeking out the cheapest, safest and most comfortable means of travel to a holiday destination, where one will experience sun, sea and good food, involves cost/benefit strategies that correspond closely to those involved when politicians aim to provide social welfare in an efficient way. So in a sense, what makes this an ethical theory is the object of *maximizing* the perceived good, so that as many people as possible benefit. Distinctive features of utilitarianism are:

- The object is to maximize achievement of the good, while minimizing any harms.
- The end result is the driving force, so that in contrast to deontology strict adherence to categorical principles is unimportant.
- Some harms may be considered acceptable in the interests of achieving the best outcome.

However, although utilitarian ethical reasoning is, for most people, intuitive, it is difficult to define what it means in practical terms. For example, performing a cost–benefit analysis every time one has to make a decision (so-called *act* utilitarianism) would be impossibly demanding: usually, one simply would not have the time or information to conduct a sound analysis. This is why most utilitarians employ rules of thumb, acknowledging that because, for example, truth-telling is usually a sound principle it is unnecessary to weigh up whether to be honest every time you make a decision. This form is called *rule* utilitarianism. However, to complicate matters, many people would argue that telling a white lie out of compassion (e.g. in deceiving a dying person as to the imminence of their impending death) is often an ethical act – which challenges the honesty rule.

There are also other difficulties with utilitarianism. For example, who (or what) is to be included in the cost–benefit analysis, and for what time period is the analysis to apply? More specifically, in considering the impacts of a new food technology, is the scope to be assessed from global, national or regional perspectives? Does it include impacts on all possible consumers, or only some; are the interests of animals part of the analysis; are impacts on the physical environment to be considered; and what attention should be paid to future generations of people, fauna and flora? It is evident that, like deontology, in the real world any notion that utilitarianism is a straightforward guide to ethical action is simplistic.

3.3.3 Justice as fairness

As noted, the above two theories were specifically formulated in the 18th and 19th centuries, but a third theory of ethics, that has become prominent in more recent times, lays emphasis on fairness. While fairness was clearly not invented in the 20th century, the political scientist John Rawls made it a central plank of the philosophy described in his highly influential book *A Theory of Justice* (Rawls, 1972). In a world in which citizens are motivated by both egoism and altruism, Rawls argued for a principled reconciliation of the demands of liberty and equality, so that limited resources might be allocated in ways that are mutually acceptable (i.e. fair).

Rawls suggested that his notion of *justice as fairness* would be acceptable to free and rational people concerned to further their own interests (i.e. it is assumed that most of us are, to some degree, egoistic). Thus, according to Rawls, fairness in the allocation of scarce resources would be facilitated if group decisions were made when personal characteristics (age, gender, race, intelligence, social status, wealth, physique, etc.) were screened out of the deliberative process by conducting it behind a notional *veil of ignorance*, in what he designated the *original position*. Rawls claimed that in trying to work out what would be a fair way of behaving in a liberal democracy, if rational people forgot who they actually were and imagined themselves to be members of the most disadvantaged group, they would be likely to arrive at the following two principles:

- *Equal liberties for all*: i.e. each person should have as much liberty as is consistent with other people having the same amount.
- *The difference principle*: this ensures equality of opportunity, but restricts social and economic inequalities to those that would benefit the least advantaged members of society. (For example, while a brain surgeon might earn a very high salary, this might be justified if her services were made readily available to impoverished members of society.)

The theory is a form of social contract, other forms of which were discussed by several earlier philosophers: but here it assumes a more egalitarian nature consistent with the emergence of Western democratic values. Over the last 50 years or so, rapid social changes have occurred throughout the nations of the world, albeit very unevenly, which have profoundly affected the common morality. For example, there have been substantial changes in attitudes to race, gender and sexual orientation: and these have led in most western democracies to laws prohibiting adverse discrimination on such grounds. The numerous revolutions in Arabian states which began in early 2011 (constituting the so-called Arab Spring) are a graphic demonstration of the public demand for equivalent reforms in these countries.

While the above discussion has focused on personal behavior, in his later work Rawls emphasized the point that:

> Justice is the first virtue of social institutions as truth is of systems of thought. A (scientific) theory, however elegant and economical, must be rejected if it is untrue; likewise laws and institutions, no matter how efficient or well arranged, must be reformed or abolished if they are unjust (Rawls, 1972, p. 3).

Conforming to this standard is increasingly challenging as societies become more diverse and multicultural, in circumstances in which "a plurality of incompatible and irreconcilable doctrines – religious, philosophical, and moral – coexist within the framework of democratic institutions" (Rawls, 1993).

3.3.4 Prima facie principles

It is clear that despite all attempts to produce a universally applicable, comprehensive theory of ethics, no one has so far devised a theory that has found widespread support: in practice, all theories sometimes appear inconsistent, incomplete or impracticable. So what is the best strategy? How do we make the best of a bad job?

One promising approach is to acknowledge that all the above theories (and others not discussed here, but in some cases referred to in Chapters 1 and 2) might well make important contributions; and that a wise combination of their insights might provide a useful framework for ethical decision-making. Thus, in the 1930s the Oxford philosopher David Ross noted that an effective way of dealing with the problems inherent in classical deontology and utilitarianism was to conceive of ethical judgments as based on prima facie principles ("at first sight"), which allow a stronger case to overrule weaker ones in particular circumstances (Ross, 1930). So this an acknowledgment that even if we possessed the moral fiber that supposedly characterizes sainthood, aiming to live according to ethical principles in a modern, multicultural society necessarily entails making compromises, and that well-meaning people will differ in both their assessment of the nature of the problem to be addressed and the weight to be assigned to different ethical principles in addressing it. Behaving ethically might then be defined as seeking to do the "right" or "ethically best" thing, *all things considered*. It also needs to be appreciated that, to some degree, ethical judgments are often subject to personal intuitions, or to express that point differently, they are influenced by tacit knowledge (i.e. understanding which it is difficult to express verbally; Polanyi, 1969).

3.4 ETHICS AND THE FOOD INDUSTRY

The principal aim of this chapter is to suggest an approach to arriving at justifiable ethical decisions by committees concerned with the prospective employment of new food technologies, a term which is intended to apply to all aspects of the food chain, from "field to fork." Perhaps the most distinctive feature of this approach is that instead of focusing on interpersonal matters, it casts the net much more widely by seeking to assess the effects of the prospective technology on different categories of person (such as producers and consumers), on farm animals and on wildlife (the biota).

Ross's prima facie principles have been adapted by Beauchamp and Childress in a principled approach to medical ethics, the development of which over a period of thirty years is described in successive editions of their *Principles of Biomedical Ethics* (as outlined in Chapter 2). In essence, they argue that ethical deliberations on specific cases of medical practice are facilitated by systematically assessing the prospective impacts of different decisions on four prima facie principles (Beauchamp and Childress, 2008) viz:

- *Nonmaleficence*: cause no harm (essentially the Hippocratic Oath of 4th century bce Greece).
- *Beneficence*: effect a cure (or provide palliative relief).
- Respect for *autonomy*: treat the patient as a person, and not just as a medical case.
- *Justice*: treat patients fairly (e.g. without racial, sexual or age discrimination).

These principles need to be specified for each particular case, and almost inevitably each decision will entail *weighing* the significance and/or likely impact of respecting each principle fully. Unsurprisingly, this analysis will invariably involve assigning unequal weights to the different principles, and there is no guarantee that members of a committee will concur in their assessments. Even so, the conscientious conduct of this structured deliberative process should guarantee a transparent procedure and an explicit rationale for decisions which are arrived at.

In the simplest case, decisions in medicine entail consideration of the interests of only two people – the patient and doctor. But this scenario is hypothetical in the extreme, because, for example, effects on patients' relatives, other patients, other medical staff, and the alternative uses of scarce resources are all important considerations that should be factored into ethical decision making. (This is evident in the illustrative case study deployed in Chapter 2.) If, however, we want to apply this principled approach to the field of food and agriculture, the interest groups, and the number of individuals in each, are greatly increased.

3.5 THE ETHICAL MATRIX

For this reason, I proposed some years ago that displaying the relevant principles (in appropriately specified forms) and the relevant interest groups in a table (an *ethical matrix*) would facilitate the deliberative process. In the ethical matrix (hereafter EM), three principles are employed: these are "respect for":

- wellbeing
- autonomy
- fairness.

These are applied to the interests of the different groups relevant to the issue being analyzed. The three principles were chosen to represent the major traditional ethical theories: i.e. respect for wellbeing represents the major utilitarian principle; respect for autonomy represents the major deontological principle; and respect for fairness is important to both the utilitarian and deontological traditions, but also encompasses the fundamental tenet of modern social contract theory.

While it is possible to discuss the EM in generalized terms, it is more effective to apply it to a particular case: and the case chosen here is the use in dairying of bovine somatotrophin (bST), which is also called bovine growth hormone (BGH), especially in the USA. This genetically engineered pituitary gland hormone was the first to be used commercially in animal agriculture, but the contrasting ways in which different governments have reacted to it provide a graphic illustration of the relationships which exist between bioethics and international politics.

In essence, the EM seeks to take account of two essential factors – the relevant prima facie principles and identification of the agents that have "interests." Appeal to principles reminds us of the overarching considerations that need to be taken into account. The appropriate list of agents with interests depends on the nature of the issue to be analyzed, but in this case it will certainly include different human interest groups (e.g. consumers and farmers); dairy animals, and the regional biota (flora and fauna) – because bST use can have significant social and environmental consequences. A practicable framework for ethical analysis is inevitably a compromise between competing requirements, because it needs to:

- be based in established ethical theory to give it authenticity
- be sufficiently comprehensive to capture the main ethical concerns
- employ user-friendly language as far as possible.

Inclusion of all three prima facie principles in the framework acknowledges the plurality of perspectives that sincere people bring to an ethical analysis, and provides a means of registering the importance of each principle in any particular context. (It should be noted that in this scheme *wellbeing* combines respect for beneficence and non-maleficence, which are given separate identity by Beauchamp and Childress.)

Some people, explicitly or implicitly, appeal to another theory, *virtue theory*, which is described in detail in Chapter 1. This emphasizes the special place of observing virtues in moral life, an approach which puts emphasis on the *person* making ethical choices rather than on the *situations* in which choices have to be made. But it is arguable that those whose primary motive is to live a life of virtue still have to decide what to do to in order to act virtuously, and it is here that a principled approach is often valuable.

3.5.1 The case of bST

We need first to consider some facts. Injecting cows every two weeks with bST generally stimulates an average increase in milk yields of 12–15%; and, although slight changes in nutrient content can be produced, their overall concentrations in bulked milk are probably largely unaffected. However, because higher metabolic demands may lead to increased rates of illness, there is an increased risk that the welfare of injected cattle will be diminished. The treatment also leads to an increase in the concentration in the milk of insulin-like growth factor 1 (IGF-1), which is a potent mitogen. If the increased milk concentration of IGF-1 was physiologically significant and if it were to remain biologically active at the level of the gut mucosa (a claim contested by some scientists), it might pose a public health threat to people consuming the milk or dairy products.

Table 3.1 shows how the use of an EM can help to summarize the ethical issues raised by this technology in a systematic way, which is based on the principles that comprise the common morality. Box 3.1 describes in more detail the ways in which the different principles are specified for each of the four identified interest groups. A more extensive analysis is provided elsewhere (Mepham, 2008), but it is important to stress that this is employed as an illustrative example of the use of the EM, which refers to the situation which existed (in the late 1990s) at the time decisions were being made on whether or not to license bST for commercial use. Consequently, more recent developments, which may or may not have influenced the ethical assessments, are not relevant to the issues under discussion here.

Table 3.1 The ethical matrix applied to use of bST in dairy farming. For details see text. For dairy farmers and consumers, both consequences and duties need to be considered, whereas for dairy cows and the biota only consequences are involved.

Respect for:	Wellbeing	Autonomy	Fairness
Dairy farmers	Satisfactory income and working conditions	Managerial freedom of action	Fair trade laws and practices
Consumers	Food safety and acceptability. Quality of life	Democratic, informed choice e.g. of food	Availability of affordable food
Dairy cows	Animal welfare	Behavioral freedom	Intrinsic value
The biota	Conservation	Biodiversity	Sustainability

Box 3.1   More detailed specification of the principles in an ethical matrix for bST use (see Table 3.1)

Dairy farmers

Wellbeing: satisfactory incomes and working conditions for farmers and farm workers: (*satisfactory* is obviously debatable, but it is a better word than *adequate*, which might imply "just enough to meet bare necessities")

Autonomy: allowing farmers to use their skills and judgment in making managerial decisions, e.g. in choosing a farming system

Fairness: farmers and farm workers receiving a fair price for their work and produce, and being treated fairly by trade laws and practices

Consumers

Wellbeing: protection from food poisoning (and harmful agents e.g. residues of veterinary drugs); this also refers to the quality of life citizens enjoy as a consequence of a productive and profitable farming industry

Autonomy: a good choice of foods, which are appropriately labeled, together with adequate knowledge to make wise food choices; this principle also encompasses the citizen's democratic choice of how agriculture should be practiced

Fairness: an adequate supply of affordable food for all, ensuring that poverty does not cause hunger

Dairy cows

Wellbeing: prevention of animal suffering; improving animal health; avoiding risks to animal welfare

Autonomy: ability to express normal patterns of instinctive behavior, e.g. grazing and mating

Fairness: treated with respect for their intrinsic value as sentient beings rather than just as useful possessions (instrumentally)

> ## The biota
>
> *Wellbeing:* protection of wildlife from harm (e.g. by pollution), with remedial
>   measures taken when harm has been caused
> *Autonomy:* protection of biodiversity and preservation of threatened species (and
>   rare breeds)
> *Fairness:* ensuring sustainability of life-supporting systems (e.g. soil and water) by
>   responsible use of non-renewable (e.g. fossil fuels) and renewable (e.g. wood)
>   resources; cutting greenhouse gas emissions

The first thing to appreciate is that the specifications in the cells (Table 3.1 and Box 3.1) set criteria which would be met if the principles concerned were *respected* by a proposed action. In one of the commonest ways in which the EM is used, the impacts of the action, in this case injecting cows with bST to increase their milk yields, are compared with the conditions when bST is not used – so the status quo represents the baseline condition. Because some ethical impacts might be *positive* (e.g. an increase in the incomes of dairy farmers using bST, so respecting their wellbeing) they could be 'scored' positively (e.g. +1 on a scale of +2 to –2). On the other hand, some impacts might be negative (e.g. the cows' welfare would be *infringed* if the additional metabolic load led to more cases of lameness), so that they could be scored accordingly (e.g. –1); while some impacts might be insignificant (and so recorded as 0). A fully scored EM would thus show a total of 12 scores indicating the perceived ethical impacts of bST use on the three principles applied to the four interest groups. This assessment is referred to as an *ethical analysis*.

It must be stressed that numerical scoring means no more than use of the adjectives *very* (for +2 and –2) and *quite* (for +1 and –1) – and if it were thought that numbers give a deceptive impression, those or similar words could be used instead. But some means of grading responses seems necessary. It is also important to emphasize that it is not possible from this analysis to directly deduce the ethical acceptability, or otherwise, of bST use, for at least two reasons.

First, the different principles will have different degrees of significance for each assessor, i.e. the different factors carry different *weights*. The next step in the process, *ethical evaluation*, involves subjectively weighing the different impacts, allowing one to reach an ethical judgment on the acceptability of bST use. A second reason why the analysis does not assess overall ethical acceptability is because it compares the impacts of two situations, neither of which might be ethically acceptable by comparison with an option which differs from both. In other words, a system adjudged marginally more ethically acceptable than another according to the analysis might still fall far short of a (third) system which has not been investigated. For example, arguably, a system of dairying that prioritized animal welfare, such as organic farming, would be a better baseline against which to compare bST use than conventional dairying systems, which already experience significant problems with animal diseases related to high productivity. What you get out of the EM is totally dependent on what you put in.

Many of the specifications in Table 3.1 and Box 3.1 are self-explanatory, but additional comment will be helpful in some cases, as follows.

Respect and infringement of principles

In theory, for any proposed action, all the principles specified in the individual cells of the EM might either be respected or infringed. But positive and negative scores do not necessarily balance each other, even for a single specification. Thus, the duty "not to harm" (non-maleficence in the terminology of Beauchamp and Childress) might be thought to be more compelling than the duty to "'do good" (beneficence). For example, in the case of bST, the duty not to harm cattle is often considered much more important than the duty to improve their lot. In some cases, it might be thought preferable to maintain the distinction between between the two (Mepham *et al.*, 1996), but this has the practical disadvantage of complicating the EM by doubling the number of columns. So it is important to bear in mind in using the EM, as shown in Table 3.1, that just as different principles often carry different weights, so also can positive and negative effects for a single principle.

Clarification of consumer autonomy

Autonomy in this context is about liberty – being able to choose the sort of food you eat and how it was produced. In many respects these are citizens' concerns and not just consumers' concerns, because you might have legitimate views on how a food is produced irrespective of whether you consume that particular food.

But the liberty of individuals, like most things, must have limits. For example, in a shrinking world, how free should the *haves* in rich countries be to appropriate the Earth's resources, pollute the environment, and exploit cheap labor overseas to the detriment of the *have-nots* in both developed and developing countries? Decisions about consumer autonomy may thus have important political consequences, and are rarely simply confined to consideration of whether or not a food bears an informative label.

Clarification of an animal's intrinsic value

The principle of fairness applied to farm animals (here, dairy cows) is specified as respect for *intrinsic value*, a term meriting further explanation. Some things (e.g. stethoscopes and taxis), that are valuable because of their usefulness, have *instrumental value*. By contrast, intrinsic value is assigned where it is possessed irrespective of any usefulness; and most of us share the fundamental belief, stressed by Kant, that all humans have intrinsic value. But most people sometimes *also* have instrumental value, so that possession of the two types of value is not mutually exclusive. For example, doctors, taxi drivers and street cleaners all perform useful tasks, making them of instrumental value. This does not raise an ethical concern if they do their jobs by choice, and receive a fair income.

Attributing intrinsic value to dairy cows makes the assumption that in addition to their instrumental value in providing milk, they are also 'subjects of a life' which we have a duty to respect (Regan, 1983). That is, they have *ethical standing*. Given all that we now know about the sentience, sensibilities, and even "personalities" of cows, it would be unfair to regard them simply as useful objects. Recent legislation gives official recognition to this concept. For example, the 1999 Treaty of Amsterdam requires that animal sentience and welfare are recognized in the implementation of EU legislation and this book contains a chapter on "The Humane Treatment of Livestock" (Chapter 7). An earlier idea that animals have merely instrumental value now seems totally discredited.

Conservation, biodiversity and sustainability of the biota

When the principles of wellbeing, autonomy and fairness are applied to the biota, a different way of assessing the impacts is adopted to that used when considering individual farm animals. The reasoning here is that ethical impacts on the biota are concerned with life in the wild and on the collective scale – as populations, species and breeds. Consequently, the principles need to be specified quite differently to be at all consistent with the common morality – for it is commonly recognized that Nature does not respect human assessments of "fairness;" and we cannot protect the interests of prey without at the same time endangering the lives of predators. The point has been expressed by saying that:

> Our relationship with wild animals arises out of an environmental ethic, which ... can only be 'eco-centric', that is, it must not assign value to natural beings themselves but rather to their diversity and to the ecological systems on which they depend" (Larrére and Larrére, 2000).

This suggests that the appropriate specifications of the principles for the biota are conservation, biodiversity and sustainability – which are all prominent environmental concerns. The rationale for translating respect for autonomy as biodiversity is that it may be seen as permitting the natural ecological interplay of the biota. Sustainability represents fairness in an intergenerational sense, by respecting the biotic impetus for survival. Unavoidably, there is often overlap between the specified principles, but this seems less important than risking gaps in the issues that need consideration. Such designations might also be regarded as rather imaginative or figurative; but the more important question would seem to be 'do they address the crucial ethical concerns?'

3.5.2 The content of the cells

The factors in each cell of the EM which are relevant to performing an ethical analysis of the impacts of bST are forms of *evidence*, that are of two major types. In some cases factual (quantifiable) evidence is required, for example, the need to know what increases in milk yield are obtained when bST is injected, whether any effects on the chemical nature of the milk have implications for consumer health, and whether the welfare of the injected animals is affected. But the nature of the "facts," whether they were obtained reliably, and whether they are relevant to the question in hand, are all matters over which there is sometimes disagreement. Even the scientific theory considered by some to justify the particular data examined may be questioned, and if the source of the data is thought to be biased (e.g. if a commercial company was relied on to produce the key data supporting their own product, or if the data were produced by a pressure group known to be ideologically opposed to the product) neutral observers might suspect that the evidence is unreliable. Assessing evidence may thus entail examining different versions of the facts where there is controversy.

In contrast to factual data, other cells of the EM require a judgment that is not dependent on the quantifiable consequences of bST use but instead concerns evidence that affects human *values*. For example, in the pursuit of economic objectives, is it right to treat animals instrumentally by chemically altering their metabolism; or is it right to take risks with human health when appropriate scientific evidence is unavailable? These considerations are necessarily more subjective, but probably no less significant.

3.5.3 The ethical matrix as an ethical map

It is important to appreciate that the aim of the EM is to facilitate rational decision making but not to determine any particular decision. To avoid confusion it might be preferable to regard the EM more as a *map* than a framework. It is, after all, a pluralist tool, which seeks to identify society's whole ethical terrain. So, far from constraining ethical reasoning, the EM provides a vehicle for the expression of the full range of ethical perspectives. Evidence in support of this claim is that both people approving of bST use and people opposing its use can use the EM to justify their differing opinions, as has been demonstrated in workshops conducted with experts (Mepham and Millar, 2001). This indicates two important points about the EM:

- It provides a means of explaining and justifying different ethical positions.
- It facilitates identification of the areas of agreement and disagreement.

3.5.4 Ethical evaluations of bST use

A summary of the lines of evidence (facts and values) which have been presented for the different cells of the EM is shown in Box 3.2. According to different interpretations of the importance to be attached to this evidence, the governments of the USA and the EU reached opposing decisions on the acceptability of licencing bST for commercial use. Although in neither case were the decisions expressed explicitly in terms of ethical acceptability, it is clear that each would be justified, if it was requested of their supporters, in ethical terms: hardly anybody admits to acting unethically. We can thus summarize the two positions, according to the ethical criteria that have been defined.

Box 3.2   A brief analysis of bST use in dairying with reference to Table 3.1 and Box 3.1

Dairy farmers

Wellbeing: Some USA farmers using bST increased their profits but economic data suggested others used it at a loss

Autonomy: Farmers in the USA had an opportunity to increase productivity, but some might have felt economically obliged to use bST (exemplifying the so-called *technological treadmill*)

Fairness: Farmers in the USA were given the option of using a productivity-boosting technology. Farmers not using bST were permitted to label milk accordingly, but generally at their own expense

Consumers

Wellbeing: An EU report by public health experts suggested possible (but poorly defined) health risks of consuming increased amounts of IGF-1 (whose concentration increases in milk of treated cows). An FAO/; committee denied any significant health risk

Autonomy: The fact that in the USA most milk was unlabeled denied consumers a choice on whether to purchase milk from treated cows

Fairness: There appeared to be no clear evidence of an impact on milk prices

Dairy cows

Wellbeing: Evidence suggested that cattle would suffer increased disease rates (such as mastitis, lameness, metabolic and digestive disorders), as noted on the product label, which listed 21 possible adverse side effects. The EU banned bST largely on animal welfare grounds, but according to the manufacturers the diseases were treatable by medication

Autonomy: Behavior may have been adversely affected by lameness, by reduced grazing opportunities due to increased concentrate feeding, and by decreased fertility

Fairness: Some people claimed that the excessively instrumental use of cows was an infringement of their intrinsic value. Others claimed that the technology accorded with accepted social norms

Biota

As *quantitative data were lacking, claims were largely speculative.*

Claimed *positive* features of bST use: reduced cow numbers (because fewer cows were needed to produce a required milk yield) would reduce both environmental pollution (e.g. reduced fertilizer use for forage growth and silage run off) and greenhouse gas emissions (methane is exhaled by ruminants)

Claimed *negative* features of bST: mergers in the dairy industry (as non-user farmers left the industry) would result in fewer but much larger dairy farms, increased point-source pollution (e.g. excess fertilizer use, silage run off) and threats to biodiversity and sustainability by reliance on fossil fuels for fertilizer production etc and routine veterinary medication.

Note: In the USA, bST was licenced for commercial use in 1994. In the EU in 1999, an earlier moratorium on its use was extended indefinitely. These comments refer to the situation at these times. FAO, Food and Agriculture Organization; WHO, World Health Organization.

Approval of bST

The ethical acceptability of bST use for those who have licensed it (e.g. the USA) would probably cite the need to respect farmers' freedom to innovate; and the economic benefits to the manufacturers of bST, to the economies of countries producing it, to the farmers using it, and, were prices to fall, to consumers of dairy products. Moreover, if its use led to reduced cow numbers it might result in marginally reduced emissions of methane. This case also rests on perceptions that the welfare of treated cows is not affected significantly (or that increased disease can be effectively treated) and that there are no risks to human safety, so that labeling is unnecessary. Job losses in the dairy industry would not be seen as an ethical issue, being merely a feature of market economies, in which competition guarantees efficient production.

Rejection of bST use

The ethical case of those who have banned bST use (e.g. the EU) would probably focus on respects in which it appears to infringe commonly accepted ethical principles. They would point to authoritative reports suggesting that bST use substantially increases the risk of pain and disease in dairy cows, and that in view of scientific uncertainties it might present a risk to human safety through ingestion of increased IGF-1 in milk. Moreover, they might consider that bST use would reduce farmers' autonomy; undermine consumer choice if milk products from treated cattle were not labeled; jeopardize public health if rejection of dairy products followed the licencing of bST (because milk is a valuable source of dietary nutrients); and increase local pollution through the intensification of dairying.

3.5.5 The ethical matrix in decision making

The above account provides a guide to identifying relevant issues in reaching a judgment on a matter of bioethical concern. But employing a suitable tool for ethical analysis does not guarantee a genuine ethical evaluation. If users adopt a partisan position on the issue, for example, allowing bias to influence the choice of scientific data, then the tool is unlikely to prove of any real value.

At its simplest, the EM is merely a checklist of concerns, which happen to be based on ethical theory. But it can play a more important role by serving as a stimulus to ethical deliberation, and as the basis of ethical decision-making. It seeks to do this by establishing a coherent and consistent approach, that gives due attention both to objective facts and human values.

The main aim of the EM is to encourage, in the phrase trumpeted by contemporary politicians, *joined-up thinking.* The necessity to consider how narrowly-focused interests interact with a wide range of other factors which are considered of value in society can only have beneficial effects. It should be no more acceptable to 'fudge' an ethical valuation than it is to fabricate experimental data.

However, it is important to emphasize that:

- The EM is not prescriptive: the fact that different people weigh the cells differently precludes its providing a definitive decision on ethical acceptability.
- Very few, if any, decisions people might reach using the EM could afford equal respect to all the ethical principles, so that in most cases some may need to be overridden by others, or respect for some only partially discharged.
- The EM is designed to facilitate, but not determine, ethical decision making by making explicit the relevant ethical concerns and providing a reasoned justification for any decisions made.
- Contrary to a (perhaps common initial) suspicion that the EM necessarily complicates decision-making (with so many issues to consider), it can often simplify matters, because when *all* the important factors are brought into the frame a *single* ethical decision might suggest itself as inevitable.

3.5.6   Other examples of EM use

The EM has been used to address a number of issues concerning food. For example, Kaiser and Forsberg have used it in public consultations about the future of the Norwegian fishing industry. Such exercises were organized to include representatives of the major stakeholder groups, or – when the interest groups were non-human – those with particular expertise and/or commitment to their cause, such as animal welfare and environmental groups. Kaiser and Forsberg (2001) ascribed the value of the EM in these exercises to the following features: namely:

- It is liberal regarding the approach to be adopted, enabling it to be read equally as a utilitarian or a deontological approach.
- It provides substance for ethical deliberation, guiding participants so that they do not stray into irrelevant paths.
- It translates abstract principles into concrete issues of direct concern to participants who may have little acquaintance with, or interest in, ethical theory *per se.*
- It facilitates extension of ethical concerns into fields benefiting from debate, such as democratic decision making.
- It captures the basic fact that because different stakeholders will be affected differently by a decision their ethical evaluations may well differ. The object is not to downplay these differences but to search for an optimal solution in the light of the conflicts).

Some other examples of employment of the EM are:

- The European Academy's investigation into the ethical and social issues associated with the development of functional foods employed the EM to structure its deliberations (Chadwick *et al.*, 2003).
- Use by the Food Ethics Council in several of its reports (e.g. FEC, 2001).
- An interactive web-based program concerning three animal production systems, designed for student use (Mepham and Tomkins, 2003).
- Use in two recent encyclopedia articles on agricultural ethics (Mepham, 2012a) and food ethics (Mepham, 2012b), respectively.
- Chapter 10 of this book contains a partial application of it to worker exploitation.

It is apparent that the adoption of the EM in ethical deliberation has now become quite common. Thus, a recent internet (Google) search (in December 2012) recorded over 20 000 hits for the term "Mepham's ethical matrix," although because of a degree of replication this undoubtedly exaggerates the actual number of different uses.

3.6 SUMMARY

The following bullet points summarize the argument advanced in this chapter.

- The complexity of the current global food system, at all stages from agricultural production to food consumption and assimilation, demands that the food industry pays due attention to the implementation of ethically sound practices.
- Despite the persistence of widespread cynicism and misunderstandings about the merits of applying ethical theory to practical concerns, there is evidence that facilitating ethical deliberation by employing the ethical matrix can prove highly effective in many contexts.
- Based on an approach that evaluates the roles of prima facie ethical principles applied to the interests of relevant groups (e.g. consumers, producers, retailers, farm animals and wildlife), the ethical matrix seeks to arrive at ethical judgments that are transparent, comprehensive and readily comprehensible to people who, although without formal training in philosophy, are willing to engage with the issues with an open mind.
- The way the ethical matrix has been used is exemplified in this chapter by a case study, the use of a milk production stimulant (bST), which illustrates the different ways ethical assessments have been evaluated in the USA and EU, respectively.
- The increasing use of the matrix approach, based on evidence from internet searches, suggests that it is proving to be an adaptable and effective conceptual tool in addressing a wide range of issues.

REFERENCES

Beauchamp, T & Childress, J (2008) *Principles of Biomedical Ethics*. 6th edn. Oxford University Press, New York and Oxford.

Bentham, J (1948) [1823] *A Fragment on Government and Principles of Morals and Legislation*. Blackwell, Oxford.

Chadwick, R, Henson S, & Moseley B (2003) *Functional Foods*. Springer-Verlag, Berlin.

Darwin, C (1883) *The Descent of Man*. Appleton, New York. p. 97.

Food Ethics Council (2001) *After FMD: aiming for a values-driven agriculture*. FEC, Southwell, UK.

Kaiser, M & Forsberg, E-M (2001) Assessing fisheries – using an ethical matrix in a participatory process. *Journal of Agricultural and Environmental Ethics* 14, 1912–00.

Kant, I (1932) [1785] *Fundamental Principles of the Metaphysic of Ethics*. Longmans, Green and Co London.

Larrére, C & Larrére, R (2000) Animal rearing as a contract? *Journal of Agricultural and Environmental Ethics* 12, 515–8.

Mepham, B (2008) *Bioethics: an introduction for the biosciences*. 2nd edn. Oxford University Press, Oxford, pp. 456–6.

Mepham, B (2012a) Agricultural ethics. In; Chadwick, R. *Encyclopedia of Applied Ethics*. 2nd edn. Vol 1. Academic Press, San Diego, pp. 869–6.

Mepham, B (2012b) Food ethics. In; Chadwick, R. *Encyclopedia of Applied Ethics*. 2nd edn. Vol 2. Academic Press, San Diego, pp. 3223–30.

Mepham, TB & Millar, KM (2001) The ethical matrix in practice: application to the case of bovine somatotrophin. *EurSafe 2000: Food safety, Food Quality and Food Ethics*. Preprints. pp. 317–319 A & Q, Polo per la Qualificatzioni del Sistema Agroalimentaire, University of Milan.

Mepham, TB, Moore CJ & Crilly RE (1996) An ethical analysis of the use of xenografts in human transplant surgery. *Bulletin of Medical Ethics* 116, 131–8.

Mepham, B & Tomkins, S (2003) *Ethics and Animal Farming*. Compassion in World Farming, Petersfield. [Online] Availablefrom www.ethicalmatrix.net (accessed 11 December 2012).

Mill, JS (1910) [1863] *Utilitarianism, Liberty and Representative Government*. JM Dent, London.

Polanyi, M (1969) *Knowing and Being*. Routledge and Kegan Paul, London.

Rawls, J (1972) *A Theory of Justice*. Oxford University Press, Oxford.

Rawls, J (1993) *Political Liberalism*. Columbia University Press, New York, p xvi.

Regan, T (1983) *The Case for Animal Rights*. University of California Press, Berkeley.

Ross, WD (1930) *The Right and the Good*. Oxford University Press, Oxford.

4 An East Asian perspective on food ethics: implications for childhood obesity in mainland China

Vinh Sum Chau

4.1 INTRODUCTION

This chapter offers a view of food ethics in East Asia, as premised on its dominant cultural values and religious influences. The ethical issue of concern relates to the marketing of junk food (from the perspective of the fast-food industry) leading to obesity, particularly childhood obesity in mainland China. In particular, this chapter determines a marketing strategy framework to provide insights for managing a range of objectives to which human beings are cognitively acquainted. The chapter begins with an explanation of what these cultures and religious influences are, and then discusses how the practice of food marketing is affected by them. The chapter then explains the growing problem of childhood obesity and its causes in China, and then models this from the perspective of potential human behaviour. Drawing from the food marketing management literature, how Chinese cultural values can lead to purchases by (or for) children is then discussed. Lastly, the chapter ends with a discussion of the potential exploitation of the dominance of Chinese cultural values by fast-food businesses, and argues that this is one reason for the trend in childhood obesity.

4.2 CULTURAL VALUES (AND RELIGIOUS INFLUENCES) IN EAST ASIA

A conventional and dominant way by which to examine the national culture of a country is through Hofstede's (1980) dimensions of national culture. In this approach, it is argued that five dimensions characterize a country: (1) power distance (the degree of inequality national culture considers normal); (2) individualism vs collectivism (the extent a culture thinks it is appropriate for people to look after themselves); (3) masculinity vs feminism (the acceptable balance between dominance, assertiveness and acquisition, and regard for people's feelings and quality); (4) uncertainty avoidance (the degree of preference for structured vs unstructured situations); and (5) long-term vs short-term orientation (saving/persistent to reach a future vs present, traditional and other social

Practical Ethics for Food Professionals: Ethics in Research, Education and the Workplace, First Edition.
Edited by J. Peter Clark and Christopher Ritson.
© 2013 John Wiley & Sons, Ltd. Published 2013 by John Wiley & Sons, Ltd.

obligations). China is diagnosed as generally being high on power distance, moderate/low in uncertainty avoidance, average in masculinity, low in individualism and high in long-term orientation. The last two can be explained by the country's deep-rooted and long-established strong set of cultural values. These are so strong that it has influenced the way food is marketed in China and a customer's general reception of, and response to, it. Interestingly, Hofstede's category of long-term orientation was only added to his model after the investigation and application of it to greater China (see Hofstede and Bond, 1988). The next two sub-sections briefly explain how religion and specific cultural values (which have come out of religion) remain to dominate behaviour today.

4.2.1 Influence of religion

Modern China, following the Chinese Communist Party's strict rule, in principle forbids the explicit belief, practice and promotion of religion, claiming unfaithfulness to, and distrust of, the Party. However, the main religions, which can be traced back thousands of years have become so deep-rooted that they have proved difficult to restrain, by what in terms of history is "relatively new" country rule. While the West's influence on China, and East Asia in general, is vast and their religious influences have also grown, such as the growth of Christianity and other smaller religions, three main religious beliefs (or philosophical underpinnings) still dominate: Buddhism, Taoism and Confucianism.

Buddhism dates back to the 5th century BCE in India, with accounts of the life of Buddha. It is based on the views and teachings of Siddhartha Gautama (known as Buddha) recorded in a scroll, with the overall mission of achieving calmness and ending suffering, known as 'nirvana' which is considered the highest order of happiness. The ways to achieve nirvana constitute the different branches of Buddhism, one of which has travelled through to China and was particularly dominant during the 17th century in the Tang Dynasty through the establishment of Buddhist temples. There are estimates of about 700 million Buddhists in the world today (although no figure is accurate), many of whom are in China. While the religion does have formal principles, such as the Eightfold Path, and other variations on them, one dominating principle is to exhibit kindness – the idea that 'good' things should be done and therefore 'goodness' should be received (i.e. a form of reciprocity).

Taoism dates back to about the same time, around the 6th–5th century BCE, and has similarities to Buddhism. It boasts a membership of around 1.3 million worldwide, mostly in China. Quite often people are confused between the two as they share operational characteristics, such as the worshipping of many Buddhas (for different things and not just the main Buddha) in Buddhism, and the worship of many gods at various times of need in Taoism (there is no overall God). The practice of Taoism has two branches: one that is purely religious (known as religious Taoism) is focused on perpetuity and immortality, and one that is philosophical (known as philosophical Taoism). The latter is more prominent, and began with the teachings of a text from Tao Te Ching and it is believed that Lao Tze founded the religion, as a personal attempt to constrain the feudal warfare during his lifetime. The overall principle is to achieve spiritual harmony with one-another. The term *Tao* means "path," and it relates to the harmony of all living and non-living things, and a balance of needs to be achieved for all things. Hence, one of the famous principles from the Chinese is the idea of yin and yang (i.e. the balance of the negative and the positive of all things must be struck, and one could not exist if it was not for the existence of the other). The operation of Taoism is based also on the "three jewels": compassion, moderation, and humility, and from these, other principles operate, particularly those relating to achieving harmony with all.

Confucianism, on the other hand, is purely a philosophical concept and not a religion *per se*. It dates back to around the 5th–4th century BCE from the Chinese philosopher, Confucius (more accurately pronounced Kong Fu Zi), and is premised on his teachings of humaneness, benevolence and reciprocity, which are extended to the family for harmony. This is known as upholding the six explicit cardinal values of moral cultivation, interpersonal relationships, family orientation, respect for age and hierarchy, avoidance of conflict and need for harmony, and face-losing (see Bond and Hwang, 1986). It was most dominant during the Han Dynasty of China (2nd century BCE to the 2nd century CE), but is still followed strictly by around 7 million people today, while its principles are applicable implicitly in more far reaching contexts of modern Chinese daily life.

4.2.2 Influence of cultural values

The cultural values of China that exist prominently today have derived in one way or another from all or any of the aforementioned three religious underpinnings. There are various views on which are the most dominant cultural values, how they are categorized, or how exactly they are relevant to different situations, but this chapter presents the cultural value of *guanxi* as the overarching principle, which is composed of other cultural values, *ganqing* (sentiment, emotions), *renqing* (human feelings and humanity), *mianzi* (loss-of-face), *bao* (reciprocity), *hé* (harmony) and *xinren* (trustworthiness).

Guanxi in the Chinese language has numerous meanings, from linguistically a colloquial saying for "no problem, it's alright" to a complex social and business phenomenon concerning networks for business strategy. In this chapter, the socio-economic context of *guanxi* is used, and this refers to "the concept of drawing on a web of connections to secure favors in personal and organizational relations . . . [and is] an intricate and pervasive relational network that contains implicit and explicit mutual obligations, assurances and understanding" (Park and Luo, 2001, p. 455). Simply put, *guanxi* in this context can be grouped broadly in three contexts: business *guanxi* (where business solutions are sought through utilizing this network), helper *guanxi* (where favors are exchanged for any reason), and family *guanxi* (where special relationships matter more over substance). In Western eyes, this practice can even be considered unethical as it carries connotations of corruption. However, the underlying principle, which was derived from religion, offers a good justification for its acceptability. Therefore, even situations as seemingly distant from one another as the marketing of food, and the exchange of family ties, can involve the utilization of *guanxi*.

Ganqing, simply put, is the sentiment and emotions of human beings for any relationship or situation. While there has been the development of scientific management and quantitative techniques to monitor human performance, human beings are still subject to emotions. It is therefore the strength of any relationship: the phrase *ganqing hao* means having good *ganqing* which is the extent of emotional attachment to something (Chen and Chen, 2004). A business or management practice is therefore more likely to be effective or successful if *ganqing* can be maximized, increased, or at least not lost or reduced.

Renqing relates to the extent of humanity associated with human beings; simply put, it is the degree of empathy a person possesses. It is the social obligation that one feels is attached to a situation, particularly as a return of favour. It is like a charitable doing that just seems right, regardless of logic or reasoning behind it. Unlike in Western cultures where charitable doing is best done privately, or even secretly, the giving and receiving of *renqing* in Chinese tradition is considered an art, and the better that one is at so doing the more successful the transaction and value created from it; hence it is explicit and flamboyant. It is also associated with the perception of immorality or wrong-doing, that if there was no intent or action of favoritism in situations that ought to bestow it, this immediately creates a negative impression that will eventually lead to a permanent negative relationship.

Mianzi directly translated means "face" and can be used to mean that literally, but *lian* is better used in the language to refer to the physical look of a person. As *lian* represents the integrity of a person's character (traditionally some Chinese judge people on their physical bone structure), it cannot change or be used. *Mianzi*, however, is therefore more associated with the "loss-of-face" for doing or not doing something, such as the return of a favor within the *guanxi* etiquette; so in essence, the act of doing something becomes the act of face-giving, and the greater the extent of face-giving the stronger the *guanxi* on which a person can later draw. More tangibly, *mianzi* is associated with a person's image, reputation and prestige: the better this can be practiced, the greater the harmonization of human relationships in society. Hwang (1987) illustrates the *mianzi* further by suggesting the difference between horizontal *mianzi* and vertical *mianzi*; the former refers to acts of face-giving, face-saving and the avoidance of losing face, while the latter refers to the projection of self-image to elevate one's status in society.

Bao is reciprocity – simply the return of a favor or any act. It can be a good thing to return or a bad thing in real terms, but the act of not considering the need to reciprocate is considered a bad thing in Chinese culture. This is supported by nearly all major religions – doing something that one would respect, or not doing something that one does not expect to have done to oneself by others. In the Chinese religions, especially in Buddhism, *bao* is particularly important as it represents reciprocation in both the present and the next life (after reincarnation), and therefore "good-doing" that enhances the *guanxi* is aspired to.

Hé is quite prominent in many East-Asian cultures. It translates directly as harmony, and is used to mean peace, calmness, unity, kindness and connectedness. For the Chinese, this is particularly so within the family context, so often members of the family live within close proximity and have regular family reunions and even evening meals together. Children quite often grow up and still live with their parents for a long time, or even permanently. Family relationships are also made clear, and so in the Chinese language there is a specific vocabulary that describes exactly all the distant cousins. This can possibly be explained by the tradition of inclusiveness and the identification of the extent of close family membership for *guanxi* and favoritism purposes.

Xinren means to trust, confide in, or have reliance on, somebody or something. It is similar to another word, *xinyong*, which is more about credibility and trustworthiness, which is less relevant to *guanxi per se* (although related) because it is a fixed status that cannot be easily manipulated and therefore used. *Xinren*, on the other hand, is inclusive of *xinyong* as one has more *xinren* when one's credibility (*xinyong*) is naturally much higher. This means that *guanxi* is higher if *xinren* is higher, but *xinren* is difficult to control as it is subject to external factors: the job of businesspersons is therefore to engage in anything that enhances *xinren* to secure the long-term survival of the business.

It is believed that these cultural values apply in numerous daily life contexts of China. These are later referred to in Section 4.6 as they relate to the marketing management of junk food and its impact on childhood obesity in China. In general though, food has always been an important part of daily life (not least the obvious reason of survival!). For example, one way of asking, "how are you?" in a few Chinese dialects is, "have you eaten your rice yet?," with the reply, "yes, I have eaten my rice" (even if the person has not eaten rice or has not eaten at all) to mean, "I'm fine, thank you." This is because rice is the staple food of the Chinese and the key source of energy, and which therefore constitutes perceived good health to the Chinese. Following this age-old tradition, the enjoyment, consumption, and hence the advertising and consequential problems of the over-consumption, of food, are particularly prominent in China. Coupled with the significance placed on cultural values, the remainder of this chapter discusses how this has resulted in the current pandemic of childhood obesity in China, and the ethical concerns raised by the way businesses try to make sense of the numerous opportunistic situations these result in.

4.3 CHILDHOOD OBESITY

The issue of obesity may be frowned upon in the modern society as a social bad, representing poor health, (mostly) associated with low-income earners who are unable to afford healthier food products that contain lower calorific content; and even to some extent the less educated who are unaware of the problems obesity can cause. However, this has not historically always been the case, and can therefore be traced back through different eras of China. In so doing, clearer light may be thrown upon the long-standing causes of childhood obesity, in order to enable a more meaningful consideration of business management opportunities and implications in these two different contexts. The term obesity is used in this paper in a neutral way, representing a state of being firmly (indisputably, by ordinary standards) overweight, and does not carry any bad connotations with it for the people it concerns.

Obesity in China can be traced as far back to the Tang Dynasty (the 6th century CE), which is considered as one of the most prosperous periods in the history of China. Men of that period found women attractive who were full-bodied, assertive and active, and being obese was a symbol of wealth. Princess Yang GuiFei, belonging to the superseding dynasty of Empress Wu, is known in ancient Chinese history for being (beautifully) full-bodied (not overweight or obese as such by present-day standards), in which period the poor were relatively and distinctively thinner (see Wu, 1924). This has meant that a long-standing preference for the full-bodied (being comparatively healthier) was thus established.

During the 18th and 19th centuries of more modern history, China had suffered many wars and natural calamities, and the population experienced insufficient food supply.

This was more recently highlighted by the Great Leap Forward period of China (1958–1961) when Communist China's plans to excel from an agricultural society to one of modernization failed, leading to a widespread famine and the deaths of tens of millions of people. The issue was therefore not one of being obese/overweight, but one of not being overly thin and malnutritioned. Later, as the Chinese population grew rapidly, which then led to the Chinese One Child Policy (in 1979), limiting only one child per family (with some exceptions to the law), this led to families focusing resources on the sole offspring for filial respect. This may be one reason why rich families may be contributing to childhood obesity. However, since this time, medical knowledge has also recognized that obesity is a cause of numerous diseases and long-term health complications, and the influence of the West has instilled an alternative mindset into the Chinese – that being thin is beautiful and therefore has many social advantages. As China is booming economically in the 21st century, creating many rich people, a mix of mindsets still dwells with Chinese families in respect of how to expend this wealth; the business management implications for childhood obesity are therefore wide-ranging.

4.3.1 Causes and problems of (childhood) obesity

The universal measure of obesity is the body mass index (BMI), calculated as body weight (in kilograms) divided by height squared (in meters). A BMI of less than 18.5 denotes being underweight, a BMI between 18.5 and 24.9 denotes normality, a BMI between 25 and 29.9 denotes being overweight, and a BMI of greater than 30 denotes being obese. The BMI does not *per se* measure the content of body fat; it provides only a crude estimate, without consideration of ethnic origin or genetic make-up, or how fat is distributed in the body, such as visceral fat (fat surrounding vital organs being more of a health risk than elsewhere), but it has become the gold-standard and has been used since the 1800s (Campbell and Haslam, 2005). More recent research has recognized that the health risks associated with the different levels of BMI vary with the ethnic origin of the individual, with those of Asian origin being more at risk of health problems at lower BMI values than their Western counterparts (Brimelow, 2009; WHO, 2003); the accuracy of such a claim will not be considered in the present chapter.

Obesity (or the state of being overweight in general) stems from four main causes: genetics (the human genetic make-up explains about 50% of the individual's state of being obese), dietary intake (knowledge of how to control this has the largest impact on controlling obesity), exercise (this is also influenced by an individual's state of mind), and society (whether there is a social advantage in being recognized as either thin or full-figured). Other less-prominent theories have also tried to explain the cause and trends of obesity, but these will be ignored so as not to complicate the kernel of the purpose, which is to examine the food marketing aspects, not the detailed medical explanations. However, medical research has linked obesity to a range of illnesses, such as heart disease, various cancers, diabetes, osteoarthritis, and a whole host of other less serious prolonged health complications, on average reducing an obese individual's longevity by about 9 years. The World Health Organization now recognizes that about 1 billion people are overweight, of whom about 300 million are classed as clinically obese (WHO, 2003). For most governments, particularly those with a responsibility to provide healthcare to the nation, this Herculean task becomes even more burdensome when the next generation that are to finance it are themselves equally or more prone to the same illnesses. To prevent the situation from becoming worse, action is necessitated by governmental policy, in relation to all the causes of obesity; action should also be addressed at the childhood stage before the problem is escalated at adolescence. The lack of so doing, or the increase in childhood obesity, is the ethical issue of concern in this chapter.

The story is somewhat different from a business opportunity and marketing point of view, and targeting children provides the best chance to pass on future generations of customer loyalty. While businesses indeed could benefit greatly from the promotion of products and services that cause or deal with obesity (such as health food products, fitness gymnasia, diet pills, etc.), there is an equivalent business opportunity for those who do not wish to, or do not see a social advantage in being thin, such as the promotion of unhealthy fast-food outlets, tailor-made extra-large sized clothing, etc. The decision of individuals of how they regard the issue of obesity is therefore not straightforward and this is dependent on the way they see a benefit in society: there are therefore a number of cognitive conditions through which these beliefs may be grouped.

4.4 COGNITIVE CONDITIONS AND SCOPE OF BUSINESS OPPORTUNITIES

In research on this subject (Chau and Tang, 2010) three possible cognitive conditions as a typology through which the issue of being overweight/obese may be regarded by individuals are identified.

The first condition recognizes persons who are not interested in change; in other words, they do not see any form of advantage in moving away from being overweight/obese. This is not to say that they prefer to be overweight/obese, but rather they are indifferent and are therefore likely naturally to become overweight/obese over time: this is therefore called the condition of "passivity" because they lack the desire to change their status. This is premised on the view that the pressures of modern society make people lazy (Ritzer, 1996). The business opportunities for such persons tend to focus on products or services that may generally be regarded as unhealthy and a major cause of obesity, such as fast-food outlets. The second condition recognizes that some people believe that the state of good health (not being obese/overweight) has some advantageous utility both for themselves by having a lower risk of health problems and that having a slimmer physique gives them a social advantage over the obese/overweight. This group of people begin with the default position of being slim, and have never been overweight/obese: this condition is termed "utilitarianism," denoting the belief of utility and need to utilize. The range of business opportunities for persons falling within this category may include fitness gymnasia, health-foods (such as fruit smoothies and low-fat products), health-food outlets (such as healthy salad and sandwich bars), and specific food supplement pills (such as omega-3 rich oils and vitamin supplements). The third condition has the same cognitive belief as utilitarianism (the existence of a social advantage in being slim), but individuals begin with the default status of being overweight/obese (and regard this as a bad status, whether self-inflicted or born with such genetic make-up) and desire, and

PRODUCT

		Current (P_0)	New (P_N)
MARKET	Current (μ_0)	(1) Market Penetration *(Passivity)*	(2) Product Development *(Rehabilitation & Utilitarianism)*
	New (μ_N)	(3) Market Development *(Rehabilitation & Utilitarianism)*	(4) Diversification *(Rehabilitation)*

Figure 4.1 Potential of business opportunity corresponding to the cognitive conditions.

make a significant effort, to put right any wrong-doing that had contributed to that in the past, to move to a different status of not being overweight/obese: this condition is coined 'rehabilitation' (to signify the need of putting right). The range of business opportunities here includes all those for utilitarianism, plus additional ones including a range of surgical procedures, dieting medication, and specific intensive fitness programmes.

Indeed, these three conditions cannot represent all psychological and behavioural possibilities of individuals, but these at least provide the most probable conditions under which business activity may be premised (for example, we ignore the condition that some persons may begin with the default position of being slim and wish to become overweight/obese for the purpose of this research, but do not deny its existence).

A basic way to look at whether a business should penetrate the existing market and business activities or pursue new activities or new businesses, or both, is through the typical two-by-two Ansoff (1957) matrix (for a typical textbook explanation, see Witcher and Chau, 2010). Figure 4.1 presents the matrix to include the three cognitive conditions within the four quadrants of the Ansoff matrix. Quadrant 1 (market penetration: p_0, μ_0) represents the condition of passivity because such persons are unlikely to be interested in products/services other than those that are the predominant cause of obesity. Quadrant 2 (product development: p_N, μ_0) represents the conditions of utilitarianism and rehabilitation because the quadrant covers business opportunities for new products within the same penetrated market; we interpret this to include dietary and health food-stuffs that are standard to sustaining good health. Quadrant 3 (market development: p_0, μ_N) also represents the conditions of utilitarianism and rehabilitation as the quadrant covers business opportunities for existing penetrated product ranges in new markets; we interpret this to mean new ways to diet and sustain good health, including new dietary and fitness programmes. Quadrant 4 (diversification: p_N, μ_N) is loosely borrowed from the original model (but does not mean having no relation with the original product range or market as such) covers only the condition of rehabilitation; this is because rehabilitation implies the need to do more than an ordinary amount of effort to ensure a healthy living. Products and services in this "diversified" segment would include the non-ordinary commodities for healthy living, and may include extreme forms of dietary foods (such as food supplements and medicines), extreme and unorthodox fitness programs, and even surgical procedures to remove flesh directly from the human body. As the model shows, the size of business opportunity is smallest in quadrant 1 and largest in quadrant 4; the size of each of the quadrants also represents the complexity and risk associated with the business activity.

4.5   CHINESE CULTURAL VALUES AND THE MARKETING OF OBESITY-RELATED FOOD PRODUCTS

This section offers a view on how the Chinese cultural values described above relate to the marketing of obesity-related food products.

4.5.1   Addiction to unhealthy products (underpinned by *hé*)

There is a cognitive view of human activity that suggests that consumers knowingly consume products for pleasure even if they are aware of their harmful consequences (Goldbaum, 2000); this phenomenon is characterised as an addiction. Examples of such products include alcohol and tobacco, and in the case of the latter, the number of smokers worldwide has not fallen despite the inclusion of warning messages on the cigarette packaging (Viscusi, 1992). Recently, excessive food consumption has also been conceptualised as an addictive behaviour (Gearhardt and Corbin, 2009), and obesity in particular is not found to be simply an over-consumption of food, but rather the consumption of specific foods that are high in saturated fat and sugar content, otherwise known as "junk food" (Flegal *et al.*, 2002). Compounded with a lack of exercise, this is the main cause of childhood obesity, particularly when the parents of the children do not see a need to instil a healthy mindset into them. When asked what their favorite food is, a large proportion of children commented that they like such food in the giant fast-food outlet chains, such as McDonalds, Burger King and KFC, etc., particularly the promotional meals for children, attracted mostly by the toys that come with them (Stanley, 2009).

The Chinese cultural value of *hé* is exemplified by the desire to ensure members of the family are close together and that leisure activities, such as Sunday meals, are done so with the family. This is a way of ensuring unity. For marketers, this poses an opportunity to take advantage of this cultural value by making the child (as well as parents and other family members) addicted to the particular food as early in their lives as possible. In essence, addiction to fast-food for the Chinese is premised on exploiting the importance in maintaining the cultural value of *hé*. This raises a major ethical issue in the sense that it provides an increased opportunity for businesses of products associated with continued bad health, which is the cause of childhood obesity which can in turn lead to adolescent obesity due to the addiction.

4.5.2 The influence of advertisements (underpinned by *ganqing*)

Sales promotions can influence consumption patterns of consumers by influencing their purchasing choices (Hawkes, 2009), generally encouraging them to buy more of something if they are more receptive to a particular advertisement (MacKenzie *et al.*, 1986; Machleit and Wilson, 1988). Advertisement specifically of children's food has long been a controversial issue, and it is generally found that such food promotion has an effect on children's preference of foods, thereby affecting purchase behaviour (of the parents) and consumption of them (Livingstone, 2006). Televised adverts seem to have the greatest impact (Kelly *et al.*, 2007). The general Chinese cultural value of *ganqing* has a similar effect here as when an advertisement is staged and received well by the potential consumer, he feels a degree of compassion and the need to reciprocate. This is not to say that adverts work particular well with the Chinese, but it is likely that the way adverts work for the Chinese is because of an internal drive at least to acknowledge an "invitation to treat," and when this is accepted as a satisfactory offering it leads to a confirmed consumption.

Advertisements for addictive products seem to work differently for three types of customers: advertisements for those who have never tried a product and know it is harmful to health have only a very little effect; for those who have tried a product, the advertisement is most important in attempting to lure them into becoming addicted to it; and for established addicts, the role of advertising is only to convince them to consume a particular brand over a competitor's brand. No matter at which stage of addiction, advertisements play a crucial role in positively influencing a customer (Hamilton, 1972; Davis, 1987), particularly as there is a degree of compassion in all human beings. Fast-food addiction, for those who have been immersed with it since childhood, is likely to fall within the third category, loosely implying that businesses advertise with the purpose to compete against each other than to attract more customers to the product (Young, 2003; Barlovic, 2006). This is likely to be the reason for the success of fast-food outlets (Teinowitz, 2005). As compassion (i.e. *ganqing*) is an emotional quality, it is likely to work differently under different advertising campaigns; it is likely that advertising does not create any increase in the addition of fast-food *per se* but it does influence a customer's preference for a particularly brand.

4.5.3 Allurement into bad health products (underpinned by *mianzi* and *renqing*)

Allurement implies a strong attraction to a particular product/service offering; it is not the same as addiction as allurement has the customer's belief that there are benefits in taking up the offering. Philipson and Posner (2008) claim that about 11% of advertisements geared to children are related to fast-food, but they do not mention the problems associated with, or state the volume of, its high fat and sugar content. The allurement is induced by the attraction of companies providing a quality service, convenience of the product, low price, good taste, and even a place for children to play in; of all these, price and convenience have long been argued to have the strongest impact (e.g. Ho and Cho, 1995). The Chinese cultural values of *mianzi* and *renqing* underpin this concept as they concern the way Chinese respond to the invitation to an offering (the perceived value in the product) with at least the same level of prestige (in the form of the consumption).

Pricing strategies are the most common and effective way to increase the general sales of a product (Milyo and Waldfogel, 1989; Spulber, 1999). Lyles *et al.* (2007) suggest that there is an inverse relationship between price changes and sales (in other words, a decrease in price will lead to an increase in sales). However, its overall impact on obesity is small, as Powell *et al.* (2007) find that there is only a very small statistically significant association between a reduction in price and its ultimate contribution to obesity. Powell and Bao (2009) further add that the association between reduction in fast-food prices and BMI values in children and adolescence is very weak (with an estimated price elasticity of about -0.12). For the Chinese, this may be because of *mianzi* – that is to say, *mianzi* ensures a consistency in the amount of perceived value offered is met by the response of the customer to the product offering; in practical terms, this may mean that the perception in the price reduction of a bad health product overrides any knowledge of the harmfulness of it, and this in turn may be a cause of obesity in China.

Similarly, convenience has become a favorable advantage in the 21st century as the pressures of modern life require people to work both at specific and flexible times, particularly as China is growing at an incredible rate, making food consumption in particular a difficult activity to plan in advance for busy working professionals. Such people have tended to turn to fast-food outlets that remain open at a wider range of hours than traditional restaurants and expect to be able to order and buy food in much shorter transaction times. Over time, their assessment of the benefits of convenience has outweighed their assessment of the health risks associated with such food (Dunn *et al.*, 2008). This is elevated by the fact that the convenience of purchasing snacks (rather than meals) can now be achieved through a number of means, including vending machines and even Internet sites (Kinsey and Ashman, 2000). *Renqing* is the Chinese cultural value that supports convenience: this is not obvious, as *renqing* does not mean convenience at all, but *renqing* discerns the level of respect that must be returned to the advertiser who tenders the (possibly non-monetary) value to the potential customer by the reciprocation of ultimately purchasing the product.

The attraction to convenience is heightened furthermore by the wider availability of that convenience, thereby elevating *renqing* to a higher level – in other words, improving the density of such available fast-food outlets. Austin *et al.* (2005) warn that the density of fast-food outlets within a close vicinity of schools may be one of the key reasons for the increase in childhood obesity, and that schools should do more to minimize children's attraction to them. On the contrary, research by Powell *et al.* (2007) and Sturm and Datar (2005) suggest that the density of such outlets does not have an impact upon children's weight. On balance, it is likely that the perception of the benefits of the convenience of consuming a bad health product overrides the perceived harmfulness of it, and therefore influences the extent of obesity, and *renqing* plays a supportive role in accepting convenience.

4.5.4 New product development and management (underpinned by *xinren*)

Business opportunities for those falling within either the utilitarianism or rehabilitation categories (as described earlier) warrant the understanding of new product development and management. This is in part at least a responsibility of a new product either from the customers' point of view as the experience is new to them or from the producers' point of view as they are to offer a product/service that is at least perceived to be new or different from what customers of bad health products were previously consuming. The Chinese cultural value of *xinren* is relevant here because of the amount of trust that must be present in the worth and usefulness of any new product. This is because the business environment influences the organizational behavior in a new product development (Gupta *et al.*, 1986), and a thorough understanding of such customer demand is important if it is to be met (Annacchino, 2003). However, there is also an equal pressure from the producer's side that influences what customers are to purchase (Liao *et al.*, 2009). For *xinren* to be achieved for the Chinese, knowledge of health matters (due to the general increase in obesity) must provide an orientation of new product development for good health products.

Nemarkommula *et al.* (2003) outline in particular that the increase in worldwide obesity levels is likely to lead to more business opportunities, with a view that there is greater inclination to favor products that combat it, such as the need for healthier food products. A common concern of food manufacturers is that reducing fat, sugar and salt content in food is likely to make them less tasty, and could lead to decline in sales. Wyman (2004) finds that this is unlikely to be the case if manufacturers make it clear that the product is a new innovation. For existing products stating that a product maintains the same amount of taste but contains a lower fat, sugar and salt content is also unlikely to lead to a fall in the sales of the product, and it may even lead to customers favoring it over the original product (Seiders and Berry, 2007). Customers seem to be satisfied if nutritional information is given to them on packaging, as a way of showing that the producers care about their well-being (Allen, 2003). While in most countries there are regulations regarding how nutritional information must be displayed on packaging, it is not necessarily an effective mode of communication. Viswanathan (1996) notes that such information is not useful to customers because not all customers actually understand it, and therefore they discard it entirely. Instead, customers' purchasing choices seem to be based on a predisposition of nutritional knowledge, or they decide in advance what they intend to purchase irrespective of what the information states (Pitts and Phillips, 1998). While the Chinese are, and have always been, interested in good health and longevity, their likely disinterest in reading nutritional information on packaging is possibly due to the cultural value of *xinyong* – that is that an established product has credibility and is safe (while not necessarily healthy) to eat and does not require further investigation of its nutritional information. To the extent that *xinyong* exists for the Chinese, the specification of nutritional information on food packaging is likely to be an ineffective way of influencing what customers choose to purchase, and does not seem to have impact over the control of obesity.

4.5.5 Product life cycles (underpinned by *bao*)

Kassirer and Angell (1998) note, particularly in the American context, a lot of people do recognize that the need to be thin is a good thing and take the necessary actions to pursue it, such as through healthy eating and exercise. The Chinese cultural value of *bao*, at least from the religious point of view, suggests that a bad outcome will result from a bad action(i.e, poor health from poor diet is a certainty). However this is not to suggest that good health can only come from a good diet. Many of those who give up healthy eating and exercise look for "quick-fix" alternatives, such as through taking dieting pills or surgical procedures, and *bao* also works here in believing that investing (a lot of money on) these quick-fixes is also a good deed that will also result in the return of good health. From the producers' point of view, the issue is how to satisfy customers' expectations of what is perceived as a quick-fix – in other words, an innovative solution/product – in a continued manner, and therefore the innovation requires to be renewed regularly for it to be effective. This is referred to as a "conceptual leap" (Christensen, 1997), which requires a close interrelation between customer and producer to ensure a good understanding (Tollim and Caru, 2008), and so that decisions about whether to introduce new products entirely as an issue of new product development or to reintroduce existing products by renewing the product life cycle (Wahlers and Cox, 1995) can be made.

Product life cycles traditionally last 15 to 20 years (Ghadar and Adler, 2003) in general, but nowadays they are much shorter, lasting about 5 years before action is required as to whether to end it entirely or renew the product; this is because of faster applied new knowledge, more products being introduced between innovations, and the time between innovations is decreasing, but the most fundamental reason is because customers nowadays expect more frequent innovations (Rifkin, 1994). *Bao* assists the reception of these innovations and hastens their pace with the belief that good health is likely to come to them if they themselves do some good, and innovations are a way of easier so doing; further, such innovations are likely the close requirements of customers of good health products and less so those relating to poor health.

4.6 CHINESE CULTURAL VALUES AND THE THREE COGNITIVE STATES

Research undertaken by Chau and Tang (2010) has already explored the extent of the Mainland Chinese population falling within each of the cognitive states. These key implications are now highlighted and discussed in relation to how they may be underpinned by one or more of the Chinese cultural values discussed earlier in this chapter (see Figure 4.2). These are also the likely ways food producers and food marketers have exploited the prevalence of these values to succeed in their business ventures.

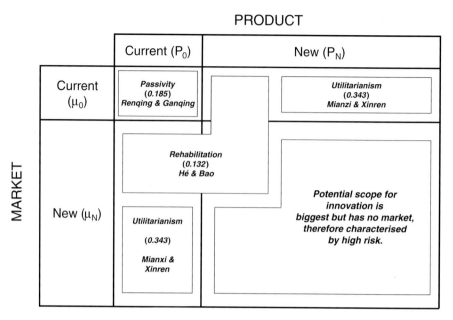

Figure 4.2 Distribution of scopes of business opportunity for the market groups.

4.6.1 Relationship of Chinese cultural values to "passivity" customers

Chinese passivity customers, based on the empirical research, make-up the second largest market group (18.5%) and have the highest BMI, which may only be served by market penetration strategies (positioned in quartile $p_0{:}\mu_0$ in Figure 4.2). Innovation for this group is likely to be effective in the form of renewing depleted life cycles of well-known products, particularly those which have become a norm in society. As these consumers are likely to be existing addicts of the product, general regular advertising, with a view to attracting generations of the same family group to the product, seems the most obvious business strategy. The Chinese cultural value related to the renewal of existing perceived successful commodities is mainly *renqing*; that is, the repeat consumption of something when it is being further developed, as an emotional acknowledgement of its former appreciation. This is an extended way to consider loyalty and improvement of customer relations; hence, *ganqing* is also relevant here but to a smaller extent. Indeed, this group of consumers is most at risk of the health complications (as outlined in Section 4.3 of this chapter) and is most likely to exacerbate the obesity pandemic. While this is a long-term social and health problem, fast-food outlets do have the most obvious advantage of sustaining employment, even in times of hard economic recession, and serve society's lower income earners, which may have positively shaped today's society.

4.6.2 Relationship of Chinese cultural values to "utilitarianism" customers

Utilitarianism customers, based on the empirical research, make up the largest market group (68.3%) but have the lowest BMI, which may be served by product development or market development. Innovation for this group refers to (i) healthy products that are an improvement on the existing range of established bad health product brands (for example, the salad option or "healthy-eating" range of fast-food outlets), or (ii) healthier products operated in a slightly different market from the well-known bad health product brand (such as, for example, if a mainstream fast-food outlet diversifies into specialist coffee-houses or healthy bakeries). The former is supported by the Chinese cultural value of *mianzi* in that the purchase is of a different product but has the same roots as the original; this means there is a notion of respect for the original but recognizing the need for the different product, as a way of 'losing face' by discharging the value of the original in the first place. The latter relates to the *xinren* cultural value as it implies the acceptance of the quality of the original product, thereby trusting it, and exerting that trusted quality in a different product/market context (Figure 4.2 indicates that utilitarianism exists either at quartiles $p_N:\mu_0$ or $p_0:\mu_N$). The combination of these two kinds of innovation is probably the reason for the low BMI for this group of customers, who are unlikely to be addicts of any product, and so marketing effort should initially be focused on targeted advertising and then move on to general ongoing advertising when the product has matured.

4.6.3   Relationship of Chinese cultural values to "rehabilitation" customers

Rehabilitation customers, based on the empirical research, make up the smallest market group (13.2%) and have the second highest BMI, which may be served by product development, market development or diversification (as positioned in Figure 4.2 at $p_0:\mu_N$ $p_N:\mu_0$ or $p_N:\mu_N$). While they make up the smallest group of customers, the scope of possible innovation/diversification strategies is the largest that are focused on health products only. Following the fast-food outlet example, complete diversifications may include vitamin supplements, fitness gymnasia or even private cosmetic surgery. Hence, as the strategies used are likely to be uncertain, a number of Chinese cultural values may play a role in this. However, most related are probably *hé* and *bao*: these work in different ways for rehabilitation. *Hé* accepts the positive side to rehabilitatory behavior; this is the need to strike a balance in the human being to achieve harmony in all aspects of life, and so customers will naturally purchase something different if this message can somehow be implicated. *Bao*, on the other hand, identifies the need to 'do a good thing' in general in fear that a regular bad happening will recur. Rehabilitators naturally accept that wrong has been done in the past, and will therefore make up for it by choosing something entirely new or different. The consequence is that marketing effort would be general, as while customers of these products/services may not be regarded as addicts, the cognitive condition to which they belong suggests a particularly earnest interest in these products.

4.7   CONCLUSION

While this chapter has covered a number of topics, it has brought specifically into discussion the relation of Chinese cultural values to marketing of food and its effect on (childhood) obesity in China, which is the current pandemic concern. It has discussed this from the perspectives of medical science, marketing and business strategies.

The unique contribution of this chapter is the identification of three prominent cognitive conditions – passivity, utilitarianism, and rehabilitation – to represent the market, using the Ansoff two-by-two matrix to map out the extent of business innovation, and applying this to the role of Chinese cultural values. Drawing on existing research in this area, it has identified utilitarianism as the largest market potential for business opportunity for the Chinese market. The fast-food industry, which is probably one of the key culprits of the present day fast increasing obesity rate, has for decades tried to fulfill consumer needs which are positioned within the utilitarianism space. The present research finds that the suitability of marketing strategies is based on the extent to which customers are addicts of the product, and the extent of their addiction is affected by the cognitive condition group to which they belong.

The premise of this chapter is that the exploitation of these cultural values leads to childhood obesity, which is the unethical aspect. For example, exploiting *renqing* and *ganqing* within passivity space is likely to exacerbate the obesity pandemic. However, such business expansion at least encourages employment and economic activity. On the contrary, the reliance of *mianzi* and *xinren* within utilitarianism space, and *hé* and *bao* within rehabilitatory space, will help to alleviate the obesity problem. Marketing strategies should be used differently for the various cognitive condition groups, and they in turn are underpinned by a particular Chinese cultural value. In this way, these values not only operate at self or close community levels between friends and family, but they can extend to wider networks, such as the relationship between retailer/food producer and customer. Understanding these should facilitate marketing effort, for objectives that are deemed either ethical or unethical.

REFERENCES

Allen, R.L. (2003), "More education needed to help consumers digest mouthful of food nutrition information", *Nation's Restaurant News*, Vol. 37 No 29, p. 24.

Annacchino, M.A. (2003), *New Product Development: Form Initial Idea to Product Management*, Elsevier Butterworth Heinemann, USA.

Ansoff, H.I. (1957), "Strategies for diversification", *Harvard Business Review*, Vol. 35 No. 5, pp. 1131–24.

Austin, S.B., Melly, S.J., Sanchez, B.N., Patel, A., Buka, S. and Gortmaker, S.L. (2005), "Clustering of fast-food restaurants around schools: a novel application of spatial statistics to the study of food environments", *American Journal of Public Health*; Vol. 9 No. 9, pp. 1575–1581.

Barlovic, I. (2006), "Obesity, advertising to kids, and social marketing", *Young Consumers*, Vol. 7 No. 4, pp. 26–34.

Bond, M.H. and Hwang, K.K. (1986), "The social psychology of Chinese people", in Bond, M.H. (Ed), *The Psychology of Chinese People*, Oxford University Press, Oxford.

Brimelow, A. (2009), "Lower Asian obesity threshold", *BBC News Channel*, http://news.bbc.co.uk/1/hi/health/8141335.stm, 9 July 2009; accessed 11 December 2012.

Campbell, I. and Haslam, D. (2005), *Obesity: Your Questions Answered*, Churchill Livingstone, Leicester.

Chau, V.S. and Tang, A.R. (2010), "It doesn't really matter if you're fat or light! Implications of Child Obesity in China and UK for Social Marketing Strategy", *Proceedings of the 24th Annual Conference of the British Academy of Management*, University of Sheffield, Sheffield, UK.

Chen, X.P. and Chen, C.C. (2004), "On the intricacies of the Chinese Guanxi: a process model of Guanxi development", *Asia Pacific Journal of Management*, 21 (3), pp. 3053–24.

Christensen, C.M. (1997), *The Innovator's Dilemma*, Harvard Business School Press, Boston, MA.

Davis, R.M. (1987), "Current trend in cigarette advertising", *New England Journal of Medicine*, Vol. 316 No. 12, pp. 725–732.

Dunn, K.I., Mohr, P.B., Wilson, C.J., Wittert, G.A. (2008), "Beliefs about fast food in Australia: a qualitative analysis", *Appetite*, Vol. 51 No. 2, pp. 331–334.

Flegal, K.M., Carroll, M.D., Ogden, C.L. and Johnson, C.L. (2002), "Prevalence and trends in obesity among US adults, 19992–000", *Journal of American Medical Association*, Vol. 288, pp. 1723–1727.

Gearhardt, A.N. and Corbin, W.R. (2009), "Body mass index and alcohol consumption: family history of alcoholism as a moderator", *Psychology of Addictive Behaviours*, Vol. 23 No. 2, pp. 216–225.

Ghadar, F. and Adler, N. (2003), "Management culture and the accelerated product life cycle", *Human Resource Planning*, Vol. 12 No. 1, pp. 37–42.

Goldbaum, D. (2000), "Life cycle consumption of harmful and addictive good", *Economic Inquiry*, Vol. 38 No. 3, pp. 458–469.

Gupta, A., Raj. P. and Wilemon, D. (1986), "A model for studying R&D-marketing interface in the product innovation process", *Journal of Marketing*, Vol. 50, pp. 7–17.

Hamilton, J. (1972), "The demand for cigarettes: advertising, the health care and the cigarette advertising ban", *Review of Economics and Statistics*, Vol. 54, pp. 401–411.

Hawkes, C. (2009), "Sales promotion and food consumption", *Nutrition Reviews*, Vol. 67 No. 6, pp. 333–342.

Ho, S.K.M. and Cho, W.K. (1995), "Manufacturing excellence in fast-food chains", *Total Quality Management*, Vol. 6 No. 2, pp. 1231–34.

Hofstede, G. (1980), *Culture's Consequences: International Differences in Work-related Values*, Sage Publications, London.

Hofstede, G. and Bond, M.H. (1988), "The Confucius connection: From cultural roots to economic growth", *Organizational Dynamics*, Vol. 16 No. 4, pp. 5–21.

Hwang, K.K. (1987), "Face and favour: the Chinese power game", *American Journal of Sociology*, 92 (4), pp. 944–974.

Kassirer, J.P. and Angell, M. (1998), "Losing weight: an ill-fated new year's resolution", *New England Journal of Medicine*, Vol. 338, pp. 52–54.

Kelly, B., Smith, B., King, L., Flood, V. and Bauman, A. (2007), "Television food advertising to children: the extent and nature of exposure", *Public Health Nutrition*, Vol. 10 No. 11, pp. 1234–1240.

Kinsey, J. and Ashman, S. (2000), "Information technology in the retail food industry", *Technology in Society*, Vol. 22 No. 1, pp. 83–96.

Liao, S., Chen, Y. and Tseng, Y. (2009), "Mining demands chain knowledge of life insurance market for new product development", *Expert Systems with Applications*, Vol. 36 No. 5, pp. 9422–9437.

Livingstone, S. (2006), "Does TV make children fat? What the evidence tells us", *Public Policy Research*, March-May, pp. 54–61.

Lyles, A., Stanton, K.R. and Acs, Z.J. (2007), *Obesity, Business and Public Policy*, Edward Elgar Publishing, London.

Machleit, K.A. and Wilson, R.W. (1988), "Emotional feelings and attitude toward the advertisement: the roles of brand familiarity and repetition", *Journal of Advertising*, Vol. 17 No. 3, pp. 27–35.

MacKenzie, S.B., Lutz, R.J. and Belch, G.E. (1986), "The role of attitude toward the ad as a mediator of advertising effectiveness: a test of competing explanations", *Journal of Marketing Research*, Vol. 23 No. 2, pp. 130–143.

Meier, G.M. and Rauch, J. (2000), *Leading Issues in Economic Development*, Oxford University Press, Oxford.

Milyo, J. and Waldfogel, J. (1999), "The effect of price advertising on prices: evidence in the wake of 44 Liquormart", *The American Economic Review*. Vol. 89 No. 5, pp. 1081–1096.

Nemarkommula, A.R., Singh, K., Lykens, K. and Hilsenarath, P. (2003), "A growing market: as obesity rates rise, so do the opportunities for marketers of specialized services", *Marketing Health Services*, Vol. 23 No. 4, pp. 34–38.

Park, S.H. and Luo, Y. (2001), "Guanxi and organizational dynamics: organizational networking in Chinese Firms", *Strategic Management Journal*, 22, pp. 4554–77.

Philipson, T. J. and Posner, R. (2008), "Is the obesity epidemic a public health problem? A review of Zoltan J. Acs and Alan Lyles's Obesity, Business and Public Policy". *Journal of Economic Literature*, Vol. 46 No. 4, pp. 974–982.

Pitts, M. and Phillips, K. (1998), *The Psychology of Health: An Introduction*. 2nd edition, Routledge, London.

Powell, L.M. and Bao, Y.J. (2009), "Food prices, access to food outlets and child weight", *Economics & Human Biology*, Vol. 7 No. 1, pp. 64–72.

Powell, L.M., Auld, M.C., Chaloupka, F.J., O'Malley, P.M. and Johnston, L.D. (2007), "Associations between access to food stores and adolescent body mass index", *American Journal of Preventive Medicine*, Vol. 33, pp. 301–307.

Rifkin, G. (1994), "When is a fad not a fad?", *Harvard Business Review*, Vol. 72 No. 5, p. 11.

Ritzer, G. (1996), *The McDonaldization of Society: An Investigation into the Changing Character of Contemporary Social Life* (revised edition), Pine Forge Press, Thousand Oaks CA.

Seiders, K. and Berry, L. (2007), "Should business care about obesity?", *MIT Sloan Management Review*, Vol. 48 No. 2, pp. 15–17.

Spulber, D. (1999), *Market Microstructure-Intermediaries and the Theory of the Firm*, Cambridge University Press, Cambridge.

Stanley, T.L. (2009), "Hollywood continues its fast-food binge", *Brandweek*, Vol. 50 No. 23, p. 6.

Sturm, R. and Datar, A. (2005), "Body mass index in elementary school children, metropolitan area food price and food outlet density", *Public Health*, Vol. 119, pp. 1059–1068.

Teinowitz, I. (2005), "4A's goads nets to enter food fight: Drake blasts 'lack of awareness' of TV's role in screening process", *Advertising Age*, Vol. 76 No. 15, p. 8.

Tollim, K. and Caru, A. (2008), *Strategic Market Creation: A New Perspective on Marketing and Innovation Management*, John Wiley & Sons, Chichester.

Viscusi, K. (1992), *Smoking: Making the Risky Decision*, Oxford University Press, New York.

Viswanathan, M. (1996), "A comparison of usage of numerical versus verbal nutrition information by customers", *Advances in Consumer Research*, Vol. 23 No. 1, pp. 277–231.

Wahlers, J.L. and Cox, I.J.F. (1995), "Competitive factors and performance measurement: applying the theory of constraints to meet customer needs", *International Journal of Production Economics*, Vol. 37 Nos. 2/3, pp. 299–240.

WHO (2003), "Obesity and overweight", *World Health Organization Global Strategy on Activity and Health*, World Health Organization, Geneva.

Witcher, B.J. and Chau, V.S. (2010), "Corporate-level strategy", in *Strategic Management: Principles and Practice*, South-Western Cengage Learning, Andover, Ch. 7.

Wu, H. (1924), *Yang Kuei-Fei, the Most Famous Beauty of China*, Appleton and Co Publishers, New York.

Wyman, V. (2004), "Demand for healthy products seen as opportunity for firms", *Food Manufacture*, Vol. 79 No. 6, p. 4.

Young, B. (2003), "Does food advertising influence children's food choice?", *International Journal of Advertising*, Vol. 22, pp. 441–459.

II Issues in food industry ethics

5 Ethics in business

Timothy F. Bednarz

5.1 INTRODUCTION

America is in a leadership crisis. Many of our prominent leaders in all walks of life view their positions of leadership as a right to power, money, and privilege. They perceive their status as a mandate to loot, pillage and abuse those who they are chosen to lead. They enjoy their wealth and the perks of power, while miserably failing to meet their professional responsibilities. In the myopic lust for profits and short-term gains, they maintain their positions as long as they deliver, often at the long-term detriment of organizations they are appointed to lead.

One may say this does not matter, but this crisis in leadership has contributed to chaos and has thrust us into the throes of our current economic crisis. It has thrown millions out of work, eliminated countless dollars of investor wealth and touched the lives of virtually every American. While these leaders' actions created chaos and havoc on America, they carefully secured and protected their personal wealth, so not to suffer the consequences of what they wrought on the rest of the country.

They have the façade of leadership but are leaders in name only. As James Burke, former CEO of Johnson & Johnson stated, "being a business leader is about giving-not taking." Jon Huntsman, founder and CEO of Huntsman Chemical identifies core elements of leadership as "talent, integrity, courage, vision, commitment, empathy, humility, and confidence. The greater these attributes, the stronger the leadership." In a world where ethically neutral leaders thrive and prosper, where are these qualities in leaders which are needed to overcome our present crisis of leadership? Have these been subverted to the quest for power and money?

There is a need to restore sound and ethical leadership. The mantle of leadership must be re-established. Customers, boards, stakeholders and employees must clamber for it and demand full accountability. This is accomplished one individual at a time, by electing or appointing the ones who possess the vision, integrity, courage, empathy, humility, commitment and confidence needed to transform their organizations.

Practical Ethics for Food Professionals: Ethics in Research, Education and the Workplace, First Edition. Edited by J. Peter Clark and Christopher Ritson.

5.2 WHY THE NEED FOR ETHICS IS JUSTIFIED?

For the past 12 years Richard Edelman, CEO of the public relations firm, Edelman has conducted a comprehensive survey of trust levels around the globe. He has documented the rise and fall of trust levels of business, government and non-profit organizations. He underscores the need for increased business ethics with his first recommendation published in his 2012 Edelman Trust Barometer Survey that companies need to:

> Exercise principles-based leadership instead of rules-based strategy. Business should not go to the edge of what is legally permissible but rather stay focused on what is beneficial both to shareholders and society (Edelman, 2012, p. 1).

He states:

> Listening to customer needs, treating employees well, placing customers ahead of profits, and having ethical business practices are all considered more important than delivering consistent financial returns – and indicate that the path forward entails continuing to do the basics well while also adopting shared values (Edelman, 2012, p. 10).

Edelman bases his conclusions in the gap between organizational performance and customer expectations in 16 attributes, which constitute trust. The top seven attributes are:

Attribute	Customer expectation (%)	Company performance (%)	Gap (%)
Listens to customer needs and feedback	67	36	−31
Offers high quality products or services	67	48	−19
Treats employees well	64	27	−37
Places customers ahead of profits	62	26	−36
Takes responsible actions to address an issue or a crisis	62	28	−34
Has ethical business practices	61	32	−29
Has transparent and open business Practices	60	27	−33

Source: 2012 Edelman Trust Barometer© Annual Global Study Executive Summary.

Eldeman's research clearly points to the demand for dramatic improvement in the practice and application of strong ethical policies. However it should be noted that an improvement in these seven key attributes builds the levels of trust that will also result in increased sales and profitability. The bottom line is that exceptional ethical behavior is good business.

Since the mid-1980s corporate America has been consumed with the consistent delivery of shareholder value, often at the expense of the long-term interests of the organization. Edelman's conclusion points to the importance of a restoration of ethical values over the consistent delivery of financial returns. The conclusion one can take away from Edelman's study is that companies need not sacrifice financial returns by emphasizing the ethical values that build customer trust. Both are achievable.

5.3 WHAT ARE ETHICS?[1]

An organization and each of its employees, wherever they may be located, must conduct their affairs with uncompromising honesty and integrity. Business ethics are no different than personal ethics and the same high standard applies to both. As a representative of their company all employees are required to adhere to the highest standard, regardless of local custom.

Everyone is responsible for his or her own behavior. We live in a culture where responsibility and accountability are minimized, with individuals hiding behind the label of "victim" as an excuse for their actions. There is right and wrong, black and white, but many would prefer to operate in shades of gray. As long as they do not cross the line, they feel that they are fine. As long as no one catches them, their behavior is acceptable.

Individuals operating in shades of gray feel ethics are not as important as the legality of their actions and think the ends justify the means. After all it is a results-driven environment and it is the results that matter.

While certain actions might be legal, they may also be unethical and reflect poorly on an organization as well as the individuals responsible for them. If these actions are tolerated and allowed, an organizational culture is created that undermines the customer's confidence in the company, as well as its products and services and ultimately destroys its reputation in the marketplace.

Allowing even a single unethical activity can pull a thread that ultimately unravels the cloth of an organization. Actions have consequences and unethical actions and their consequences can have a rippling effect within a company. If all employees understand this and apply it to their actions and the actions of their colleagues, it will result in a stronger company. Both the company and an employees' ongoing employment within it require compliance to this philosophy.

Ethical behavior cannot be legislated. It is a combination of strong values and the impact of the example set by peers and superiors. To better appreciate ethics, individuals must understand how the following factors interact with each other to impact their actions, behaviors and decisions.

5.4 THE CORRELATION BETWEEN ETHICS AND INTEGRITY

Integrity incorporates a personal affirmation of ethics. Ethics is definitely correlated with integrity. Both go hand in hand with one another.

So what is ethical integrity? The definition from the dictionary broke down the words into their category. We know honesty is the core root of principles that we live by. The key really is in the morality of principles that we use (our conscience mind is pretty unique and valuable to our inner most selves). To be kind and giving, to be true to ourselves and to others.

---

[1]Most of the issues raised in this chapter are discussed in more detail in Bednarz, 2011.

Ethics is basically the moral part of the principle in the way we act, what we say, and do as to how we listen and use our conscience for the common sense. As said already when it comes to teaching others, we have a code that is lived by, whether it is at work or at home it is for the living day-to-day basis as life brings many challenges. To be responsible and utilize the tools given and with that keep learning more skills for life as each day is lived. As an individual we are liable for our behavior and our actions. The wisdom gathered can be from many sources. There are those who live by the Ten Commandments, there are those who have a set of principles that they live by through an organization or non-profit organization. There are those that have a different religion with their beliefs. The good does not have to come as a book or a piece of paper, but it does come through the inner self of a person and by the laws of nature, we are entitled to use respect, to keep compassion, to seek rather than be the one to get, do rather than be lazy, and so on. The wisdom gathered as mentioned endures with experience. We use ethics as an external overview while integrity is from within, this keeps the compatibility as open as our mind (Source: *How Do I Maintain My Integrity and Remain Competitive in the Workplace?* By Lesley Ni, CPD – Raffles Institute Shanghai [Scribbler's Ink, 2005]).

But just what is actual integrity?

Integrity… is defined as an internal system of principles which guides our behavior. The rewards are intrinsic. Integrity is a choice and conveys a sense of wholeness. It may be influenced, but cannot be forced. To sum it up is basically just doing what is right (Source: *How Do I Maintain My Integrity and Remain Competitive in the Workplace?* By Lesley Ni, CPD – Raffles Institute Shanghai [Scribbler's Ink, 2005]).

Joseph Wilson (Xerox) stated:

Integrity is so much more than not doing wrong: Business and personal ethics mean always doing right – to the highest possible level – not just in response to issues presented, but by actively looking for issues and then flooding the system with affirmative actions on core values (Ellis, 2006).

The true test of any leader's integrity occurs in the face of adversity, when he/she is under intense and unrelenting pressure. When revenues and profits are endangered, as well as one's position, individuals are often tempted to take actions that are either unethical or illegal. After all, a sense of self-preservation can easily become stronger than one's integrity. This is the true test of great leadership, and many have failed it.

James Burke [Johnson & Johnson] demonstrated his integrity when he put the interests of consumers ahead of company profit. In 1982, after seven people in the Chicago area died after ingesting cyanide-laced Extra-Strength Tylenol capsules, Burke pulled all forms of Tylenol from every store in the country. Although the cost to the company was $100 million, Burke recognized that the public's trust was more important (Source: *How Do I Maintain My Integrity and Remain Competitive in the Workplace?* By Lesley Ni, CPD – Raffles Institute Shanghai [Scribbler's Ink, 2005]).

Mary Kay Ash (Mary Kay) continually stressed integrity to all of her associates as the foundation of her business, as well as theirs as independent contractors.

Integrity is the ingredient that will enable you to forge rapidly ahead on the highway that leads to success. It advertises you as being an individual who will always come through.

Whatever you say you will do. Do it even if you have to move heaven and earth. Each of us should have a philosophy about how we conduct ourselves with others. A long time ago, I chose as my standard the Golden Rule: 'Do unto others as you would have them do unto you.' Some might consider the Golden Rule corny and old-fashioned, but no one can deny its simple truth. Imagine how much better our world would be if everyone lived by this creed (Kay, 2010).

5.5 THREE CLASSES OF ETHICAL BEHAVIOR[2]

There are three classes of individual leaders: ethically challenged, ethically neutral and ethical. Obviously leaders who are ethically challenged, or ones who are devoid of any ethical standards are dangerous to their organizations. Many charismatic leaders fall into one of these categories and may pose a danger to the companies they lead.

Charisma is value neutral, it does not distinguish between good or moral and evil or immoral charismatic leadership. This means the risks involved in charismatic leadership are at least as large as the promises. Charisma can lead to blind fanaticism in the service of megalomaniacs and dangerous values, or to heroic self-sacrifice in the service of a beneficial cause (Gibson *et al.*, 1998).

When unethical or immoral behavior occurs within an organization . . . it greatly impacts everyone, similar to the effect a prodigal son or an unfaithful spouse has on a family . . . If top executives fail to follow their moral compasses, how can one expect those they lead to adhere to moral values? And if employees in the workplace do not care about ethics or morality, how can they expect their children to be any different? Everyone loses (Huntsman, 2008).

If ethics are poor at the top, that behavior is copied down through the organization (Guzzardi, 1989).

When leaders are ethically neutral, many of their decisions are based upon what is most advantageous, rather than what is the best or right thing to do. These leaders can also pose a danger to their organizations due to their ethical compromising. On this topic, Curtis Carlson (Carlson Companies) remarked, "Companies lose because they start making compromises, and they keep compromising until the company goes down (Carlock, 1999)".

What characterizes the great leaders is their level of principled and uncompromising ethics. Thomas Watson Jr. (IBM) noted:

You cannot treat management differently from the employees. If a manager does something unethical, he should be fired just as surely as a factory worker. It took me a number of years to realize that a CEO has to spot-check decisions made by his subordinates (Fortune Magazine, 1987).

---

[2]This section is adapted from Bednarz (2012: pp. 251–252).

5.6 ETHICAL CONDUCT IN THE WORKPLACE

Employees are expected to be honest and ethical in their dealings with each other, clients, vendors and all other third parties. In addition, doing the right thing means doing it right every time.

Ethics in the workplace means more than words on a card or in a company code of ethical behavior. Ethics is the creation of a corporate culture where every employee must not only "talk the talk," they must "walk the walk." This means that words have meaning and that all employees are expected to put the words into action. While ethical direction comes from the top of the organization, for a culture to be effective, every employee must assume a leadership role in defining and communicating the company's values that shape and guide its decisions and actions.

Every employee assumes a role in building a shared understanding of what an organization is about and how it should operate as an ethical company. In this way each employee is responsible for building and maintaining an ethical culture. Each should model the ethical behavior expected of them and make a habit of "catching someone doing the right thing" as an example for all other employees. This reinforces ethical behavior as the overall corporate culture. These behaviors should be consistent in all actions and should not only be maintained when others are observing employees.

The practical application of ethical behavior implies observing norms or standards that govern the conduct of each employee in the workplace.

5.6.1 Integrity

Employees should not place themselves under any financial or other obligation to other employees, clients, vendors or third parties that might influence them in the performance of their jobs and the fulfillment of their responsibilities.

5.6.2 Respect

Honesty, integrity and ethical behavior are based upon an employee's respect for an organization, and for its decisions, activities and actions, as well as those of other employees, clients, vendors and third parties. A lack of respect for any of these undermines an employee's personal attitudes and affects his or her ethical behavior.

5.6.3 Behavior

An employee's behavior is a reflection of their personal attitudes, values and ethical beliefs. While behavior cannot be legislated, it can be guided by ethical norms and standards. Rotary International adopted the Four-Way Test as an ethical decision-making tool that guides and directs an individual's behavior. It simply states:

Of the things we think, say or do ...

Is it the TRUTH?

Is it FAIR to all concerned?

Will it build GOOD WILL and BETTER FRIENDSHIPS?

Will it be BENEFICIAL to all concerned?

5.6.4 Choices

Ethical behavior is defined by the choices one makes in his or her personal or business activities. Applying ethical norms and standards requires that the choices employees make be both fair and just. The application of Rotary's Four-Way Test to each decision or choice provides employees with an ethical yardstick to measure their choices.

5.6.5 Consistency

Consistency is the ability to maintain a particular standard with minimal variation. In practical terms it means constantly attempting to do the right thing. However, this does not demand perfection. Since all individuals are human, mistakes will be made. Yet, consistency requires that when mistakes are discovered, they be quickly resolved.

5.6.6 Objectivity

Objectivity requires that employees who are responsible to hire or promote, award contracts or recommend individuals for rewards and benefits make their choices based solely upon merit and not be influenced by outside interests of other employees, managers, clients, vendors or third parties.

5.6.7 Openness

Employees should be as open as possible about all of the decisions and actions they take and should give reasons for their choices. There may be a legitimate justification for restricting information, but employees should be able to provide it to their managers or committees they report to.

5.6.8 Accountability

An open and ethical organizational culture requires all employees be held accountable to the company for their decisions and actions. Each must subject him or herself to whatever scrutiny is appropriate to their position within the company. Employees should remember that with rights come responsibilities.

5.6.9 Example

Employees should be examples of high standards of ethical behavior to all other employees, clients, vendors and third parties without question or repudiation. This also means not tolerating the unethical or questionable behaviors of other employees. Tolerance of unethical behavior by other employees undermines the organizational culture and will lead to increased problems and internal conflict.

5.7 ORGANIZATIONAL CULTURE, LAWS AND REGULATIONS

Business practices, customs and laws differ from country to country. An organization and its employees need to conduct their affairs in a manner consistent with the applicable laws and regulations of the countries where they do business. When conflicts arise between an organization's ethical practices and the practices, customs, and laws of a country, employees need to resolve them consistently with the ethical beliefs and standards of conduct of their company. If the conflict cannot be resolved in a manner consistent with its ethical beliefs and standards, the organization might consider not proceeding with the proposed action that is causing the conflict.

There are a number of reasons why companies are compelled to create codes of business conduct and ethical behavior. Treating employees, customers and vendors in a fair and consistent manner is good business. It enhances a company's reputation and overall value in the marketplace. However, one of the most compelling reasons is to be in compliance with the laws and regulations that govern business activities.

All ethics programs are built upon the interconnection of personal values and beliefs, corporate policies and applicable laws. However, ethical behavior cannot be legislated for. Organizations and the employees that comprise them are either ethical or they are not. The ethical corporation is defined by a combination of strong values and leadership by example at all levels. These are supported by both formal and informal norms and standards defining acceptable behavior, including fair and equitable treatment of employees, customers and vendors.

An organization that stresses an ethical culture reduces both internal and external conflict within the company and provides the means to settle any conflicts in a fair and efficient manner. In the absence of an ethical framework, conflict arises from varying personal values and beliefs, corporate policies or applicable laws.

The meshing of an organization's ethical conduct with applicable laws and regulations requires consideration of the following norms and boundaries.

5.7.1 Business practices

Employees need to be sensitive to the fact that formal business standards can change from country to country and within specific cultures. There are also informal business practices that may be regionally and culturally specific and may encourage unethical means of conducting business. These include conflicts of interest, the use of bribes, kickbacks, and gifts, as well as the unethical collection of competitive information and violations of fair competition and antitrust laws. While it may be expedient to tolerate and overlook these activities as a perceived necessity to successfully conducting business in these regions, this behavior is inconsistent with an ethically minded organization.

5.7.2 Ethical vs. legal

Many individuals in business walk a fine line between what is legal and what is ethical. They may be aware that their actions are unethical, but also know that they still fall within the boundaries of the law. Since they are not legally liable for their actions, the behavior is considered acceptable. This is the ultimate in gamesmanship, and a losing strategy.

What these individuals fail to understand or choose to ignore is the ramifications of their actions. While they may win in the short term, they and the organization will lose in the long term. Nothing will outrage employees, customers, vendors, and society as a whole more than individuals and companies operating in this manner. They feel used and will quickly leave the relationship, harming the business and its reputation. Customers and vendors will turn from advocates into activists, looking for every opportunity to damage the company and employees playing these games. The long-term ramifications of these actions include greater customer dissatisfaction and potential for legal action against the company.

5.7.3 Consistent practices and standards

Ethical practices and standards are clear and in black and white, not cloudy and in shades in gray. Cultural differences between countries and geographic regions are a continual source of conflict. If an organization hopes to maintain multinational status, it must be consistent in its ethical practices and standards so that situations and conflicts that do arise will be handled fairly and consistently. This is both the most expedient and practical way to manage ethical business practices on a global scale as one standard applies to all employees, clients, vendors and third parties regardless of their geographic location.

5.7.4 Cultural differences and attitudes

Multinational organizations and their employees must respect cultural differences that may surface in the course of their business activities. As previously stated, cultural differences are a continual source of conflict. The norms and standards governing ethical conduct and relationships should also govern employees' behavior when dealing with employees, clients, vendors and third parties who share their own (and especially different) geographic cultural values and beliefs.

5.7.5 Objectivity

Ethical situations and conflicts will arise when organizations interact with other government agencies, entities and cultures. Continually applying an ethical code of conduct and demanding that employees always "do the right thing," creates an atmosphere where objectivity prevails and subjectivity, surrounded by external agendas and emotional responses, is minimized. This is not to say these conditions will not exist, but always walking an ethical path provides all employees with a moral compass to follow and keeps them centered in their thinking and objective in their decision making.

5.8 ETHICAL ACTIVITIES AND EXPECTATIONS

The following requirements concern frequently raised ethical questions.

5.8.1 Conflicts of interest

Employees must avoid any personal activity, investment or association, which could appear to interfere with good judgment concerning an organization's best interests. Employees may not exploit their position or relationship with a company for personal gain.

5.8.2 Gifts, bribes and kickbacks

Other than modest gifts given or received in the normal course of business (including travel or entertainment), neither employees nor their relatives may give gifts to, or receive gifts from, an organization's clients and vendors.

Dealing with government employees is often different than dealing with private persons. Many governmental bodies strictly prohibit the receipt of any gratuities by their employees, including meals and entertainment. Any employee who pays or receives bribes or kickbacks may be immediately terminated and reported, as warranted, to the appropriate authorities. A kickback or bribe includes any item given to improperly obtain any type of favorable treatment.

5.8.3 Improper use or theft of company property

Every employee must safeguard company property from loss or theft, and may not take such property for personal use. Usually an organization will have in place a policy regarding employee use of its electronic communications systems which prohibits using such systems to access or post material that is:

- Pornographic
- Obscene
- Sexually-related
- Profane or otherwise offensive
- Intimidating or hostile
- In violation of company policies or any laws or regulations.

5.8.4 Covering up mistakes, falsifying records

Mistakes should never be covered up, but should be immediately and fully disclosed and corrected. Employees that falsify company, client or third-party records may be faced with dismissal as well as legal ramifications.

5.8.5 Protection of company, client or vendor information

Employees should not use or reveal company, client or vendor confidential or proprietary information to others. They must take appropriate steps, including securing documents, limiting access to computers and electronic media and utilizing proper disposal methods, to prevent unauthorized access to such information.

5.8.6 Gathering competitive information

Employees should not accept, use or disclose the confidential information of competitors. When obtaining competitive information, they need to keep in mind and not violate competitors' rights. Particular care must be taken when dealing with competitors' clients, ex-clients and former employees. Employees should never ask for confidential or proprietary information or ask a person to violate a non-compete or non-disclosure agreement.

5.8.7 Sales: defamation and misrepresentation

Aggressive selling should not include misstatements, innuendos or rumors about the competition or their products and financial condition. Employees should not make unsupportable promises concerning an organization's products.

5.8.8 Use of company and third-party software

Company and third-party software is usually distributed and disclosed only to employees authorized to use it, and to clients in accordance with the terms of an agreement. Company and third-party software should not be copied without specific authorization and may only be used to perform assigned responsibilities. All third-party software must be properly licensed.

5.8.9 Developing software

Employees involved in the design, development, testing, modification or maintenance of company software should not tarnish or undermine the legitimacy and "cleanliness" of an organization's products by copying or using unauthorized third-party software or confidential information. Employees should not possess, use or discuss proprietary computer code, output, documentation or trade secrets of another party, unless authorized by such party. Intentional duplication or emulation of the "look and feel" of others' software should also be avoided.

5.8.10 Fair dealing

No employee should take unfair advantage of anyone through:

- Manipulation
- Concealment
- Abuse of privileged information
- Misrepresentation of material facts
- Any other unfair-dealing practice.

5.8.11 Fair competition and antitrust laws

Every organization and its employees must comply with all applicable fair competition and antitrust laws concerning the following.

Securities trading

It is illegal to buy or sell securities using material information not available to the public. Persons who give such undisclosed "inside" information to others might be as liable as persons who trade securities while possessing such information. Securities laws may be violated if an employee, any relatives or friends trade company securities, or those of any of its clients or vendors, while possessing "inside" information.

Political contributions

Usually, company funds may not be given directly to political candidates.

5.9 SHADES OF GRAY

The list of ethical expectations above details a number of situations that would be considered major violations. Each presents a situation that is either black or white, right or wrong. In the majority of situations on this list, the right choice is straightforward and it is a conscious decision to do something obviously unethical.

Keeping in mind the personal choices on the major items detailed above, an employee is often confronted with situations that fall within less definitive areas or "shades of gray":

- Altering a financial report at the request of a superior
- Protecting a friend and coworker whose drinking is causing productivity problems
- Taking credit for work on a report that was prepared by someone else
- Put off correcting a safety situation because the cost will decrease profitability
- Changing a performance appraisal to reflect more positively on an individual whose advancement is important to a supervisor
- Giving a bad review to save money on a raise
- Justifying an unethical action because it was requested by a superior
- Overcharging a client if revenue growth was down
- Selling used equipment as new
- Using copied software to save money for an organization or business unit.

When comparing the above ethical expectations with the list of situations outlined earlier, most individuals would find this list poses more difficult ethical decisions. In many cases these choices may impact personal performance or the department or group they are responsible for. Often, individuals make unethical decisions without first thinking about what they are about to do. There are numerous ethical dilemmas that all individuals are faced with on a daily basis and that collectively impact the success or failure of an organization.

The practical application of ethics in the workplace creates dilemmas for the average individual, whether they realize it or not. Some people can be impervious to these decisions and act in accordance to their personal beliefs and values without consideration of the impact on an organization. The practical daily application of ethics encompasses the following personal considerations:

5.10 VALUES

Ethics are ethics, whether in the workplace or in one's personal life. They are all based upon an individual's values. While the major decisions are easy, it is the day-to-day activities where an individual's ethics are put to the test, especially where a friendship, one's job, personal gain or money is involved.

The small, everyday decisions are where an individual's ethical character is defined. Most individuals will not cross the line when major forms of misconduct or unethical behavior is involved. The stakes are too high; the decisions are too high profile, and the risks and consequences enormous. However, the small, everyday decisions make it easier to cross the line. Many are made in private and most often will not be discovered. Most can justify these choices if it gives them "the edge." They often ask themselves: Who is harmed and who is to know?

What most people do not realize is that these poor everyday decisions can lead to a deterioration of personal values. Each unethical choice erodes one's values slowly and often without them even realizing what is happening. Step-by-step it becomes easier to expand one's unethical choices. When no one notices or they are not held accountable, it is easier to make poor choices. As an individual rises within an organization, each decision has more of an impact within it. Without realizing it, the major choice an individual would never make in the past, due to its consequence and risk, is suddenly easy to make.

5.10.1 Integrity

Personal levels of integrity protect one's values. When an individual acts with integrity, they are not putting their own interests before those of an organization. Integrity especially comes to the forefront when individuals make ethical decisions in private, when they are not being observed and they know that no one will be aware of their choice. Integrity means always "doing the right thing" no matter what the circumstances or the scale of the decision. Individuals with integrity and personal character will do the right thing in all circumstances, no matter what and no matter who knows, even if it is only themselves.

5.10.2 Courage

The final element of the application of practical ethics is courage. The dilemma surrounding most ethical decisions is the impact that they will have on others or the company. Most often the tough decisions involve some form of pain and discomfort that will be experienced by friends, employees, clients, vendors or third parties. These decisions may personally impact promotions, raises and bonuses. Individuals with strong personal values and integrity will have the courage to "do the right thing."

Personal courage can be painful to develop, but it is an asset organizations highly value as it builds character and molds individuals into leaders.

5.11 TESTING ETHICAL BEHAVIOR

In the final analysis, an employee is the guardian of their organization's ethics. While there are no universal rules, when in doubt ask:

- Will this action be ethical in every respect and fully comply with the law and with company policies?
- Will this action have the appearance of impropriety?
- Will supervisors, employees, clients, family and the general public question this action?
- Is the employee trying to fool anyone, including themselves, as to the propriety of their actions?

If uncomfortable with an answer to any of the above, the employee should not take the contemplated actions without first discussing them with superiors.

The focal point of this training program is the premise that there are "rights" and "wrongs." These create choices that are either black or white. Operating between the boundaries of black or white is the gray areas of ethics. The majority of people are most comfortable in this range as they feel it gives them the most latitude in their choices or decisions.

In reality, gray areas are often defined by the following five characteristics:

- A lack of understanding of the full nature of the problem
- A lack of critical analysis
- A lack of effort in discovering all the facts and a lack of understanding regarding ethical alternatives
- A lack of desire to take a certain path
- The competition between two rights, two wrongs or a right and a wrong.

An examination of the five characteristics reveals that four of the five reflect either ineffective or lazy thought processes. Ethical choices and decision making requires forethought and contemplation about the options. Functioning in the gray areas is easy because most people do not have to think or work at what they are doing.

The final characteristic defining gray areas – an ethical dilemma or competition between two rights, two wrongs or a right and a wrong – is one that most individuals choose not to deal with. Functioning in the gray areas allows them to avoid dealing with these issues. In reality, they are turning a "blind eye" to this decision. This is an unethical choice.

5.12 CONCLUSION

This chapter has examined many facets of organizational ethics and conduct. In essence, ethics comes down to the personal choices that employees make and that apply to their professional life within a company.

Virtually every individual knows the inherent difference between right and wrong. This is where all ethical choices and decisions are made, whether they are personal or professional. As was stated previously, the big ethical choices are usually easy to make due to the clear risks and consequences, but it is the small, lower risk daily decisions that trip up many individuals.

If an individual makes a firm personal commitment to operate in the areas of right and wrong, black or white, the choices are straightforward. He or she will always opt for "the right thing." It becomes second nature. These decisions are made without thought or consideration. This individual automatically defaults in their thinking to the right choice.

However many individuals do not possess the internal fortitude and courage to take such a firm stand. There are too many pressures that are too difficult to deal with. If an employee feels this way they must continually measure their choices and decisions against an ethical benchmark.

REFERENCES

Bednarz, T, *Business Ethics* (Majorium Business Press, Stevens Point, WI, 2011).

Bednarz, T, *Great! What Makes Leaders Great* (Majorium Business Press, Stevens Point, WI, 2012).

Carlock, R, 1999 *Minnesota Business Hall of Fame: A Conversation with Curt Carlson* (Twin Cities Business, July 1999) Online. Available from http://tcbmag.com/Honors-and-Events/Minnesota-Business-Hall-of-Fame/1999-Minnesota-Business-Hall-of-Fame/Curtis-L-Carlson.

Eldelman, R, 2012 *Edelman Trust Barometer© Annual Global Study Executive Summary*. Edelman, New York, p. 1.

Ellis CD, *Joe Wilson and the Creation of XEROX* (John Wiley & Sons, Inc., Hoboken, NJ, 2006) p. 362.

Gibson JW, Hannon JC, Blackwell CW., Charismatic leadership: the hidden controversy (*Journal of Leadership Studies*, v 5, issue 4).

Guzzardi W, *Wisdom from the Giants of Business* (Fortune Magazine, July 3, 1989).

Huntsman, JM, *Winners Never Cheat Even in Difficult Times* (Wharton School Publishing, Upper Saddle River, New Jersey 2008) pp. 166–167.

Mary Kay Inc. Corporate Website, 2010 http://www.marykay.com/en-US/About-Mary-Kay/CompanyFounder/Pages/default.aspx (accessed 2 January 2013).

Ni, L, *How Do I Maintain My Integrity and Remain Competitive in the Workplace?* (Scribbler's Ink, 2005). Online. Available from http://www.raffles-cpd.com/CPD/News/ArticlesDetails.aspx?ItemID=d82e8d42-ba1b-4e56-9506-91b343357b58&WebResTypeID=659de181-494e-49f1-bb88-d50a464e1e9e&privs=0 (accessed 2 January 2013).

The Greatest Capitalist in History (Fortune Magazine, August 31, 1987).

6 Ethics in publishing/reporting food science and technology research

Daryl Lund

6.1 INTRODUCTION

Are ethics in publishing still important? Are they important in writing research reports? Some would argue that ethics are more important today than ever before. Frequently there are exposés about scientific reports that have been fabricated, used false data, included authors who made no contribution to the work, were ghost written by a non-scientist not involved in the research, and included materials previously published by someone else verbatim without proper reference and attribution. The importance of ethics in food science publishing and reporting has elevated to the point where several university food science departments are including ethics of publishing and writing reports in the topics presented to graduate students. Increasingly, there are letters to the editor or a letter by the editor announcing an impropriety in something that was published in the journal. Publishers have taken greater pains to ensure that what they publish contains appropriate and properly attributed material. Journals have rewritten their guidelines for authors to emphasize the importance of "truth in publishing," and with electronic submission and review of manuscripts, journal editors have increasingly used computer programs that search for previously published work within the manuscript so they can check that proper attribution has been cited. In this chapter, several critical elements of ethics in publishing and report writing are shared. Most of these ideas can be found in the "Guidelines for Authors" that journals post on their webpage.

6.2 WHO SHOULD BE LISTED AS AN AUTHOR ON A PUBLICATION OR A REPORT?

This seems like an innocuous question but can present a real dilemma depending on culture and autocratic bosses. In addition, some universities require that candidates for promotion identify his or her role in every publication in which they are named as a co-author. For scientific journals published by the Institute of Food Technologists (*Journal of Food Science*, *Comprehensive Reviews in Food Science and Food Safety*, and *Journal*

Practical Ethics for Food Professionals: Ethics in Research, Education and the Workplace, First Edition.
Edited by J. Peter Clark and Christopher Ritson.
© 2013 John Wiley & Sons, Ltd. Published 2013 by John Wiley & Sons, Ltd.

of Food Science Education), co-authorship is restricted to those who have "contributed substantially to one or more of the following aspects of the work: conception, planning, execution, writing, interpretation, or statistical analysis." (IFT Scientific Journals, 2011). The decision as to who is included in authorship, of course, ultimately lies with the person submitting the manuscript or report. Journals have no intention of checking the authenticity of authorship nor should they be required to. Hence, this is a matter of ethical consideration. Equally important is the requirement that those listed as authors must be willing to vouch for the authenticity and validity of the research.

Probably every editor has had situations arise in which the authorship of a published paper has been questioned. When this occurs, the process for ascertaining the truth is easy to follow (just ask the author in question and the other co-authors the role each played in the publication) but knowing the "truth" is not so easily obtained. This is where culture and personal relationships come into the situation. In some cultures it is perfectly acceptable to list the boss, center director, dean or other person of authority as a co-author. Generally this practice is frowned upon and should not be the norm. When this does occur, situations can arise wherein the results in a publication can be challenged and the person who really played no role in the publication has to defend the results. Obviously this could become a "sticky wicket" since the person may have no knowledge of the work or insufficient knowledge to defend the research. The bottom line is that each person listed as a co-author should know what they contributed to the publication and be able to defend the results.

Another circumstance that has arisen and was publicized recently is the use of ghost writers. For researchers whose native language is not English, once the research work has been completed and the results and discussion taken place, a person with excellent command of the English language is hired to write the manuscript. Wong (2010) described such occurrences in China where researchers have used English teachers to write the manuscript. She implied this was academic cheating. At a minimum, ethically, ghost writing would be considered inappropriate if the writer is not included in the list of authors.

6.3 IS THE RESEARCH CONDUCTED IN AN ETHICAL MANNER?

All research organizations today have guidelines that must be followed when animals or humans are used in carrying out the research. The National Institutes of Health has a website devoted to proper use of animals in research and alternatives to using animals (http://bioethics.od.nih.gov/animals.html). Their guidelines are nearly universally followed. To ensure that the guidelines are strictly followed, all research universities have established review committees to which protocols for use of animals and humans in research projects must be subjected.

Currently less than half the journals that have published papers in which animals are used in the research have a policy on proper use or ethics of using animals in research and use of human subjects (Osborne, 2011). Osborne points out that these journals and in fact the review committees on campuses rarely consider if it is necessary to use animals or human subjects in the research. Their only evaluation is on the procedures for using animals or human subjects.

As a result of this apparent lack of journal attention to ethics in using animals and human subjects in research, Osborne and the Royal Society for the Prevention of Cruelty to Animals (RSPCA) have initiated a program to alert biomedical and biosciences journals to establish guidelines on ethical use of animals in research submitted to their journals, and specific guidelines are currently under development.

In spite of lack of more specific guidelines, most journals have some statement on ethical use of animals and humans in research. For example, the *Journal of Food Science*, published by the Institute of Food Technologists, has the following in their guidelines for authors (IFT, 2011):

4. Ethical issues:

If the work involves experimentation on living animals, authors must provide evidence that it was performed in accordance with local ethical guidelines. In the case of work involving human beings, evidence must be provided that it was performed with the approval of the local ethics committee.

6.4 IS THERE A CONFLICT OF INTEREST BY THE RESEARCHERS?

Recently there has been increased interest in requiring scientists to disclose potential conflicts of interest in the research they are conducting. This stems in part from the relationship of doctors to drug companies when the doctors are part of the evaluation team assessing efficacy of drugs. Campbell and Zinner (2010) called for a re-examination of the federal policy on disclosure of the relationship between researchers and industry. For the food and nutrition community, the International Life Sciences Institute of North America (ILSI-NA) assembles a distinguished team of researchers and public policy experts to develop guidelines for researchers in the field (Rowe *et al.*, 2009). Although the primary focus of the guidelines is for researchers and the conduct of the research, the principles also apply to publishing the results of the study.

As an example of conflicts of interest policies and guidelines, since IFT's scientific journals are actually printed and distributed through Wiley-Blackwell, the journals also subscribe to the code of ethics for Wiley-Blackwell (http://www.blackwell publishing.com/Publicationethics/). These are very extensive and by association with Wiley-Blackwell, IFT's standards include requiring the authors to disclose conflicts of interest that would compromise the integrity of the research. In peer-reviewed journals such as IFT's there is often a check list for reviewers to identify if they believe there is a bias in the research or a conflict of interest for the authors. By the same token, the reviewer is also asked if he/she has a conflict of interest that would influence his/her ability to fairly and impartially review the manuscript. Biases and conflicts of interest by reviewers are an issue not only in publications but also in reviewing research proposals. As a result, the first requirement of a reviewer is to declare his/her potential biases and conflicts of interest.

6.5 IS THERE BIAS BASED ON FUNDING SOURCE?

One of the major criticisms of food science research particularly from those who purport to be protecting the consumers' interest is that the study was funded by industry. Given the lack of public funding of food science and technology research, industry funding is essential to make advances in knowledge and application of scientific principles to improvement of food, food ingredients, and food processing.

Most peer-reviewed journals require that the authors disclose all sources of funding for the research. In addition if the editor feels there may be a bias based on funding source he/she has the authority to query the authors in the interest of full disclosure and transparency. This transparency of investigator and relationship to the funding source is essential to quell the critics and move the science and knowledge forward. Generally, the funding source is in the acknowledgement of the manuscript and published with the paper.

Aside from full disclosure and transparency, the question remains, "Is there a bias in research based on funding source?" Unfortunately there is very little published research on this topic. The major reason for this is that there is no large database that can be dissected to analyze results coupled with funding source. Such a study would require an analysis of the results by experts to determine if the results were indeed influenced by funding source.

Although many societies and journals do not maintain such a database, the American Dietetic Association does maintain a database on nutrition research. This allowed Myers *et al.* (2011) to use the American Dietetic Association's Evidence Analysis Library database of 2539 peer-reviewed research papers to examine if there is a link between funding source and quality of the research paper reporting.

Myers *et al.* (2011) found that 43.3% of the research reports were rated positive, 50.1% neutral and 6.6% negative. The majority of those rated negative were reports in which the funding source was not stated. They concluded that industry funded research was no more likely to be rated neutral or negative than those funded by federal government sources.

6.6 WHAT IS PLAGIARISM AND WHAT ARE ITS CONSEQUENCES?

The plague of the publishing business today is plagiarism. With the internet, it is so easy to manufacture a manuscript by piecing together bits and pieces from one's own or others' previously published work and call it a new contribution. Although plagiarism will be briefly discussed here, it clearly is a matter of ethics since it is illegal. In fact, journals have really tough penalties if plagiarism is found in manuscripts. To assist journals in making decisions about plagiarism, the Committee on Publishing Ethics (COPE; http://publicationethics.org/) has developed roadmaps for editors to use. COPE was founded in 1997 by journals in medicine and currently has over 7000 members worldwide. All the major publishing houses including Elsevier, John Wiley & Sons, Springer and Taylor-Francis, the major publishers of food science and technology books

Table 6.1 Terms in use for plagiarism

Term	Definition
Copy-paste writing, or cut-paste writing	The reuse of text published by others in one's own manuscript – usually for the sake of using "good, already-published English" or of producing a manuscript faster. The reused text may be substantial strings of words that may be sentence fragments, sentences, several sentences or whole paragraphs. Authors might do this with or without attribution.
Micro-plagiarism	A form of copy-paste writing in which the copied texts are consistently small (a clause or a sentence or two) but frequent in one or more sections.
Patch writing, or mosaic writing	The end result of copy-paste writing. These terms convey the choppiness a text can have when copy-paste writing strategies are used.
Plagiarism	Copying of substantial amounts of text with an intent to deceive the reader into assuming that the writing and ideas belong to the author.
Self-plagiarism	Reuse of substantial portions of text from one's own previous work.
Duplicate or redundant publication	Reuse of one's own previous work that goes beyond text (i.e. the use of wholly or substantially overlapping data).
Translated plagiarism	The use, after translation, of strings of sentences, paragraphs, or even larger blocks of prose, with or without attribution, keeping the informational structure of the original intact.

Adapted from Kerans and de Jager (2010). Reproduced by permission of the European Association of Science Editors.

and journals, are members. COPE provides best practice guidelines, flowcharts, sample letters, and a database of all cases used as examples.

Karens and de Jager (2010) provided a definition of terms related to plagiarism (Table 6.1), and these are useful in thinking about plagiarism. If an author uses any one of these forms of plagiarism, he/she is subject to the penalties inflicted by the journal. To assist journal editors in assessing if plagiarism has occurred, Cross Check was developed, and most journals subscribe to a program for checking manuscripts against other previously published work. The program is called iThenticate and, for example, it is used by the Institute of Food Technologists' peer- reviewed journals.

The way the system works is that the editor, either on his/her own or as suggested by an Associate Editor, subjects the manuscript to iThenticate. The report that comes back (after a few minutes) reports the percent similarity to other published works (including information on internet sites). The information can be filtered to exclude references (since one would expect references to be similar in manuscripts or published works on the same topic). The report provides the number of similar words between the manuscript and the previously published work, and the editor can then examine the previously published work and the manuscript and determine how egregious the overlap is. If it is sufficiently apparent that plagiarism, deliberate copying of materials from a previously published work (either one's own or another's), has occurred then the process moves to making a decision of what action to take against the author.

COPE has produced a roadmap for dealing with instances of alleged plagiarism, and these are widely used by editors. In summary (since there are a number of different paths depending on severity of the offense – see types in Table 6.1), in most cases the editor will contact the author of the manuscript asking for an explanation and the author(s) whose work has been plagiarized will also receive notice of the alleged plagiarism. The penalty will then depend on the outcome of the correspondence between the editor and the author of the manuscript. As an example, the *Journal of Food Science* has experienced several instances of plagiarism (and one of these instances actually prompted the journal to incorporate iThenticate in its subscription to ScholarOne, the automated manuscript handling system that the journal uses uses). The most severe penalty imposed to date is to (1) print a retraction of the paper in the journal citing the papers that were plagiarized, (2) barring all the plagiarizing authors from submitting any manuscripts to the Journal for 5 years, and (3) notifying the immediate supervisor of the authors that this incident of plagiarism has occurred. This may seem harsh, but committing plagiarism is a very serious matter and journals must take strong action to curb it.

The easiest way to avoid plagiarism is to make sure that proper attribution is given to previous published work. Use of quotation marks with identification of the source is perfectly legal and should be used when a direct quote is used from another publication. Rewording previously published work without direct copying is also appropriate, provided recognition is given to the origin of the thought or idea.

6.7 SUMMARY

Lest you think that this chapter is written only for those readers who are going to author manuscripts for publication, think again. Authors of internal reports for industry and government laboratories need to keep these same principles in mind as they prepare their reports. To assist in making good decisions as writing begins for reports and manuscripts, consider the questions in Table 6.2. The following are the principles of making sure your publications/reports are ethical: (1) recognition of authorship, (2) ethical conduct of research involving animals and human subjects, (3) ensuring that conflict of interest is transparent, (4) removing bias in conclusions of the results as a result of funding source or personal financial interest, and (5) making sure that others previously published work is properly cited and attributed.

Table 6.2   Checklist question for authors of manuscripts/reports

Consideration	Questions to ask
Recognition of authorship	Have each of the authors made substantiated contributions to the research/report?
Ethical conduct of research involving animals and human subjects	Have the experiments using animals or humans been conducted within the national recognized guidelines?
Ensuring that conflict of interest is transparent	Have all of the authors revealed all conflicts of interest in the results of the research?
Removing bias in conclusions of the results as a result of funding source or personal financial interest	Has the appearance of or actual bias based on funding source been explored and explained?
Making sure that others previously published work is properly cited and attributed	Has this manuscript/report properly attributed all ideas and previous investigations to the appropriate sources/authors?

REFERENCES

Campbell, E.G. and D.E. Zinner. 2010. Disclosing industry relationships-toward an improved federal research policy. *N Engl J Med* 36: 604–606.

Committee on Publishing Ethics. http://publicationethics.org/ (accessed 12 December 2012).

Institute of Food Technologists. 2011. IFT Scientific Journals: Author Guidelines. http://www.ift.org/Knowledge-Center/Read-IFT-Publications/Journal-of-Food-Science/Authors-Corner/Author-Guidelines.aspx (accessed 12 December 2012).

Kerans, M.E. and M. de Jager. 2010. Handling plagiarism at the manuscript editor's desk. *Eur Sci Editing* 36(3):62–65.

Myers, E.F., J.S. Parrott, D.C. Cummins, and P. Splett. 2011. Funding source and research report quality in nutrition practice-related research. *PLos One* 6(12): e28437.

National Institutes of Health. 2012. URL for Bioethics of Animal Use in Research. http://bioethics.od.nih.gov/animals.html (accessed 12 December 2012).

Osborne. N. 2011. Point of view: reporting animal research-worthy of a rethink. *Learned Publishing* 24(1): 5–7.

Rowe, S., N. Alexander, F.M. Clydesdale, *et al*. 2009. Funding food science and nutrition research: financial conflicts and scientific integrity. *Nutr Rev* 67:264–272.

Wong, G. 2010. In China, academic cheating is rampant: Some say practice harmful to nation. *Associated Press*. April 11, 2010.

7 Humane treatment of livestock

Temple Grandin

7.1 INTRODUCTION

Humane treatment of livestock is good for business for many reasons. First of all, it is the right thing to do. Today everybody has a mobile phone with a camera. The last thing that any business wants is bad publicity. Pictures of people abusing animals on one of your supplier's farms is going to damage the reputation of your company. There are many videos on the internet showing animal abuse and some have been shown on national news. The videos have caused some people to stop eating meat. There has been an unfortunate tendency for many companies to react to bad press by trying harder to keep cameras out. Shortsighted pig and poultry producers in several states have passed laws making it a felony to photograph an animal operation without permission. This has caused a big backlash and has reduced consumer trust. The *New York Times* wrote a scathing editorial that the livestock industry has lots of bad things to hide. However, a few progressive slaughter plants and farms have done the opposite. They have invited the press in. The Cargill Corporation allows press in their beef slaughter plants and they had one of their plants featured on the nationally televised Oprah Winfrey Show. The best approach is to open the door. This can be done electronically. J.S. West, a poultry farm and Fair Oaks dairy have live videos that can be viewed by the public. A video I did on methods to humanely stun and slaughter pigs has over 800 000 views on YouTube.

7.2 IMPROVE MEAT QUALITY

There are many situations where maintaining high standards of animal welfare will improve the quality of the meat. Careful handing of livestock during transport and handling at the slaughter plant will reduce bruises. Bruised meat cannot be used for human consumption. It must be cut out and put into rendering. Poor handling methods will double the amount of bruising (Grandin, 1981). Bruising can occur right up until the moment of slaughter (Meischke and Horder, 1976). The most common cause of bruising are excited cattle jamming in a truck door or hitting gate latches and other objects that

Practical Ethics for Food Professionals: Ethics in Research, Education and the Workplace, First Edition.
Edited by J. Peter Clark and Christopher Ritson.
© 2013 John Wiley & Sons, Ltd. Published 2013 by John Wiley & Sons, Ltd.

have sharp edges. On cattle, the muscle can have huge bruises, but the hide may not appear damaged.

Many research projects show that handling during the last few minutes before slaughter can have a big effect on meat quality. Cattle that have been prodded repeatedly with electric prods within 10 minutes before slaughter will have tougher meat (Warner *et al.*, 2007). Excitement, squealing, and poking with electric prods within five minutes before slaughter will raise lactate levels and increase the incidence of PSE (pale soft exudative) pork (Edwards *et al.*, 2010; Hambrecht *et al.*, 2005a,b).

7.2.1 Industry programs

Some corporations in the livestock and poultry industry have greatly improved the treatment of the animals that are raised for food. In some cases, a company had used good animal welfare as a marketing tool, but in other cases, real improvements have been made but the company had failed to communicate them with the public. They had an excellent program that could have been used to improve public perception, but they had failed to take advantage of it. In many companies, pressure from animal activists and undercover videos has served as an impetus to make them improve the care and treatment of animals. The author has worked with many large corporations on implementing their animal welfare programs. Some of the companies I have worked with are McDonald's Corporation, Wendy's International, Safeway, Whole Foods, KFC, Tesco, and others. In many organizations, I have observed the animal welfare issue changing from an abstract nuisance that is delegated to the legal and PR department to a reality. I call this opening up the executive's eyes. I have taken many high-level executives from many large restaurants and retailers on their first trips to farms and slaughter plants. I have seen their attitude change. At first, their attitude was "Why do we have to do this," and then their attitude changes and they understand that welfare is a real issue. Their company needs to do something about the treatment of animals. When conditions on a farm or slaughter plant were good, the executives were pleased. But when conditions were bad, they were motivated to use their huge buying power to bring about reforms.

I remember the day when a vice president watched an emaciated, half dead, dairy cow walk up the chute and into his company's hamburger. He was appalled. In several big food companies, the supply chain management staff got motivated and brought about huge improvements, but the rest of the company did not communicate them to the public. For example, in 1999, when the McDonald's program first started, maintenance of stunning equipment was so poor that only 30% of the plants rendered 95% of the cattle insensible with one shot from a captive bolt gun (Grandin, 1998a). Seven years later, a typical plant was stunning 97% of the cattle correctly (Grandin, 2005). When large meat buyers use their huge purchasing power wisely, great improvements can happen. Taking a few plants off the approved supplier list for McDonald's made it clear to the meat industry that McDonald's was serious about animal welfare in slaughter plants. Both McDonald's and Wendy's have great programs but they have not communicated effectively with the public. When I travel, and my seatmate on an airplane asks me what I do, I tell them that I worked as a consultant for McDonald's and I implemented their animal welfare program. Most fellow passengers are surprised to learn that McDonald's has an animal welfare program.

7.2.2 Welfare used in branding

There are many different labels in the market place for welfare-branded meat and poultry products. Some of these are Whole Foods, Gap program, RSPCA in the UK, Humane Certified, and others. These programs cater mainly to the niche markets and most of the animals are raised on pasture or straw-based systems instead of more intensive confinement. These labels cater to a high-end market. However, there is a basic level of animal welfare that should apply to everybody. Abuse, neglect, or deliberate cruelty should never be tolerated. For example, all cattle and other livestock, regardless of the type of production system they are raised in, should NEVER be beaten, have an eye poked out or be starved.

7.2.3 Motivation for the farm and food industry to change

It is unfortunate that many needed reforms that have occurred in animal welfare were motivated by pressure from activist groups. I have a saying, "heat softens steel." Pressure from animal activist groups and undercover videos forced many companies to address the welfare issue and implement programs. When the executives were taken out of the office to see what their suppliers were doing, they were motivated to make change. Over 30 years I have worked to improve animal handling and transport. "Heat" from activist groups has "softened the steel" and I have been able to bring about improvement. I have another saying. When heat softens steel, I can bend it into pretty grillwork.

7.2.4 Motivation for activist groups

When I first started working with livestock in the 1970s, most activist groups and the people in them were motivated to reform the meat industry. There is a good review of different moral views on the use of animals in Palmer and Sandøe (2011). In the 2000's a new younger generation of activists has taken a more abolitionist approach. People who believe in the abolitionist approach are motivated to eliminate animal agriculture instead of reforming it. In some cases, instead of doing things that would improve animal agriculture, they file lawsuits to make raising animals more and more difficult. Staunch abolitionists do not approve of reformers like me. If you are trying to eliminate meat eating then making animal treatment better is wrong. Some abolitionists state that animals should not be kept as property, therefore raising animals for food is wrong.

People who are animal welfarists, such as myself, believe that using animals for food can be done in an ethical manner (Grandin, 2010b; Rollin, 2010). The Farm Animal Welfare Council (FAWC) in the United Kingdom has stated that an animal that is raised for food should have "a life worth living." Bernard Rollin stresses the need for good stockmanship and animal husbandry. I often get asked, why do you continue to work in the animal industry when it has so many problems? I have seen many ranchers, cattle feeders, and pork producers who treat animals really well. There are people who do things right. It is the ethical, good people in the animal industry who have motivated me to continue to work to improve livestock production and slaughter. Many research studies have shown that good stockmanship and gentle, considerate treatment of farm animals improves weight gain, reproduction, and milk production (Hemsworth *et al.*, 1981, 1986; Rushen *et al.*, 1999).

7.2.5 What does the public think?

When I was doing book tours for *Animals in Translation, Animals Make us Human*, and the HBO movie *Temple Grandin*, I had extensive contact with the public. The general public are the people that animal agriculture needs to communicate with. Animal agriculture will never be able to satisfy all the activists. One of the problems with the internet is that it enables extremists on both sides of an issue to have a much bigger voice. The internet can turn into a screaming match between radicals on both sides. The book and movie tour was an eye-opening experience on how the general public feels.

1. *Curiosity*: Most people are extremely curious about how animals are raised. At a movie press conference, all the press wanted to talk about was how things were done in animal agriculture. They asked, "What is a CAFO (confined animal feeding operation) and how are cattle killed at the slaughter plants?"
2. *Urge to reconnect with food*: Many people want to learn more about where food comes from. This has motivated many people to plant a garden, have a few backyard chickens, or buy local food. In 2011, a large percentage of the public is totally removed from where their food comes from. One lady attending a recent livestock exhibition thought that piglets were puppies.
3. *The industry needs to open the door*: Some progressive companies have installed video cameras so that the public can tune in and see how the animals are produced. When free video services such as YouTube (www.YouTube.com) first became available on the internet, they were mostly filled with ghastly, cruel undercover videos. In 2011, when I typed "cattle feedlot", "chicken farm", and "pig farm" into YouTube, the top videos showed normal procedures such as caring for piglets or feeding cattle. Most people want to look at normal stuff. Progressive farms are proud of what they do and they are really maintaining high standards.

7.2.6 Lip service is not enough

During my career, I have seen many companies make a pretty manual of animal care standards but they are never really implemented. In 2004, Timothy Pachirat worked in a dreadful beef slaughter plant as a quality assurance manager. The plant management was not serious about either food safety or humane treatment of animals (Pachirat, 2012). This book really upset me because they tortured cattle in a piece of equipment I had designed. Employees were told to poke electric prods up a steer's anus but to never do it when the US Department of Agriculture (USDA) inspector was around. This plant had lots of paperwork that they faked and they were a good example of a place that paid "lip service" to animal welfare. Mr. Pachirat quit his job as a quality assurance person because his managers asked him to ignore obvious welfare and food safety violations. In 2012, conditions are better than in 2004. Additional improvements have been brought about by increased USDA enforcement, pressure from meat buying customers, and some really horrendous undercover videos. On one of the worst videos, workers deliberately drove a forklift into an old dairy cow, and knocked her over. A USDA official called this video a "policy changing event" and enforcement was increased.

7.2.7 Equipment does not replace management

Some of the worst animal abuse I ever witnessed was in the 1980s and 1990s when I was supervising installation of the center track conveyor system and other equipment I had designed for meat plants. Since I was part of the construction and engineering staff, I would have been fired if I had reported it. In every case, I did talk to the plant manager. Too often people buy nice equipment that they can brag about but they do not operate correctly. If I had a choice, I would rather have superb management with older, adequate equipment than a brand new state-of-the art plant with terrible management. The plant that Mr. Pachirat worked in was a brand new, state-of-the art independent plant with old-fashioned management that cut every corner that could be cut.

During my career, I have worked on cattle and pig handling consulting, equipment design and supervision of start-up of the equipment. I have witnessed every type of management from the very best to the absolute worst. When a good manager hires a quality assurance person to monitor both animal welfare and food safety, that person is given the power to implement real changes. In a poorly run plant, the person is not able to make real improvements because management ignores their recommendations. Instead, they sit in the office and fabricate paperwork. In well-managed plants, this does not happen, because upper management believes in the importance of both food safety and animal welfare. When I worked on implementing the McDonald's auditing program there were about 75 plants on the pork and beef supplier list. Out of the 75 plants, three plant managers had to be either fired or replaced. This may sound harsh but this was the reality. Practices in the three plants changed from terrible to really good when top management was replaced.

7.2.8 Scientific animal welfare principles

Scientific studies clearly show that animals can suffer. Mammals, such as cattle, pigs, and birds feel fear and pain. To determine that animals feel pain, a self-medication experiment was done with chickens and rats. When a leg is injured, the bird or animal will drink or eat a bitter tasting fast acting painkiller (Colpaert *et al.*, 2001). When the leg heals, the animal will stop ingesting the bitter painkiller. The self-medication experiment shows very clearly that animals will seek relief for pain. One of the most severe animal welfare problems in intensively housed animals is lameness. In dairy cows, lameness is the number one welfare issue and research shows very clearly that it causes pain (Rushen *et al.*, 2006; Flowers *et al.*, 2007). Other species that have high levels of lameness are pigs and poultry. Lameness can be easily measured by watching animals walk and scoring them as lame or not lame (Grandin, 2010a).

7.2.9 The importance of reducing fear

The fear circuits in the brains of animals have been fully mapped (Rogan and LeDoux, 1996; Grandin, 1997). In some situations, fear can be worse than pain. If you have ever been in a car accident or in a natural disaster such as a tornado, you will know how terrifying fear can be. Neuroscience research shows very clearly that the fear circuits in both human and animal brains are very similar (Setckleiv *et al.*, 1961; Matheson *et al.*, 1971; Redgate and Faringer, 1973; Davis, 1992; Rogan and LeDoux, 1996).

One of the most fear-provoking situations is handling animals for veterinary procedures, transport, and slaughter. When animals are handled quietly and calmly, fear stress can be greatly reduced. Reducing fear stress also has the advantage of improving animal productivity. Sows, dairy cows, and other animals that fear people, run away and avoid human contact will have less piglets born, lower weight gain, and less milk production (Hemsworth *et al.*, 1981; Hemsworth and Coleman, 1998). Many research studies have shown good stockmanship improves productivity. When people remain calm, animals remain calm. Screaming and hitting animals can be very stressful. Screaming at cattle increases heart rate more than the sound of a gate slamming (Waynert *et al.*, 1999).

One question that gets asked all the time is "Do animals know that they are going to get slaughtered?" I have observed that cattle, pigs, and sheep react to handling on the farm the same way they react to walking up a chute at a slaughter plant. When animals are being handled in a strange, novel place, such as an auction or slaughter plant, they are afraid of a whole lot of little things that most people do not notice. They may refuse to walk over a shadow on the floor or walk up an alley that has a dangling loose chain (Tanida *et al.*, 1996). Animals often get scared when they go to a new place that has many novel sights and smells. Agitated behavior at a slaughterhouse is often due to the fear of novelty (Terlouw *et al.*, 2011). Many animals are fearful about walking into a dark alley. Adding additional lighting will help attract animals into a chute (Grandin, 1980, 1996). Installing solid panels to prevent animals from seeing people up ahead will improve animal movement in truck loading ramps and slaughter plants. Animals notice little sensory details that most people fail to notice. To reduce fear stress, it is essential that people working with animals are trained in low stress handling methods. There is much more information on low stress animal handling in Grandin and Deesing (2008). Stress levels will be much lower and animals will be easier to handle if animals are carefully acclimated to handling procedures on the farm (Ried and Mills, 1962; Hutson 1985; Fordyce, 1987; Geverink *et al.*, 1998). Animals need to become accustomed to people walking quietly among them and being moved through chutes before they arrive at a slaughter plant. It is best to do these acclimation procedures on young animals, and it is essential that their first experiences with new people, equipment, or places are good experiences.

7.2.10 The importance of good animal health and nutrition

Many people mistakenly think that health is the only thing that matters for good animal welfare. Health is essential to have good welfare because sick animals would have bad welfare. To maintain good health, animals need to always have access to plenty of clean water and be fed sufficient feed to prevent them from becoming too thin.

However, animals need to have more than good health to have a satisfactory level of welfare. There are situations where animals can be healthy and productive but welfare would be poor. One study showed that the highest producing dairy cows had more swellings and lesions on their legs (Fulwider *et al.*, 2007). Chickens bred for a high rate of growth also had high percentages of lame birds (Knowles *et al.*, 2008). Bored animals that constantly pace, chew bars, or perform repetitive behaviors, may also have good health.

7.2.11 Stocking density

When pigs, laying hens, and other animals are jammed too tightly in cages or pens, the productivity of each individual animal will be reduced (Bessci, 2004). Unfortunately, there is often an economic incentive to overstock animals because even though individual productivity declines, the economic return on the entire building may increase when animals are jammed in. Some producers have packed laying hens so tightly into a cage that when the birds sleep at night, they have to lie on top of each other. An absolute minimum space requirement for a housing facility for both birds and livestock is that all the animals can lie down at the same time without being on top of each other. This is a bare minimum and more space is strongly recommended (Grandin, 2010b). Low air ammonia levels are also essential. The goal should be less than 10 ppm (Kristensen and Wathes, 2000; Jones *et al.*, 2005). The maximum level for worker safety is 25 ppm.

7.3 BEHAVIORAL NEEDS

To have an acceptable level of welfare, an animal needs to have its basic behavioral needs met. The field of ethology has provided scientific documentation that shows that animals have real behavioral needs in addition to maintaining their health. Ethology is the study of an animal in a natural environment. When an animal is studied in its natural environment, it is possible to create an ethogram which catalogs all the behaviors that an animal naturally does. For example, pigs will root and nose fibrous materials and hens will lay their eggs in a secluded spot. Obviously, on a modern farm, it will not be possible to provide a pig with everything that it likes to do in the wild. What is essential is to determine which behaviors are the most important to the animals (Duncan, 1998). Behaviors that are highly motivated are the most important and they should be accommodated on a commercial farm. Scientists can measure strength motivation to perform certain species typical behaviors in a very objective manner. Below are three common tests.

- Weighted door test – an animal is taught to go through a door (similar to a doggie door) to get something it wants. To measure the strength of motivation, increasing amounts of weight are attached to the door and more and more effort is required to push it open (Duncan and Kite, 1987).
- Pushing a switch – an animal is trained to push a switch to get something it wants. Motivation is higher if the animal is willing to push a switch many times to get it.
- Time off feed – length of time that the animal will go without food in order to perform a behavior it wants to do (Duncan and Petherick, 1991).

An example of a highly motivated behavior is as follows. Hens prefer to lay their eggs in a secluded nest box. They are highly motivated to do this, because this is an instinctual fear motivated behavior to hide in a safe place to lay eggs. When hens lived in the wild, the hen that laid her eggs in the open clearing was eaten by the fox. Hens that hid when they laid their eggs survived and raised chicks. Pigs, on the other hand, are highly motivated to root and manipulate fibrous material such as straw (Van de Weerd *et al.*, 2003; Studnitz *et al.*, 2007). Below is a list of the different types of animals and some of the highly motivated behavioral needs that should be accommodated in large-scale commercial systems. These are the systems that activists call factory farms. This list also includes environmental devices, which help to improve bone health, reduce lameness, or reduce damage to the animal. For all species, ammonia levels in the building must be kept at a low level.

7.3.1   Laying hens – minimum behavioral needs

- Secluded nest box (Duncan and Kite, 1978, Freire *et al.*, 1967).
- Perches to roost on at night. All the hens should be able to fit on the perches at the same time (Hughes *et al.*, 1993).
- A place to scratch in litter.

7.3.2   Broiler chickens and turkeys – minimum behavioral needs

- Provide friable litter that has dry material for the birds to scratch in. Additional dry litter must be added if the litter becomes wet and transfers soil onto the birds.
- Provide a dark period at night (Prescott *et al.*, 2004).

7.3.3   Pigs – minimum behavioral needs

- Change individual gestation stalls to group housing because sows in gestation stalls cannot turn around or walk for their entire pregnancy.
- Provide small amounts of roughage for rooting and chewing (Studnitz *et al.*, 2007; Van de Weerd *et al.*, 2003).Providing a handful of straw prevented bar biting in sows (Fraser, 1975). Another alternative is to put additional roughage in sow feed.

7.3.4   Cattle – both beef and dairy, minimum behavioral needs

- Provide sufficient roughage in the diet to maintain normal rumen function and help prevent the development of stereotypic tongue rolling.
- House in social groups with their own kind
- A place to sleep that will not injure the legs. Poor stall design or lack of stall mainte-nance increases the incidence of swollen legs in dairy cows (Fulwider *et al.*, 2007). All animals should have a dry place to lie down.

7.4   BIOLOGICAL SYSTEM OVERLOAD

Since the 1960s, the productivity of farm animals has continued to increase. In 1970, dairy cows gave half as much milk compared to 2011. Lucy (2001) reported that milk production has been increased 2% each year since 1991. I visited a farm in 2011 and the sows weaned 10 to 11 piglets per litter. Broiler chickens and turkeys grow at an astonishing rate. There is a point where the animal's metabolism is pushed so hard for increased productivity, that problems are occurring. Lameness is a huge issue in dairy cows. Surveys of dairies showed that 24% of the national herd is lame (Espejo *et al.,* 2006). In the worst 20% of dairies, 33% to 62% of the cows are lame (Webster, 2005). In the top 5% of dairies, only 5% of the cows are lame (Espejo *et al.*, 2006). In 2011, I have observed that in well-managed dairies, 5% or less of the cows are lame.

Broiler chickens and turkeys also have had problems with lameness. One side effect of increased productivity is that animals have become weaker and they have a shorter life. Some high producing dairy cows may last for only two years of milking. In the 1980s, a dairy cow would normally last for 4 to 5 years of milking. Another problem is disease resistance. I visited Australia in the early 2000s shortly after rapidly growing American broiler chickens were introduced. When these new hybrid chickens were introduced, there was a serious outbreak of Newcastle disease. Studies on local breeds of chickens in Egypt showed that there are genetic differences in resistance to Newcastle (Hassen and Afify, 2004). When lean, rapid growing pigs were introduced in the 1980s, porcine reproductive and respiratory syndrome (PRRS) erupted. Research done by Vincent *et al.* (2006) showed that some genetic lines are more susceptible to PRRS. Since 1990, the swine industry started having problems with diseases that previously did not cause problems such as swine wasting disease, and mycoplasma arthritis. Pigs are getting so susceptible to disease that some producers are putting special HEPA filters in the ventilation system to filter out pathogens (Vansickle, 2006). This is the same type of filter that is used to protect immunocompromised people. I think this is the wrong approach. Breeding animals that are more disease susceptible is both an animal welfare and a food security issue.

High feed prices caused by large quantities of corn going into ethanol plants provides an economic incentive that may be really detrimental to animal welfare. When feed grain is expensive, this motivates producers to push the biology of animals even harder. In beef cattle, and pigs, beta agonists such as ractopamine (Paylean, Optiflex) make the animals put on muscle mass really fast. The downside of doing this is increases in lameness, heat stress, excitability and animals that are more difficult to handle. Pigs have become weak and fragile. I have observed truckloads with over 20 non-ambulatory pigs. I have also seen pigs that were so weak and excitable that they had difficulty walking off a truck and moving through the stockyards at a slaughter plant. Scientific studies have documented handling problems and hoof cracking in pigs fed ractopamine (Marchant-Forde *et al.*, 2003) and (Poletto *et al.*, 2009). There is also evidence that ractopamine increases aggression in female pigs (Poletto *et al.*, 2010). Many people working in the industry have observed problems in some groups of feedlot cattle fed either ractopamine or zilpaterol. Some of the problems are heat stress, lameness, and increased death losses during hot weather. Macias-Cruz *et al.*, (2010) reported higher skin temperatures in ewe lambs fed zilpaterol. Dairy cows that are pushed too hard with rBST growth hormone lose body condition and may become too skinny. They can also have increased mastitis (Widman, 1993; Kronfield, 1994; Collier *et al.*, 2001). Both beta agonists and rBST have to be used very carefully to avoid animal welfare problems.

7.4.1 Strive for optimum productivity

The sensible thing to do is to strive for optimum productivity instead of maximum productivity. Perhaps a cow that lasts for 4 years of milking would be a reasonable tradeoff. Biological system overload is not just an animal welfare concern; it is also a food security concern. If animals get wiped out by disease, that would be a disaster from a food security standpoint. There are always tradeoffs. Overzealous genetic selection for rapid growth or some other production trait has a price. An animal is like a country. If a country puts all its resources into the economy, then nothing is left for the military and if all the resources are put into the military, there is nothing left for the economy. The "economy" would be meat, milk or eggs and the "military" would be the immune system. Hardy wild-type animals with a strong ability to fight disease are usually not very productive. When animals are indiscriminately selected for productivity, resistance to disease or parasites may decrease. There is a need to find a balance to prevent serious welfare problems such as lameness, and heat stress, and maintain an adequate level of disease resistance.

7.5 CORE CRITERIA – CRITICAL CONTROL POINTS FOR WELFARE

There are certain core criteria that are essential for a minimum level of animal welfare. The meat market has two main sectors. These are large scale commercial and the different specialty and niche markets such as organic, pasture raised or ultra high welfare standards. There will always be a need for a large-scale commercial sector to keep food, such as eggs, affordable. CAFOS or factory farms are what activist groups call these facilities. The niche markets will cater to a wealthier clientele. There is a minimum set of standards that both sectors should adhere to. It is impossible for a farm to have a high level of welfare if 30% of the dairy cows are lame. In an auditing program, core criteria need to be weighted heavily when the audit is scored. Most of the core criteria are directly observable and are not done as a paperwork audit. There is an unfortunate tendency to turn both animal welfare and food safety audits into paperwork audits. The most important welfare indicators are directly observable.

There are three basic types of standards. They are (1) animal-based outcome standards, (2) prohibited practices, and (3) input standards (Grandin, 2010b). The trend in welfare auditing is to move away from input based or engineering standards that go into detail on how to build housing. However, there are a few input standards that are needed.

7.5.1 Core animal-based outcome standards

- *Body condition score*: Basically it means how many really skinny animals does a farm have? Low body condition score is an outcome of either insufficient feed, disease or pushing diary cows too hard to produce more milk. Body condition scoring charts are available for beef, cattle, sows, and dairy cows. Body condition scoring charts for Holstein dairy cows are available from Wildman *et al.*, (1982) and University of Wisconsin (2005). Body condition scoring is most important in breeding animals, lactating females, such as dairy cows, beef cows, sows and ewes. When feed is scarce, all animals should be scored for body condition.
- *Percentage of lame animals*: Lameness can be measured by watching how animals walk. Some common lameness scoring tools are in (Knowles *et al.*, 2008; Zinpro®

http://www.zinpro.com/lameness (accessed 13 December 2012). Good dairies have only 5% lame cows (Espejo *et al.*, 2006). I have observed that on the best broiler farms, less than 1% of the birds have difficulty walking. Dawkins *et al.* (2004) has a simple lameness scoring system for market ready broiler chickens. A normal bird walks 10 paces evenly. A lame bird walks 10 paces crooked and a bird scoring 3 is not able to walk 10 paces. Lameness is more common in lactating animals and rapidly growing animals.

- *Percentage of animals with sores, lesions, and swellings*: Scoring cards with lots of photos are available at many industry websites and European Welfare Quality Assurance for dairy cows, beef cows, sows, sheep, and poultry. For laying hens, go to LayWel (www.laywel.eu; accessed 13 December 2012). Important measures for all types of poultry are footpad lesions, breast blisters, and hockburn.
- *Coat condition and feather condition*: Important for organically raised animals because the use of anti-parasite medication is restricted. Some of the easily observed conditions are ringworm and bald spots caused by lice. In poultry, feather condition is scored. Pictures which show both good and poor feathers can be found at www.laywel.eu.
- *Hygiene score*: What percentage of the animals or birds are dirty from lying in wet manure? An animal or bird should be scored as severely soiled if the belly or breast is covered in manure.
- *Abnormal behavior*: Examples of abnormal behavior that can be readily observed are bitten tails on pigs, feather pecking in chickens, tongue rolling in dairy cattle or beef cattle, belly nosing in pigs, and wool pulling in sheep.
- *Udder condition in dairy cows*: Some of the problems are damaged suspensory ligaments and in some extreme cases, the udder may be almost dragging on the ground.
- *Animal handling scoring*: The quality of animal handling can be scored by measuring variables such as the percentage of animals falling (should be 1% or less), percentage of animals vocalizing, percentage that run into a fence or gate, and the percentage speeding (going faster than a walk or trot) (Grandin, 2010b). In a well-run slaughterhouse, less than 5% of the cattle will vocalize (moo, bellow) in the stun box (Grandin, 2005). Vocalization during handling is associated with aversive events such as electric prodding (Grandin, 1998b). Vocalization is also associated with physiological indicators of stress (Dunn, 1990; White *et al.*, 1995). Other indicators that cattle are being stressed and scared is, whites of the eye show in cattle (Sandem *et al.*, 2006; Core *et al.*, 2009), tail swishing, and the heads are up. At the slaughter plant it is easy to measure the consequences of rough handling, such as broken wings in poultry, bruises, broken legs, death losses, and the percentage of downed non-ambulatory animals. Audit information for slaughter plants can be found at Grandin (1998a, 2010a, 2010b, 2010c).
- *Thermal comfort*: Cattle that are severely heat stressed will have open mouth breathing (Mader *et al.*, 2005). Poultry, cattle, and sheep will all have open mouth breathing when they become overheated. Overheated animals must be provided with either shade, fans or water sprays. In extreme cold weather, windbreaks and other shelter must be provided. A combination of high temperature, high humidity, and no air movement is the most dangerous and can cause death losses.

7.5.2 Input or engineering measures

There are a few input measures that are important, but the trend is to move away from telling producers exactly how to design equipment.

- *Life support backup*: This is essential in any housing system where the animal's lives depend upon mechanical ventilation. It can consist of back-up generators, devices to open up the building, such as curtain drops or alarms. If the animals live in a naturally ventilated building, life support back up is not required.
- Ammonia levels in enclosed housing: High ammonia levels are detrimental to the welfare of both people and animals. For safety for people, ammonia levels should not exceed 25 ppm. Ten ppm is recommended (Jones *et al.*, 2005; Kristensen and Wathes, 2000).
- *Minimum space requirements* for housing, water trough space, and feed trough space.
- *Clean water and plenty of it.* Water consumption may double or triple during hot weather. Check to insure that hot, thirsty animals are not sucking water troughs dry.

7.5.3 Commercial audits need to be simple

An audit program for commercial use must be simpler than measuring tools that are used for research. The core criteria that can be directly observed are more important than looking at paperwork that people can fake. The core criteria also work on the HACCP principle of identifying multiple problems with a single measurement. For example, lameness in dairy cows can be caused by foot disease, poor hoof trimming, poor flooring, freestalls (cubicle) that are designed wrong, or rough handling. Lameness is the outcome of all of these bad conditions.

Three legs on the tripod

Lots of places have fancy books of standards they can show their customers, but are they really doing what they say they are doing? A good auditing program for both animal welfare and food safety has three parts. To insure you are doing what you say you are doing requires all three parts. A tripod is not stable unless it has all three legs.

1. Internal audits done by the farmer or the slaughter plant. People manage the things they measure.
2. Large buyers of meat, milk, or dairy products should have every one of the farms or slaughter plants that they buy from, audited at least once a year. This can be done either by a third party auditor or field staff that work for the company.
3. Random Check Audits – Depending on how your system is set up, a certain percentage of places need to be audited by another entity to make sure the people doing the yearly audit are doing their job. If the system relies mainly on third party auditors, then corporate staff does the check audits or a third party audit company does check audits on field staff. The places that are chosen for the check audits should be a combination of places with past problems and random choices.

Video auditing over the internet

In 2011, 23 large beef slaughter plants in the US installed video auditing, which is monitored by a third party auditing firm over the internet. This has resulted in great improvements. To be effective, the cameras must be monitored by an outside auditor. I have observed that plant management fails to monitor the cameras after the novelty wears off. A major advantage of the cameras is that they prevent the problem of people "acting good" when they see a person doing an audit. Data from the video audits has also revealed that people change their behavior when they know an auditor is watching. The video audits have also located plants where internal auditors faked records. On the positive sides, the video auditing system has been used to promote positive competition between plants owned by the same company. The plant with the best animal handling gets a reward. The employees handling the animals may get a pizza party or baseball caps when they win best plant for humane handling.

REFERENCES

Bessci, W. (2004) Stocking density, In: C.A. Weeks and A. Butterworth (editors) *Measuring and Auditing Broiler Welfare*, CABI International, Wallingford, UK, pp. 133–144.

Collier, R.J., Byatt, J.C., Denham, S.C., *et al.* (2001) Effects of sustained release bovine somatotropin (sometribove) on animal health in commercial dairy herds, Journal of Dairy Science, 84:1098–1108.

Colpaert, F.C., Taryre, J.P., Alliaga, M., and Kock, W. (2001) Opiate self administration as a measure of chronic nociceptive pain in arthritic rats, Pain 91:33–34.

Core, S., Miller, T., Widowski, T., and Mason, G. (2009) Eye white as a predictor of temperament in beef cattle, Journal of Animal Science 87:2174–2178.

Davis, M. (1992) The role of the amygdala in fear and anxiety, Annual Review of Neuroscience, 15:353–375.

Dawkins, M.S., Donnelly, C.A. and Jones, T.A. (2004) Chicken welfare is influenced more by housing conditions than stocking density, Nature 427:343–348.

Duncan, I.J.H. (1998) Behavior and behavioral needs, Poultry Science 77:1766–1772.

Duncan, I.J.H., and Petherick, J.C. (1991) The implications of cognitive processes for animal welfare, Journal of Animal Science, 69:5017–5022.

Duncan, I.J.M. and Kite, V.G. (1989) Nest for selection and nest building behavior in the domestic hens, Animal Behavior, 37:215–231.

Dunn, C.S. (1990) Stress reaction in cattle undergoing ritual slaughter using two methods of restraint, Veterinary Record, 126:522–525.

Edwards, L.N., Engle, T.E., Corca, J.A., Paridis, M.A., Grandin, T., and Anderson, D.B. (2010) The relationship between exsanguination blood lactate concentration and carcass quality in slaughter pigs, Meat Science 85:435–340.

Espejo, L.A., Endres, M.I., and Salfer, J.A. (2006) Prevalence of lameness in high-producing Holstein cows housed in freestall barns in Minnesota, Journal of Dairy Science, 89:3052–3058.

Flowers, F.C., de Passillé, A.M., Weary, D.M., Sanderson, D.J., and Rushen, J. (2007) Softer, higher friction flooring improves gait of cows with and without sole ulcers, Journal of Dairy Science, 90:1235–1242.

Fordyce, C. (1987) Weaner training, Queensland Agricultural Journal, 113:323–324.

Fraser, D. (1975) The effect of straw on the behavior of sows in tether stalls, Animal Production, 21:59–68.

Freire, R., Appleby, M.C. and Hughes, B.O. (1997) Assessment of pre-laying motivation in the domestic hen by using social interaction, Animal Behavior, 34:313–319.

Fulwider, W.K., Grandin, T., Garrick, D.J., *et al.* (2007) Influence of free-stall base on tarsal joint lesions and hygiene in dairy cows, Journal of Dairy Science, 90:3559–3566.

Geverink, N.A., Kappens, A., van de Burgwal, J.A., Lambooij, E., Blockhuis, H.J., and Wiegant, V.M. (1998). Effects of regular moving and handling on the behavioral and physiological responses of pigs to preslaughter treatment and consequences on subsequent meat quality, Journal of Animal Science, 76:2080–2085.

Grandin, T. (1980) Observations of cattle behavior applied to the design of cattle handling facilities, Applied Animal Ethology, 6:19–31.

Grandin, T. (1981) Bruises on southwestern feedlot cattle, Journal of Animal Science, 53(Supplemental 1) 213 Abstract.

Grandin, T. (1996) Factors that impeded animal movement in slaughter plants, Journal of the American Veterinary Medical Association, 209:757–759.

Grandin, T. (1997) Assessment of stress during handling and transport, Journal of Animal Science, 75:249–257.

Grandin, T. (1998a) Objective scoring of animal handling and stunning practices in slaughter plants, Journal of the American Veterinary Medical Association, 212:36–39.

Grandin, T. (1998b) The feasibility of using vocalization scoring as an indicator of poor welfare during slaughter, Applied Animal Behavior Science, 56:121–138.

Grandin, T. (2005) Maintenance of good animal welfare standards in beef slaughter plants by use of auditing programs, Journal of the American Veterinary Medical Association, 226:370–373.

Grandin, T. (ed) (2007) Livestock Handling and Transport, 3rd Ed., CABI International, Wallingford, UK.

Grandin, T. (2010a) The importance of measurement to improve the welfare of livestock, poultry and fish, In: T. Grandin (Editor) Improving Animal Welfare: A Practical Approach, CABI International, Wallingford, UK, pp. 1–20.

Grandin, T. (2010b) Implementing effective standards and scoring system or assessing animal welfare on farms and slaughter plants, In: T. Grandin (Editor) Improving Animal Welfare: A Practical Approach, CABI, International, Wallingford, UK, pp. 32–49.

Grandin, T. (2010c) Auditing animal welfare at slaughter plants, Meat Science, 86:56–65.

Grandin, T., and Deesing, M. (2008) Humane Livestock Handling, Storey Publishing, North Adams, MA, USA.

Hambrecht, E., Eissen, J.J., Newman, D.J., Smits, C.H.M., den Hartog, L.A. and Vestegen, M.W.A. (2005a) Negative effects of stress immediately before slaughter on pork quality are aggravated by suboptimal transport and lairage conditions, Journal of Animal Science, 83:440–448.

Hambrecht, E., Eissen, J.J, Newman, D.J., Verstegan, M.W. and Hartog, L.A. (2005b) Preslaughter handling affects pork quality and glycolytic potential of two muscles differing in fiber type organization, Journal of Animal Science, 83:900–907.

Hassen, M.K., and Afify, M.A. (2004) Genetic resistance of Egyptian chickens to infection bursal disease and New Castle disease, Tropical Animal Health and Production, 36:1–9.

Hemsworth, P.H., and Coleman, G.J. (1998) Human–Livestock Interactions, CABI International, Wallingford, UK.

Hemsworth, P.H., Brand, A., and Willems, P. (1981) The behavioral response of sows ot the presence of human beings and its relation to productivity, Livestock Production Science, 8:67–74.

Hemsworth, P.H., Barnett, J.L. and Hansen, C. (1986) The influence of handling by humans on the behavior, reproduction and corticosteroids of male and female pigs, Applied Animal Behaviour Science, 15:303–314.

Hughes, B.O., Wilson, S., Appleby, M.C. and Smith, S.F. (1993) Comparison of bone volume and strength as measures of skeletal integrity in caged laying hens with access to perches, Research in Veterinary Medicine, 54:202–206.

Hutson, G.D. (1985) The influence of barley feed rewards on sheep movement through a handling system, Applied Animal Behavior Science, 14:263–273.

Jones, E.K.M., Mathes, C.M. and Webster, A.J.F. (2005) Avoidance of atmospheric ammonia by domestic fowl and the effect of early experience, Applied Animal Behaviour Science, 90:293–308.

Knowles, T.G., Kestin, S.C., Hasslam, S.M., et al. (2008) Leg disorders in broiler chickens: Prevalance, risk factors and prevention, PLOS One, 3(2) available from http://www.ncbi.nlm.nih.gov/pubmed/18253493 (accessed 8 January 2013).

Kristensen, H.H and Wathes, C.W. (2000) Ammonia and poultry: A review, World Poultry Science Journal, 56:235–243.

Kronfield, D.S. (1994) Health management in dairy herds treated with bovine somatotropin, Journal of the American Veterinary Medical Association, 204:116–130.

LayWel (2009) LayWel – Periodic Final Activity report. Available at: http://ec.europa.au/food/animal/welfare/farm/laywel_final_report_en.pdf (accessed 1 February 2013).

Lucy, M.C. (2001) Reproductive loss in high producing dairy cattle: Where will it end? Journal of Dairy Science, 84:1277–1293.

Mader, T.L., Davis, M.S., and Brown-Brandl, T. (2005) Environmental factors influencing heat stress in feedlot cattle, Journal of Animal Science, 84:416–422.

Marchant-Ford, J.N., Lay, D.C., Pajor, E.A., Lunstra, B.D., Rohrer, G.A. and Ford, J.J. (2007) The effects of ractopamine on the behavior and physiology of finishing pigs, Journal of Animal Science, 81: 416–422.

Macias-Cruz, U., Alveraz-Valenzuela, F.D., Torrentera-Oliversa, N.G., *et al.* (2010) Effect of zilpaterol hydrochloride on feedlot performance and carcass characteristics of ewe lambs during heat stress conditions, Animal Production Science, 50:983–989.

Matheson, B.K., Branch, B.J., and Taylor, A.N. (1971) Effects of amygdoid stimulation on pituitary adrenal activity in conscious cats, Brain Research, 32:151.

Mieschke, H.R.C., and Horder, J.C. (1976) A knocking box effect on bruising in cattle, Food Technology in Australia, 28:369–371.

Pachirat, T. (2012) Every Twelve Seconds, Yale University Press, New Haven, CT.

Palmer, C. and Sandøe, P. (2011) Animal ethics, In: M.C. Appleby, J.A., Mench, I.A.S. Olsson, and B.O. Hughes (Editors) Animal Welfare, 2nd Edition, CABI International, Wallingford, UK, pp. 1–12.

Poletto, R., Rostagno, M.H., Richert, B.T., and Marchant-Forde, J.N. (2009) Effects of a "step up" ractopamine feeding program, sex and social rank on growth performance, hoof lesions, and Enterobacteriaceae shedding in finishing pigs, Journal of Animal Science, 87:304–311.

Poletto, R., Cheng, H.W., Meisel, R., Garner, J.P., Richert, B.T., and Marchant-Forde, J.N. (2010) Aggressiveness and grain amino concentration in dominant and subordinate finishing pigs fed B2 adrenoceptor agonist ractopamine, Journal of Animal Science, 88:3017–3120.

Prescott, N.B., Kristensen, H.H. and Wathes, C.M. (2004) Light, In: C.A. Weeks, and A. Butterworth (Editor) Measuring and Auditing Poultry Welfare, CABI International, Wallingford, UK, pp. 101–116.

Redgate, E.S., and Faringer, E.E. (1973) A comparison of pituitary adrenal activity elicited by electrical stimulation of preoptic amygdaloid and hypothalamic sites in the rat brain, Neuroendocrinology, 12:334.

Ried, R.L. and Mills, S.C. (1962) Studies in carbohydrate metabolism in sheep, XVI, The adrenal response of sheep to physiological stress, Australian Journal of Agricultural Research, 13:282–294.

Rogan, M.T. and LeDoux, J.E. (1996) Emotion systems, Cells Synaptic Plasticity Cell, 85:469–475.

Rollin, B. (2010) Why is agricultural animal welfare important? The social and ethnical context, In: T. Grandin (Ed.) Improving Animal Welfare: A Practical Approach, CABI Publishing, Wallingford, UK, pp. 21–31.

Rushen, J., Taylor, A.A., and dePassille, A.M.B. (1999) Domestic animals fear of humans and its effect on their welfare, Applied Animal Behavioral Science, 65:285–303.

Rushen, J., Pombourceq, E. and dePaisselle, A.M. (2006) Validation of two measures of lameness in dairy cows, Applied Animal Behaviour Science, 106:173–177.

Sandem, A.I., Janczak, K.M., Salie, R. and Braastad, B.O. (2006) The use of diazepam as a pharmacological validation of eye white as an indicator of emotional state in dairy cows, Applied Animal behavior, Science, 96:177–183.

Setckleiv, J., Skaug, O.E., and Kaada, B.R. (1961) Increase in plasma 17-hydroxycorticasteroids by cerebral cortical and amygdaloid stimulation in the cat, Journal of Endocrinology, 22:119.

Studnitz, M., Jenson, M.B., and Pederson, L.J. (2007) Why do pigs root and in what will they root? A review on the exploratory behavior of pigs in relation to environmental enrichment, Applied Animal Behavior, Science 107:183–197.

Tanida, H., Miura, A., Tanaka, T. and Yoshimoto, T. (1996) Behavioral responses of piglets to darkness and shadows, Applied Animal Behavior Sci., 49:173–183.

Terlouw, E.M.C., Bourguet, C., and Deiss, V. (2011) Stress at slaughter in cattle: Role of reactivity profile and environmental factors, Recent Advances in the Welfare of Livestock at Slaughter, Humane Slaughter Association, Wheathampstead, UK.

University of Wisconsin (2005) Body condition score. What is body condition score? Why does it help us manage? http://babcock.wisc.edu/sites/default/files/de/en/de_12.en.pdf (accessed 1 February 2013).

Van de Weerd, H.A. and Day, J.E.L. (2009) A review of environmental enrichment for pigs housed in intensive housing systems, Applied Animal Behavior Science, 116:1–20.

Vansickle, J. (2006) Filters first line of defense against PRRS, National Hog Farmer, www.national hogfarmer.com/mag/farming_filters_first-line/ (accessed 13 December 2012).

Vincent, A.L., Thacker, B.J., Halbur, P.G., Rothschild, M.F., and Thacker, E.L. (2006) An investigation of the susceptibility to porcine respiratory and reproductive syndrome virus between two genetically diverse commercial lines of pigs, Journal of Animal Science, 84:49–57.

Warner, R.D., Ferguson, D.M., Cottrell, J.J., and Knee, B.W. (2007) Acute stress induced by preslaughter use of electric prodders causes tougher meat, Australia Journal of Experimental Agriculture, 47:782–788.

Waynert, D.E., Stookey, J.M., Schwartzkopf-Gerwein, J.M., Watts, C.S. and Waltz, C.S. (1999) Response of beef cattle to noise during handling, Applied Animal Behaviour Science, 62:27–42.

Webster, J. (2005) The assessment and implementation of animal welfare: theory into practice. Review Scientifique et Technique (International Office of Epizootics), 24:723–734.

White, R.G., DeShazer, I.A., Tressler, C.J., et al. (1995) Vocalizations and physiological response of pigs during castration with and without anesthetic, Journal of Animal Science, 73:381–386.

Wildman, E.E., Jones, G.M., Wagner, P.E., Boman, R.L., Troutt, H.F., and Lesch, T.N. (1982) A dairy cow body condition scoring system and its relationship to selected production characteristics, Journal of Dairy Science, 65:495–501.

Widman, P. (1993) Bovine somatotropin and clinical mastitis: Epidemiological assessment and welfare risk, Livestock Production Science, 36:55–66.

8 Sustainable food production and consumption

Jeanette Longfield

8.1 INTRODUCTION

The problem of feeding a growing world population, and how that should be done, has been rarely out of the headlines in recent years. What "sustainability" means, and the extent to which we should focus our efforts more on increasing production, or more on reducing or changing what we eat, have become hotly contested issues. Given the political, economic and environmental implications of different choices, the heat of the debate is likely to rise rather than fall. This chapter aims, if not to cool the debate, at least to inject some ethical principles that are all too often lacking.

8.2 A WORKING DEFINITION OF SUSTAINABLE FOOD AND FARMING

As a UK alliance of some 90 plus national, independent organizations working on various aspects of sustainable food and farming, Sustain (the organization for which the author of this chapter works) is well placed to develop a definition of the term. The general principles set out below were drawn up in 2007 (Sustain, 2007).

1. Use local, seasonally available ingredients as standard, to minimize energy used in food production, transport and storage.
2. Specify food from farming systems that minimize harm to the environment, such as certified organic produce.
3. Limit foods of animal origin (meat, dairy products and eggs) eaten, as livestock farming is one of the most significant contributors to climate change, and eat meals rich in fruit, vegetables, pulses, wholegrains and nuts. Ensure that meat, dairy products and eggs are produced to high environmental and animal welfare standards.
4. Exclude fish species identified as most "at risk" by the Marine Conservation Society, and choose fish only from sustainable sources – such as those accredited by the Marine Stewardship Council.

Practical Ethics for Food Professionals: Ethics in Research, Education and the Workplace, First Edition. Edited by J. Peter Clark and Christopher Ritson.
© 2013 John Wiley & Sons, Ltd. Published 2013 by John Wiley & Sons, Ltd.

5. Choose Fairtrade-certified products for foods and drinks imported from poorer countries, to ensure a fair deal for disadvantaged producers.
6. Avoid bottled water and instead drink plain or filtered tap water, to minimize transport and packaging waste.
7. Protect health and well-being by cooking with generous portions of vegetables, fruit and starchy staples like wholegrains, cutting down on salt, fats and oils, and cutting out artificial additives."

As we noted at the time, these principles are – and will remain – a "work in progress." New evidence is emerging all the time on how best to improve the sustainability of our complex food and farming system. Already, if we were to update them, we would add a principle on reducing waste, and expand the principle on fair trade to incorporate good jobs in rich, as well as poor countries.[1] We could also add, among others, sections on the importance of having a variety of ways to obtain sustainable food, from growing your own to having a choice of easily accessible outlets – shops, markets, cafés and restaurants and so on – to buy food from.

In short, we do not think there is a definitive way to describe sustainable food and farming, nor do we think there should be. Indeed, the search, in some quarters, for such a definition can often be a thinly disguised excuse for postponing making an effort to improve our food. So, working on the proverb that "perfect is the enemy of good", Sustain is clear that these principles are a good enough basis for action.

We are also clear that, although the principles are listed, they are not in priority order and they should be taken in the context of the "classic" definition of sustainable development (Brundtland Commission, 1987), which has indivisible economic, environmental and social elements. Our principles of sustainable food are therefore also intertwined and should not be "cherry picked". Just as some will postpone any efforts to improve the sustainability of our food and farming system, others will focus only on one or two elements, sometimes to the detriment of other principles. This is sometimes called "greenwash" and it is not "sustainability."

Misleading marketing like "greenwash", along with many other issues, is dealt with at greater length in other chapters in this book. However, this chapter will touch on most of the elements of a sustainable food and farming system. This will not only help illustrate the interconnections in the system but also, more important, show how they can be mutually reinforcing so that taking steps to improve one aspect can and should improve others.

Throughout this chapter, from here on, sustainable food and farming will be referred to mainly as good food and farming. This is not only for the sake of brevity, important though that is. The word "sustainable" is not widely understood, in any context, including in food and farming. It is often misunderstood to mean "good for the environment" (excluding the social and economic aspects), and increasingly often misused, as in "sustainable

---

[1] This chapter will use the terms "rich" and "poor" to describe countries usually called "developed" and "developing" respectively. The reason is best described by Colin Tudge in "The greatest folly of our age", *New Statesman*, April 2002: "Then there's the broader issue – in some ways, the biggest of all: the pervasive notion that all the world's countries should be categorised as 'developed', those that in some sense have arrived, or 'developing', whose task and destiny it is to try to become more like us. This conceit is gross."

growth" (Jackson, 2009). Using "good" to summarize the food system explored in this chapter is part of an attempt to reclaim the word and reframe efforts to improve the food system as both a normal and desirable part of our society. By the same token the term bad food will be used instead of unsustainable food.

8.3 ELEMENTS OF SUSTAINABLE AGRICULTURE

This section outlines some key elements of a good farming system, specifically: land, finite agricultural inputs, biodiversity, animals and people, and how they might be combined into a resilient system that will last.

8.3.1 Land

What can be done about land in a good food and farming system given, as Mark Twain noted, "they are not making it any more?" Conventional wisdom is that, given the finite supply of land and increasing demands on it, we should intensify production and get as much food (and other products such as biofuels and cotton) from it as possible. However, mindful of the bad reputation of intensive farming, the phrase "sustainable intensification" has now entered our language.

The main problem with this approach is that it proposes to solve the problems of intensification by using slightly modified versions of the methods that have caused the problems in the first place. So, for example, slightly less toxic pesticides are suggested alongside more careful use of finite stocks of artificial fertilizer, and more sophisticated applications of equally finite water supplies. In addition, the focus continues to be on a narrow concept of efficiency: how much product will result from how much land and other inputs?

But even a passing glance at the countryside from a window will confirm that land "produces" a great many other things, not least beautiful landscapes and wildlife. That these important features of our history, culture and well-being are termed "externalities" in traditional economic literature speaks volumes about the mindset that proposes intensifying pressure on them. These "externalities" have recently been thrust into the public eye around the issue of "land-grabbing".

Put bluntly, proponents argue that when agricultural land in poor countries is bought by rich countries or global companies, this improves the efficient use of that land by increasing production. It is also sometimes argued that more and better jobs are created. While there is some debate about the accuracy or otherwise of these predictions, a more fundamental question is under what circumstances – if any – is it ethical for control of food producing land to be taken out of the hands of the people who live there?

Prior, informed consent to the sale, a fair contract and a fair price should cover most objections, and it is lack of some or all of these that is probably at the root of many campaigns against land-grabbing. However, vital though all these are, they are not enough. People's passionate attachment to land has been recently demonstrated in the UK.

At the time of writing the UK Government has just finished the consultation for the National Planning Policy Framework which, among other things, aimed to make "yes" the default answer to any application for development, unless there were good reasons to say no. The proposals have been opposed by a large and diverse coalition of groups and individuals, largely due to fears that they would lead to the destruction of the British countryside. It is testament to the powerful emotional bonds we have with land that a country of urbanites should feel so protective. Similar smaller-scale battles are played out if, for example, a local food growing site is threatened.

So, to return to the question, what can be done about land, perhaps the short answer is "protect it." A longer answer would add treating it like the precious, finite natural resource it is and enriching it so that future generations can enjoy the multiple benefits it provides. This is the approach taken by organic farmers and others who treat the land as part of an integrated ecosystem. In some ways this is also officially recognized by European governments in the rather ugly phrase "multifunctional agriculture." Unfortunately, official recognition is too rarely translated into official policy and practice, so land continues to disappear – whether under buildings or asphalt, or eroded or degraded by industrial farming methods.

However important the homes or infrastructure might be, the people living there will need to eat, and that needs land. Once lost it is very difficult to recover, so the presumption should surely be to protect food growing land.

8.3.2 Finite agricultural inputs

Organic and other ecological farmers not only give a high priority to protecting and enhancing the quality of their land, they also do their best to minimize finite agricultural inputs on their land and instead use sustainable ones. The finite inputs are not just the fossil fuel-based energy for running the farm (and transporting products to and from the farm), but also the artificial fertilizers, nitrogen,[2] phosphorus (Hislop and Hill, 2011), and potassium (NPK) that industrial farming relies on. Sustainable alternatives include animal manure, composted waste material, and a variety of plants which, as well as being renewable, can be provided from local sources (ideally on the farm), thereby reducing transport costs too.

These also have the benefit of including a wide range of micronutrients alongside providing bulk and structure to soils, which can help retain water. Given global warming, it is becoming increasingly important to conserve fresh water and agriculture, as the major user of the world's fresh water supplies, has a significant role in using water both less often and more wisely.

This is not to say that careful stewardship of water and using renewable sources of energy and fertilizers is entirely unproblematic. If this was the case, farms would already be using these methods routinely. There may be technical issues still to solve, such as application rates, and yields may fall, at least temporarily while soil fertility recovers (Rodale Institute, 2011) and farmers become more skilled in new methods. This research shows, *inter alia*, that productivity on organic farms rises, in the long term.

---

[2] The Haber–Bosch process has allowed the commercial production of ammonia fertilizer by combining nitrogen and hydrogen under high pressure, in the presence of a catalyst. The main source of hydrogen for the process is natural gas (a fossil fuel) and the process itself is very energy intensive.

However, there are more serious obstacles. Primarily, in this author's view, finite agricultural inputs (and the machinery to use them) are readily available (certainly in rich countries), relatively cheap (although there have been sharp price rises for fertilizers when oil prices spike), and energetically promoted by the multinational companies which sell them. In contrast, sustainable inputs, are not yet mainstream so may be regarded as riskier options, and they require more skilled staff (who may not be readily available). Worse still, such good farming methods not only do not get the same quantity or quality of marketing, but also are often the subject of denigration.

The most common line of attack is that, largely due to what are alleged to be lower yields, organic food cannot feed the world. One commentator even went so far as to brand organic farming immoral, on these grounds (Driver, 2011). In fact, as a large and growing body of evidence shows, ecological methods of farming are the only ways we are going to be able to feed ourselves (IAASTD, 2008). They are designed as closed loop systems and, while not all practitioners have yet achieved this ideal, they are at least travelling in that direction. Intensive farming, by contrast, relies on finite supplies of, for example, oil, artificial fertilizers and water, to fuel infinite growth. This is a physical impossibility.

8.3.3 Biodiversity

Even if, by some alchemy, it proved possible to squeeze infinite agricultural growth into the finite planetary pot, there are other problems with intensive farming. Relying on fossil fuels contributes, of course, to greenhouse gases and therefore global warming, which itself can lead to loss of biodiversity. In addition, the damage caused by pesticides to wildlife (alongside the risks from pesticides to human health) is well established and a body of European-wide legislation[3] as well as global codes[4] have developed to try to reduce this damage. In some instances this seems to be working, with some farmland bird populations for example returning to healthy levels, but in many other cases biodiversity – including bees, butterflies and other bird species – is still under threat.

Intensive farms are also usually monocultures. It is richly ironic that the industry that gave the world the phrase "don't put all your eggs in one basket" now routinely does so. It is common for farmers to specialize and produce, for example, only wheat or corn or soya, or only dairy cows. If there is a disease problem in such a farm, the effects can be catastrophic.

Good farming methods take a different approach. Rather than trying to mitigate damage, ecological farmers instead integrate diverse wildlife into the farming methods themselves and aim (successfully, according to the evidence) to increase biodiversity. Good farming methods also integrate biodiversity into the farms' own plants and animals by growing and rearing a variety of both. As well as reducing the risks associated with

---

[3]The European Commission maintains a website of information about these regulations http://ec.europa.eu/food/plant/protection/index_en.htm (accessed 13 December 2012).

[4]The Codex Alimentarius develops rules on pesticides, among other things, to facilitate food trade http://www.codexalimentarius.net/pestres/data/index.html;jsessionid=0A1229161657A7839FD83E35DD37BDA0 (accessed 13 December 2012).

monoculture this diversity could also be good for our health, by putting diversity on our plates (Pollan, 2007), as well as enhancing it in our countryside. Pollan's work is one among many pointing out the numerous dangers of humans relying on a very small number of plants (particularly corn and wheat) and animals.

8.3.4 Animals

Chapter 7 deals with this issue in more depth. Suffice to say here that the production and consumption of meat and dairy products (including eggs) is arguably the hottest topic in the debate about how best to create and maintain a good food and farming system. The same arguments – about productivity, efficiency, and feeding the world – are rehearsed but, perhaps because the subjects of the argument are sentient beings rather than plants, the temperature of the debate seems higher.

The intensive farming industry continues to argue for mega dairies and giant pig farms on these grounds, while insisting that welfare standards can be maintained or even improved. This is progress of a sort since, only a few years ago, the industry might not have felt it necessary to make the case for animal welfare standards at all. Opponents emphasize the risks of air and water pollution, the damage from increased traffic to and from the site, and the risks to human health of outbreaks of disease in these factory farms. These are, essentially, technocratic arguments about how best to produce a given amount of meat and dairy produce.

The more profound challenge is that we should eat fewer meat and dairy products and, therefore, produce less of them. There is growing acceptance of the argument that this approach would improve public health, reduce the environmental damage caused by intensive animal farming and allow smaller numbers of animals to be kept to higher welfare standards (Garnett, 2008; SDC, 2009; Macdiarmid *et al.*, 2011). Perhaps more difficult than winning the argument are the practicalities of converting our current, heavily meat and dairy focused system to one that can function well – economically as well as environmentally and socially – focusing on a variety of plants.

For proof that it can be done we need only look at human history, but that may not help us much in setting out what a good farming system, with fewer animals, should look like in future. What seems certain is that farmers who have invested heavily in intensive farming systems may need some public funding to help them diversify, and develop a wider range of "baskets" into which more diverse agricultural products (and probably not eggs) can be put.

The principle for this kind of support for agriculture is well established in the European Union's Common Agricultural Policy (CAP). The payments some farmers receive for implementing environmental measures on their farms recognizes that these are "public goods" that the current market does not pay for, either at all, or at least not adequately. We are still a long way from the CAP supporting a more diverse and sustainable, plant-based agricultural system but no laws of physics need to be broken to devise such an approach.

8.3.5 People

Similarly, the CAP recognizes (though various policies and payments) that a thriving rural society is something that the current food and farming system is unable to provide.

Other chapters examine in more detail worker exploitation, ethics at work, and fair trade. What is clear is that, except at the level of celebrity chefs, work in the food and farming system is often low paid, low status and sometimes downright dangerous. It is no surprise, then, that the sector is largely unattractive as a career.

Farming is still the most hazardous occupation, even in the UK (and in other countries it is far more dangerous). The average age of those still farming is well above 50 years, and farmers continue to leave the industry. Many in the processing, retail, catering sectors have stated how difficult it is to recruit high-caliber staff. Yet if we are to have a good food and farming system, where finite and damaging sources of energy and other inputs are replaced by renewable and sustainable ones, we are clearly going to need more and better jobs in the system.

This flies in the face of conventional "development" theory where it is axiomatic that development means moving out of agriculture into industry and services. Certainly this is the path that rich, Western countries have taken and many others are rapidly pursuing a similar path. Is it anti-development and even anti-progress to propose that more of us should work in the food and farming system?

A return to some mythical rural idyll is not being proposed. Some traditional farm labor can be back-breaking and tedious, as indeed can work in much of the rest of the food sector. However, it can be enjoyable and fulfilling work and there has been a resurgence of interest, particularly among young people, in urban agriculture, in combining farming with other rural enterprises, and in selling food directly to customers in a variety of outlets (MLFW, 2011). "Local" has become a popular shorthand[5] for food that is authentic, flavorsome, and linked to creating local jobs, having high welfare standards and a low impact on our world. As such it may be attracting a new type of entrepreneur that might pave the way for "green jobs" in food and farming, analogous to "green jobs" in home insulation and generating renewable energy.

Certainly we are going to have to think of some gainful employment opportunities (Tudge, 2003, 2007) for the world's population of 7 billion and counting. The Western model of industrial growth is not possible for the rest of the world and nor, indeed, is it possible for rich, Western countries to continue with it. If we are not going to work in factories or sell each other insurance, then decently paid and satisfying work, producing one of the essentials of life does not seem like a bad option.

8.3.6 Resilience

This author argues that a more labor intensive food and farming system would be more resilient, but does resilience mean being self-sufficient? Certainly, there are many advantages to growing perishable, high value fruit, vegetables, salad and herbs in urban areas, close to the market. It enhances freshness, reduces spoilage and waste, and reconnects urban dwellers with the source of some of their food, while creating attractive green spaces and opportunities for physical activity. However, at the risk of laboring the point,

---

[5]Sustain and its predecessors are partly responsible, given the Food Miles publications http://www.sustainweb.org/foodandclimatechange/archive_food_miles/ (accessed 13 December 2012), for popularizing local food. We have always acknowledged that food miles are not the best measure of sustainability, as has been demonstrated by a number of reports, including Watkiss, *et al.* (2005).

this is not suggesting some unattainable urban utopia. Although some small scale live-stock – for example, chickens, bees and non-carnivorous fish such as tilapia – is already being produced in towns and cities, most livestock farming (and also arable farming) should continue to be done in the countryside and on a significant scale.

This is not an argument for a siege economy, where nothing is traded and we consume only what we can produce in our respective countries. This would merely replace one set of risks – associated with global warming, exhausted natural resources, and so on – with another, where droughts, floods or other damage to agriculture in some countries could not be offset by buying surpluses from other countries not so afflicted.

The question is not whether we should trade in food, but what food should we trade, when and what should the terms of trade be? Fairly traded bananas, for example, can be promoted on the grounds that they maintain good livelihoods for people in poor countries, are transported by ship (which is less energy intensive than flying), and provide a nutritious fruit that cannot be grown in western Europe. Strawberries in winter, on the other hand, tend to be grown in wealthier countries (so alternative employment is readily available), have to be flown in (due to their perishability), and are at their best in summer (not winter), when they can and should be grown and eaten in northern European countries.

A resilient food and farming system would use benign methods (outlined above) to produce a diverse range of plants and animals that are suited to the soil and climate in a given area, while reducing the risks of relying on finite and damaging inputs by using as little of them as possible. What we cannot produce here should be imported, on fair trade terms, from elsewhere and we would expect similar, fair terms for any surplus produce we export. Global food trade would continue, but probably at much lower volumes and on an entirely different basis.

8.4 SUSTAINABLE PROCESSING AND PACKAGING

It is widely accepted that fresh food, skillfully cooked is the tastiest and most nutritious way to feed ourselves, so much so that seasonal food is currently the height of fashion for celebrity chefs and expensive restaurants. It is not likely to be feasible, though, to rely solely on seasonal food[6] for every meal for the whole year. Even the most skilled cook will struggle to make a variety of appetizing dishes from turnips and potatoes.

In addition, more than half the world's population is now urban, the proportion is considerably higher in rich countries, so trade – both within and between countries – requires safe and reliable methods of getting food from one place to another. What kind of processing, preserving and packaging should we have in a good food and farming system?

---

[6]Unless, of course, we are using the definition of seasonal developed by the UK Department of Environment Food and Rural Affairs (Defra), which includes food from anywhere in the world as long as it is grown outdoors during a natural growing or production period. See the terms of reference for research on "Understanding the Environmental Impacts of Consuming Foods that are Produced Locally In Season – FFG 0811", Defra, March 2009.

8.4.1 Safety

Perhaps the fundamental and perennial problem is that good food goes bad. Fresh fruit and vegetables turn rotten, cereals grow toxic molds, fresh meat turns rancid and so on. Since civilization began, humans have found ever more inventive ways to try to halt this process; salting, pickling, curdling, making jellies and other preserves and so on. These processes have produced some of the world's great traditional delicacies, such as fabulous cheeses, and cured meat products. More recently the food industry has developed a range of food additives, canning, freezing and modified gases in special packaging, among other processes, and the quest for new methods continues. As well as providing plentiful, year-round supplies of food, this also reduces food waste, an issue now firmly in the public eye (WRAP, 2011).

It is worth noting, however, that there is a vital economic benefit in processing food. It reduces economic losses, as well as food waste, and each process not only adds value to a food product but provides an opportunity to add a profit margin. This is the basis on which global multinational companies have been built. Consider the humble potato, for example. There are very few wealthy potato farmers, and most people will only pay a basic price for what they consider a basic product. Once the potato is processed into, say, a crisp (or chip, in the USA) by slicing, frying, salting, flavoring, and packaging the price per gram rises dramatically, alongside the profits.

Sometimes processing goes wrong, leading to food safety problems and product recall, which are covered in other chapters. But even when the process is safe, it very often has unfortunate side-effects, either making the product less nutritious, or causing environmental damage, or both.

8.4.2 Nutrition

Adding salt and sugar to products to preserve them is an ancient process and, when people are unable to afford to consume many of these products, it does not cause much of a public health problem (certainly not compared to the much more serious problems of malnutrition and poisoning). However, in rich countries we can afford to, and do consume far too much salt, sugar (and fat), and even in poor countries consumption of such food is rising.

The damage to public health caused by fatty, sugary, salty food has been well documented for at least 40 years or more (Cannon, 1992). Industry resistance to reducing levels of these elements in our food goes back almost as long (Walker and Cannon, 1985; Nestle, 2007). In the UK and in some other rich countries efforts are now, at last, being made to reduce the amount of salt, fat and – to a lesser extent – sugar in our food by reformulating product recipes, sometimes by using artificial substitutes. Examples are provided in other chapters.

While some of these new products are better than their saltier, fattier or more sugary predecessors, some of them may not be. Artificial sweeteners, for example, remain controversial, and not only due to safety concerns about some types of sweetener.[7]

---

[7]The safety of some artificial sweeteners is, at the time of writing, being reviewed by the European Food Safety Authority – for instance, for aspartame see: http://www.efsa.europa.eu/en/topics/topic/aspartame.htm (accessed 13 December 2012).

A link between artificial sweeteners and weight gain is proposed by some nutritionists (Fowler *et al.*, 2008; Yang, 2010) who suggest that, by reinforcing our taste for sweetness, artificially sweetened food and drinks simply add to, rather than replace our sugar consumption. Similarly, safety concerns have dogged some fat replacement ingredients[8] and, even if overcome, may have the same effect as artificial sweetener and simply reinforce our taste for fatty foods.

The simplest approach, of course, which requires no complex reformulation or new ingredients, is to eat fewer of the foods that contain a lot of fat, sugar and salt. This would have the added advantage of reducing the energy required, as less food would be processed.[9] However, eating less and therefore buying less is an obvious problem for the profitability of the companies that make such products, hence their resistance to any such approach.

8.4.3 "Greening" packaging

Despite its ubiquity, food packaging comes into and out of public focus seemingly almost randomly. Issues about the safety of packaging ingredients that might migrate into our food continue to surface. These can be tackled by removing the ingredient or developing a safer alternative. More challenging is the energy and materials used by making any packaging, and the waste when it is thrown away. One of campaigning charity Friends of the Earth's earliest campaigns in the UK was about wasteful food packaging, and some 40 years on there was a great public furore about plastic bags (with, of course, other issues hitting the headlines from time to time.)

Compostable packaging may have some value in some circumstances. It is hard to avoid, though, the relentless logic of the waste hierarchy where recycling (of which composting is a form) comes last, after re-using, with reducing firmly at the top of the desirability list.

The packaging industry is not slow to point out that packaging can protect food and help prevent food waste, and this is clearly important. They are less keen to acknowledge, though, that one of the main roles of packaging is to be a vehicle for marketing the product. Marketing is dealt with in more detail below, but a brief example here will illustrate the broader point.

In the long-standing competition between Coca-Cola and Pepsi Cola much was made (in Pepsi advertising) of Pepsi being preferred to its rival in blind taste tests. In his book *Blink*, Malcolm Gladwell (Gladwell, 2005) explains why this did not translate, as the company expected, into higher sales. One reason is that taste tests ask people to sip samples, but in reality people consume the whole bottle or can, and this can make a sweeter product (which is preferred at first sip) unpleasantly cloying when consumed in volume. The other reason is that most people do not buy or consume blind. Coca-Cola packaging is powerfully linked to all the other marketing imagery and brand values that

---

[8] See, for example, information on the fat replacer, Olestra, from the Center for Science in the Public Interest in the USA http://www.cspinet.org/olestra/ (accessed 13 December 2012).

[9] Provided that people eat less food overall (which would help to reduce obesity) and do not replace the fatty, sugary and salty products with food that requires the same amount or more energy to produce.

are associated with the product, which leads people to buy it, even if they prefer the taste of other products.

Similarly, much confectionary, for example, is marketed and packaged in combinations of card, plastic, foil, and paper. The materials are not usually made from recycled sources, take energy to produce, are far in excess of what is needed to protect the product, are rarely recycled (or even recyclable) and so are as far from good packaging as can be imagined.

This contrasts starkly with the classic British glass milk bottle. This shows that, where packaging is necessary, it should be simpler, more elegant, single material packaging that can be repeatedly re-used and, at the end of its life, easily recycled. With short supply chains for some local produce, food may not need to be packaged at all. Some pioneering retailers are encouraging packaging-free shopping and re-usable containers for a much wider range of products[10].

8.5 SUSTAINABLE CONSUMPTION

By far the most common policy proposal to achieve the change from bad food to good food is to educate people. This suggestion is often applied to people in general but is most energetically promoted for children's education. This section will argue that education is (probably) necessary, but far from sufficient, and that changes in the rules on marketing, along with changing the price and availability of products, is far more effective.

8.5.1 Education, education, education?

People in some poor and middle income countries are not better educated than people in rich countries, and yet their diets often promote health and damage the environment less. At the same time, people in rich countries are very well educated, and our diets tend to be nutritionally poor and environmentally costly. Clearly, the link between education and good food is not straightforward.

This has led some to conclude that educating people about good food is a waste of money, either because the educational programs are badly designed and/or because people are feckless. The latter argument is particularly common when discussing why people on low incomes in rich countries have such unhealthy diets. The notion that poor people are too uneducated or weak-willed to eat well ignores the fact that rich people's diets are just as unhealthy in many respects (Lobstein, 2007).

So is it pointless to provide people with knowledge and skills about good food? There are circumstances when this approach can be valuable. The Food for Life program of the Soil Association, the UK organic certification body, has shown (Soil Association, 2011) that providing children with tasty, healthy and sustainable produced meals is entirely feasible, and is particularly popular when children and their parents are involved in the process. Good food education and practical skills, such as cooking and growing food, are more effective when they are matched by good food on the children's plates. This ought to

---

[10]See, for example, London independent retailer, Unpackaged http://beunpackaged.com/about/ (accessed 13 December 2012).

be too obvious to state, but there are far too many instances when good education has been undermined by bad food. Yet even if this charitably-funded school program was universal, as it should be, it would still not be enough to shift consumption towards good food.

8.5.2 The "choice" myth

Outside the school gates (and in most places inside them) we have created an obesogenic environment (Gortmaker *et al.*, 2011), which is bad for our health and also for the Earth's life support systems on which we depend. Bad food is easily affordable (until recent price increases, as cheap as it has ever been, relative to income in rich countries), ubiquitously available (with 24 hour supermarket shopping and ready-to-eat food on every high street and in every shopping centre) and very attractive, thanks to a sophisticated range of marketing techniques. Good food, on the other hand, tends to be more expensive (sometimes dramatically so), not always easy to find (requiring a different shop or other outlet to where the rest of the shopping is done), and has an image which is often, at best, "worthy" and "elitist" or, at worst, downright dull and unappealing.

Given this environment it is not surprising that education often fails to produce higher sales of good food. Indeed it could be argued that, given the obstacles, it is impressive that anyone at all makes the effort to buy good food. That there is a growing market for good food is partly due to choice editing, which means that people do not have to make an effort. The Co-operative, a British consumer-owned retailer, has been at the forefront of providing, for example, only certified Fairtrade products[11] for some types of food and drink in their stores, and another UK food retailer, Sainsbury's,[12] now sells only free-range eggs in its stores. A significant number of companies are reducing salt levels in their standard product ranges,[13] so that people's salt intake can go down without them having to look for (and sometimes pay extra for) low-salt products. The School Food Trust, established by the government in England to improve school meals, has also shown that there has been a third consecutive annual increase in children eating school dinners, with an increase of 173 000 in 2010–2011 (Nelson *et al.*, 2011). These encouraging figures follow the nutritional improvements brought about by legislation (due, at least in part, to a celebrity chef, Jamie Oliver[14]).

This handful of examples exposes the myth of free choice in the food market. Major retailers, caterers and others in the industry do not – and could not – provide us with an infinite choice of products from which we, as rational actors, make choices. They choose, on our behalf, a range of products to offer us and they decide how they will be

---

[11] See the Co-operative's ethical trading policy. Information about their Fairtrade policy, including the history, is available online: http://www.co-operativefood.co.uk/ethics/Ethical-trading/Fairtrade/ (accessed 19 December 2012).

[12] See Sainsbury's corporate responsibility policies, including on animal welfare and free-range eggs. Available online: http://www.j-sainsbury.co.uk/responsibility/our-values/sourcing-with-integrity/ (accessed 19 December 2012).

[13] The Food Standards Agency's salt reduction program was taken over by the Department of Health as part of the Coalition Government's Responsibility Deal. Salt reduction targets and updates, appear in these bulletins http://www.dh.gov.uk/prod_consum_dh/groups/dh_digitalassets/documents/digitalasset/dh_129528.pdf (accessed 13 December 2012).

[14] Jamie Oliver's Jamie's School Dinners program was aired on Channel 4 TV in 2005 and started the Feed Me Better campaign which, at time of writing, is still active. http://www.jamieoliver.com/school-dinners (accessed 13 December 2012).

priced and how they will be marketed, down to where they will be placed, for example, on which shelf. Although customer preferences will be considered, this sits alongside other issues such as profitability, reliability of supply, what competitors are doing, and so forth. Whether or not the product contributes to, or undermines progress towards a good food and farming system is only now beginning to feature occasionally in that list of factors.

Given that choice editing in favor of good food has been a popular move (as the Co-operative and Sainsbury's will attest), as well as the right thing to do, it needs to be extended rapidly. Of course, while we are waiting for this to happen, we do not have to buy our food from major retailers, caterers and others that promote and sell mainly bad food. The number of alternative options, although still small, is at least growing. We have already touched on the advantages of growing some food close to the towns and cities where it will be eaten, and there has been a renaissance in the most local of local food; growing your own. In addition, farmers' markets, box schemes and a host of community supported social enterprises and co-operatives are blossoming in the UK, the USA, and other rich countries. However, it remains an open question how large this market sector can grow, given the considerable market muscle exerted, often unfairly, (Competition Commission, various) by the major retailers.

One of the factors that would help sell more good food, through whatever outlets, is making it cheaper than – or at least the same price as – bad food. Currently the costs of bad food – such as poor health, environmental damage and low quality jobs – are not reflected in the price so are termed "externalities." As a society we do pay for these costs, whether in monetary terms (for example in costs of treating diet-related diseases or cleaning up pollution) or in the suffering caused by ill-health or cruel practices in animal farming. We just do not pay for them at the same time we buy the food.

The price of good food, however, does include some of the costs of preventing damage, so it appears to be more expensive. Not using pesticides, for example, is good for biodiversity and creates jobs, but skilled labor costs more than chemicals, so the price of the food is higher. Higher welfare standards also require more skilled labor to care for the animals, and paying for Fairtrade standards, while supporting poor farmers, also increases costs. This saves us money in the long run but, at the point of purchase, this is not apparent.

For other products, such as tobacco, alcohol and fuel, where the "externalities" are not reflected in the price, governments use taxes to "internalize" the costs. This not only makes the price reflect more fairly the costs to society as a whole but also, if the tax is high enough, discourages people from buying the product and reduces the damage done by using it. An increasing number of governments are now using, or considering using a similar approach with some aspects of bad food. For several years a number of US states have taxed sugary drinks and France (NACSonline, 2011) has just introduced such a tax. Denmark (BBC, 2011) introduced a tax on fatty foods, as did Hungary (Cheney, 2011), and at time of writing, Ireland is considering a similar policy (Carbery, 2011).

It is not inconceivable that fiscal measures could be used more broadly, so that other elements of sustainability, alongside nutrition, would be considered when setting a tax. This could make good food cheaper (or at least not more expensive) than bad food and generate revenue that could be invested in much-needed research and development to support a good food and farming system.

8.5.3 **The power of persuasion**

Earlier in this chapter we referred to bad food being affordable, available and attractive and we have touched on availability (through choice editing) and affordability (in the brief exploration of externalities). What, then, of attractiveness? Other chapters examine ethical marketing (including to children), and also health and other claims, including on labeling.

It is also dealt with briefly here because, arguably, making bad food attractive is the engine that drives the bad food and farming system. Marketing can be so powerful that people can be persuaded to pay happily for, for example, bottled water when the equivalent product can be obtained almost for free from the taps in their homes and offices. Tim Jackson, among others, has noted that with marketing "... people are being persuaded to spend money we don't have, on things we don't need, to create impressions that won't last, on people we don't care about" (Jackson, 2009). An excellent explanation of, *inter alia*, why "sustainable growth" is an oxymoron can also be found in Jackson (2009). Evidence continues to accumulate to confirm what many of us will readily verify; that the constant pursuit of material goods and the status associated with them does not improve our well-being or make us happy (Jackson, 2009; Layard, various).

It is already accepted in many countries, including the UK, and increasingly so elsewhere, that children should be protected from marketing, not only for bad food but also from marketing in general. This is mainly on the grounds that it is exploitative, since children are, by definition, not mature enough to know that they are being manipulated for commercial gain. We also want to protect children from marketing for harmful products, and fatty, sugary and salty food is widely agreed to be in this category. Thus it is no longer legal, in the UK, to advertise bad food during children's TV programs, nor place (as a form of advertising) branded bad food products in any TV programs.[15]

Sadly, some in the food industry have chosen to follow the letter rather than the spirit of these rules. They continue to promote bad food to children in the growing range of other media, such as via Facebook and Twitter, and on websites (Children's Food Campaign, various), as well as through sponsorship deals with children's sporting and entertainment heroes.

While efforts will continue to close these loopholes in the laws protecting children, what controls are appropriate on marketing bad food to adults? There is already long-standing legislation that, on paper at least, protects everyone from misleading marketing in general, including for food. Alongside the laws there are also voluntary codes, devised and run by the advertising industry. Although there is room for improvement in the content of the laws and the codes, there is a serious problem due to the lack of enforcement powers.

In the UK, local authority trading standards departments are responsible for enforcing the laws on food labeling, but the staff are thinly stretched and their budgets are small and shrinking. Broadcast advertising is regulated but, although advertisements are vetted

---

[15] Sustain's Children's Food Campaign, working with a large coalition of supporting organisations and individuals, can justifiably claim to have brought about these policy changes http://www.sustainweb.org/childrensfoodcampaign/ (accessed 13 December 2012).

before being broadcast, some will slip through the net. When this happens, the complaints process runs so slowly that, even if a complaint is upheld, the advertisement will have long since run its course so the sanction of not broadcasting it is no longer relevant.

For non-broadcast advertising the situation is worse because advertisements are not vetted before being published. The sheer volume of non-broadcast advertising arguably makes pre-vetting impractical and the volume also makes it very difficult to monitor. The Advertising Standards Authority does undertake some monitoring but, even when it repeatedly finds a problem – as is the case with advertising for slimming products – it has virtually no powers to tackle it.[16]

Even if the rules were tightened, and enforcement made more effective, would it still be acceptable to market bad food to people? It could be argued that, if good food companies and bad food companies had similar amounts of money to spend on marketing, then adults could simply make up their own minds. But the balance of marketing power is a long way from fair, with the marketing budgets of multinational food companies outstripping the GDP of some small countries. So would equally large (or small) marketing budgets for good and bad food make marketing ethical?

A recently published report (PIRC & WWF-UK, 2011) has challenged the entire edifice of marketing, arguing persuasively that it contributes to the consumerist culture that is leading us to use more of the world's natural resources than can possibly be maintained and, of course, adding to potentially calamitous global warming. If, as this report proposes, marketing methods (often acting at a subconscious level) are inherently unethical, should such methods be used to help us shift towards a good food and farming system? The authors note that there is little evidence on whether the effects of such marketing are broadly positive or broadly negative, but such research as exists indicates that even "good" advertising for good products (such as food) can sometimes continue to reinforce materialism.

8.6 ASKING THE IMPOSSIBLE?

So where does this leave us? It is not possible to be perfect but, as was noted at the beginning of this chapter, striving to be good enough (while continuing to learn more) would bring significant progress towards a good food and farming system and this author's approach can be summarised as follows:

- Protect land for growing food and improve its quality.
- Do not use finite natural resources – oil, fertilizer, water – as if they were infinite. Use them sparingly to buy time to invest in renewable inputs.
- Integrate biodiversity into animals and plants that are farmed, and also into the way they are farmed.

---

[16]In a 2005 survey of slimming advertising, the Advertising Standards Authority concluded that it "was concerned at the high breach rate and will continue to monitor the slimming sector to ensure an improvement in the compliance rate" (http://www.asa.org.uk/~/media/Files/ASA/Reports/Slimming_Survey_Sep2005.ashx; accessed 19 December 2012). But in the last six years, no further ASA monitoring has been published and self-regulation of the slimming industry continues to fail (see, for instance: Slimming pills: Do the claims add up? 9 July 2010, www.bbc.co.uk/news/10241981; accessed 13 December 2012).

- Eat more, and more diverse plants, and fewer animals and animal products, while ensuring high welfare standards.
- Create and maintain decently paid, skilled and attractive jobs in the food and farming sector as a whole.
- Be resilient, by producing domestically as much as possible (given the above) and fairly trading the rest.
- Preserve food to reduce waste, but design preservation techniques to use less energy and retain nutritional value, taste and safety.
- Use the least amount of packaging necessary to protect food, and use materials and designs that can easily be re-used and, eventually, recycled.
- Provide good food in schools, along with education and practical cooking and food growing skills.
- Extend choice editing and use fiscal measures so that good food is more easily available and cheaper than bad food.
- Protect children from junk food marketing in all media, and develop tough rules and effective enforcement for marketing to adults.

None of this is technically difficult, and many – if not most – of these approaches have been proposed for years, sometimes decades. Why, then are more people not already acting ethically? The response is usually that to do so is too complicated and difficult. This author's response is that the vested interests of money and power mean that making the ethical choice is too often unavailable, unaffordable and unattractive – precisely the opposite of what is needed to make the ethical choice the easy choice.

REFERENCES

BBC (2011) *Denmark introduces the world's first food fat tax*, 1 October 2011, BBC News, http://www.bbc.co.uk/news/world-europe-15137948 (accessed 13 December 2012).

Brundtland Commission (1987) *Our Common Future*, Oxford University Press, Oxford.

Cannon, G (1992) *Food and Health: The Experts Agree. An analysis of one hundred authoritative scientific reports on food, nutrition and public health published throughout the world in thirty years, between 1961 and 1991.* Which? (formerly Consumers' Association), London.

Carbery, G (2011) *Would sweet tax leave a bitter taste?*, The Irish Times, 18 October 2011. Available online: http://www.irishtimes.com/newspaper/health/2011/1018/1224305986378.html (accessed 19 December 2012).

Cheney, C (2011) *Hungary Introduces 'Fat Tax'*, Der Spiegel, 09/01/2011, Available online: http://www.spiegel.de/international/europe/0,1518,783862,00.html (accessed 13 December 2012).

Children's Food Campaign (various) See several publications available online: http://www.sustainweb.org/childrensfoodcampaign/publications/ (accessed 13 December 2012).

Competition Commission. See a number of inquiries undertaken into the groceries market during the last decade. Available online: http://www.competition-commission.org.uk/inquiries/subjects/supermarkets.htm (accessed 13 December 2012).

Driver, A (2011) CropWorld 2011: Organic farming is 'morally indefensible. *Farmers Guardian*, 2 November 2011. The article quotes consultant Sean Rickard.

Fowler, S *et al.* (2008) 'Fueling the obesity epidemic? artificially sweetened beverage use and long-term, weight gain', *Obesity*, 168, p. 1894–1900.

Garnett, T (2008) *Cooking up a storm: Food, greenhouse gas emissions and our changing climate.* Food Climate Research Network, Centre for Environmental Strategy, University of Surrey. Available online: http://www.fcrn.org.uk/fcrn/publications/cooking-up-a-storm (accessed 13 December 2012).

Gladwell, M (2005) *Blink: The power of thinking without thinking.* Penguin, London.

Gortmaker, S. Swinburn, B, Levy, D *et al.* (2011), 'Changing the future of obesity: science, policy, and action', *The Lancet*, 378, 9793, 838–847.

Hislop, H & Hill J (2011) *Reinventing the wheel: a circular economy for resource security.* Green Alliance, London. Available online: http://www.green-alliance.org.uk/grea_p.aspx?id=6044 (accessed 13 December 2012).

International Assessment of Agricultural Knowledge, Science and Technology for Development (2008) *Agriculture at a Crossroads.* Available online: http://www.agassessment.org/ (accessed 13 December 2012).

Jackson, T (2009) *Prosperity Without Growth.* Sustainable Development Commission, London. Available online: http://www.forumforthefuture.org/project/consumer-futures-2020/overview (accessed 13 December 2012).

Layard, R. (various) Centre for Economic Performance, London School of Economics. Publications available online: http://cep.lse.ac.uk/_new/staff/person.asp?id=970 (accessed 13 December 2012).

Lobstein, T (2007) *Low income diet and health – next steps: A paper prepared to aid discussion at the FSA stakeholder meeting on diet and low income,* 5 November 2007. Sustain, London.

Macdiarmid, J, Kyle, J, Horgan, G *et al.* (2011) *Livewell: A balance of healthy and sustainable food choices.* WWF-UK and Rowett Institute of Nutrition and Health. Available online: http://www.wwf.org.uk/what_we_do/changing_the_way_we_live/food/food_publications_library.cfm?4574/Livewell-a-balance-of-healthy-and-sustainable-food-choices (accessed 13 December 2012).

Making Local Food Work (2011) See information produced by many of the organisations and initiatives involved in this programme, funded charitably by the Big Lottery. Available online: http://www.makinglocalfoodwork.co.uk/ (viewed 13 December 2012).

NACSOnline (2011) *France backs soft drinks tax,* NACSOnline, 25 October 2011. Available online: http://www.nacsonline.com/NACS/News/Daily/Pages/ND1025111.aspx (accessed 13 December 2012).

Nelson, M, Nicolas, J, Riley, K, Wood, L, Russell, S. (2011), Statistical Release – Take up of school lunches in England 2010–2011, School Food Trust, London. Available online: http://www.schoolfoodtrust.org.uk/news-events/news/more-families-choose-healthy-school-meals-despite-tightening-belts (accessed 13 December 2012).

Nestle, M (2007) *Food Politics: How the food industry influences nutrition and health.* Revised and expanded from 2002 edition. University of California Press, Berkeley, CA.

Pollan, M (2007) *The Omnivore's Dilemma: The search for a perfect meal in a fast-food world.* Bloomsbury, London.

Public Interest Research Centre & WWF-UK (2011) *Think of me as evil? Opening the ethical debates in advertising.* Available online: http://assets.wwf.org.uk/downloads/think_of_me_as_evil.pdf (accessed 13 December 2012).

Rodale Institute (2011) *The Farming Systems Trial: Celebrating 30 years.* Available online: http://www.rodaleinstitute.org/fst30years (accessed 13 December 2012). Soil Association (2011) *The impact of the Food for Life Partnership.* Available online: http://www.foodforlife.org.uk/LinkClick.aspx?fileticket=YyUBCvfUWCc%3d&tabid=310 (accessed 13 December 2012).

Sustain (2007) 7 principles of sustainable food. Available online: http://www.sustainweb.org/sustainablefood/ (accessed 13 December 2012).

Sustainable Development Commission (2009) *Setting the Table: Advice to Government on priority elements of sustainable diets.* Available online: http://www.sd-commission.org.uk/publications.php?id=1033 (accessed 13 December 2012).

Tudge, C (2003) *So Shall we Reap: What's gone wrong with the world's food – and how to fix it.* Allen Lane, London.

Tudge, C (2007) *Feeding People is Easy.* Pari Publishing, Italy.

Walker, C & Cannon, G (1985) *The Food Scandal, What's Wrong with the British Diet and How to Put it Right.* Available online: http://www.cwt.org.uk/publications.html#scandal (accessed 13 December 2012).

Watkiss, P, Smith, A, Tweddle, G *et al.* (2005) *The Validity of Food Miles as Indicator of Sustainable Development,* Defra, London. http://archive.defra.gov.uk/evidence/economics/foodfarm/reports/documents/foodmile.pdf (accessed 18 December 2012).

WRAP (2011). This organization runs the Love Food Hate Waste campaign http://www.lovefoodhatewaste.com/. It has published a series of publications on the amount of food waste in the UK. Available online: http://www.wrap.org.uk/retail_supply_chain/research_tools/research/report_new.html (accessed 13 December 2012).

Yang, Q. (2010) 'Gain weight by "going diet?" Artificial sweeteners and the neurobiology of sugar cravings', *Yale Journal of Biology and Medicine*, 83, 2, p. 101–108.

9 Good or bad foods? Responsible health and nutrition claims in Europe

Sue Davies

9.1 A DIET CRISIS

There has been an unprecedented amount of debate about diet and health in recent years. Rising rates of obesity and diet-related disease globally have focused attention on actions to improve people's diets as never before.

Around a quarter of the UK population are now obese and people who are a healthy weight are now in the minority (Department of Health, 2011a). This compares with obesity rates of over 35% for women in the United States and over 32% for men (OECD, 2012). But added to this, diet-related diseases, such as cancers, heart disease and stroke are the major killers. The overall burden falls on individuals, but also wider society, the health service, and the economy.

A lot of attention has therefore focused on what we should be eating to maintain good health and what measures need to be in place to help people to identify and choose healthier foods. Successive strategies and action plans have evolved at national, European and international level in order to try and reverse the tide of diet-related diseases.

In the UK alone, there has been a succession of strategies and initiatives over the last two decades aimed at tackling poor diet, from the Nutrition Task Force of the 1990s through to the most recent Responsibility Deal initiative under the Coalition Government. The World Health Organization (WHO) has attempted to harness action by member governments and food companies in order to respond to the burden of disease in both developed and developing countries through a Global Strategy on Diet, Physical Activity and Health (WHO, 2004) and more recently a broader United Nations Summit specifically on non-communicable diseases (NCDs), including those that are diet-related.

At the European level, WHO Europe has initiated a Ministerial Conference and Charter on Counter-acting obesity and a European Action Plan on Food and Nutrition Policy. The European Union has also developed a Strategy on nutrition, overweight and obesity, including a Platform for Action on Diet, Physical Activity and Health to encourage voluntary measures from the private sector and other stakeholders.

Practical Ethics for Food Professionals: Ethics in Research, Education and the Workplace, First Edition.
Edited by J. Peter Clark and Christopher Ritson.
© 2013 John Wiley & Sons, Ltd. Published 2013 by John Wiley & Sons, Ltd.

These strategies have all reinforced that the issue is complex, multifactorial and requires a range of interventions by a range of stakeholders if healthier choices are to become easier. The rhetoric has moved away from a focus on the individual making healthier choices in isolation to highlighting a range of actions that are needed by different stakeholders, particularly the Government and food companies, in order to create an environment that supports healthier choices.

9.2 THE RIGHT CHOICE

In terms of what we should actually be eating in order to be healthy, the advice has changed little since the 1980s, when the UK body – the National Advisory Committee on Nutrition Education – first proposed what we should be eating more of, and what less of, in order to help reduce diet-related diseases (NACNE, 1983).

This advice has evolved and been reassessed by successive advisory committees nationally and internationally. However, despite over 20 years of initiatives designed to improve our health through diet, most people still eat too much fat, sugar and salt and not enough fruit and vegetables. The balance has been taken out of our diets and needs to be re-aligned.

Research conducted by Which?, the UK consumers' association (Which?, 2009a), has found that many people actively try to choose a healthy diet, but find it difficult to do so in practice. The National Diet and Nutrition Survey (Department of Health, 2011b) shows that many people fail to meet recommended intakes. Just 30% of adults meet the "five a day" recommendation for fruit and vegetables and just 14% of boys and 7% of girls aged 11–18. Average saturated fat and non-milk extrinsic sugar (NMES) intakes also exceed the recommended levels.

It is against this backdrop that the role of health and nutrition claims on foods needs to be considered.

9.3 A GROWING MARKET

Somewhat ironically, at the same time that overweight and obesity have been on the increase, so has the market for healthier foods. The intense scrutiny that food and health has had over the last few years has meant that many people are interested in healthier eating and want to make healthier choices. A whole industry has therefore developed to respond to this demand.

A recent Mintel Oxygen report (Mintel Oxygen, 2011), for example, estimated that the UK functional food market (products claiming to offer specific health benefits) had seen a growth of 32% between 2006 and 2011, with an estimated value of £785 million.

As consumers have become more interested in health issues, therefore, foods have increasingly been promoted to them on this basis. A central issue has however been whether the types of products promoted to people as healthy really are. Can consumers trust the claims that they make and are these products really responding to the issues affecting the majority of people when it comes to poor diet? Or are they largely a marketing tool and an irrelevance designed for the "worried well"?

Alongside the greater focus on obesity and chronic diet-related diseases, nutrition science has evolved leading to a whole range of food products and food supplements claiming to offer consumers specific health benefits – from digestive health to mental health.

As well as these so-called "functional foods", which appeal mainly to a niche market, a wide range of health and nutrition claims have started to appear on more mainstream products and are being used in food advertising – from supermarket healthy eating ranges to claims about specific nutrients and endorsements from celebrities and health professionals. Whether these claims are explicit or implied (through the shape or illustrations on food packaging for example), they are a quick and easy way for consumers to identify healthier options. But they have also become an effective way to sell products – and often at a premium.

9.4   TRUST IN CLAIMS

Research conducted by Which? over the years has shown that people generally like claims on foods as they are a quick and easy way to identify healthier foods. But there has also been a lot of cynicism among consumers about health and nutrition claims (Which?, 2000). People do not always feel that they can trust them and want them to be independently verified.

Which? research (Which? 1997, 2009b, 2011) has also highlighted that some of these product claims could be misleading. Problems that Which? found included suggestions that the product was healthier than it was in reality. For example, beneficial nutrients may be highlighted (e.g. a particular vitamin), while less healthy ones (e.g. high sugar content) stay hidden. Other common issues included products highlighting benefits that are in fact a standard feature or health claims being made that are not underpinned by scientific evidence.

The responsible use of health and nutrition claims has therefore been an important issue in recent years and there has been a steady evolution of the regulatory framework that surrounds them. Labeling requirements have steadily been enhanced – initially through specific, voluntary measures dealing with claims through to a more comprehensive regulatory framework in recent years. But there is still much work to do to create a consistent and user-friendly labeling system that enables consumers to discern genuinely healthy food items from those simply claiming to have health benefits.

9.5   GREATER SCRUTINY OF NUTRITION CLAIMS

Nutrition claims highlight the levels of positive or negative nutrients in a food (e.g. low sugar, high fiber). In 1990, the UK Government's Food Advisory Committee, made up of a range of experts with different backgrounds across the food chain, produced voluntary guidance on the use of nutrition claims used by different manufacturers (Food Advisory Committee, 1990). This stated, for example, that claims such as "90% fat free" should not be used because they could mislead consumers by suggesting that a product was low in fat, when this was not the case. It also defined the meaning of terms such as "low fat" and "high fiber."

The use of claims on foods also came under scrutiny at international level. The Codex Alimentarius Commission (a joint initiative between the Food and Agriculture Organization (FAO) and WHO that sets international standards) initiated work through the Codex Committee on Food Labelling (CCFL) to provide clarity over the use of claims. In 1997, a list of definitions for nutrition claims was agreed by the member governments of Codex (CAC, 1997). While not legally binding, these standards are used as reference texts in trade disputes and governments should aim to implement them. As a result the UK Food Advisory Committee's guidance was revised to bring it in line with the Codex guidance.

9.6 RESPONSIBLE HEALTH CLAIMS

Health claims, however, were a different matter. They were also on Codex's agenda and started to be scrutinized more closely in general. Under the UK Food Safety Act 1990 (and previous Food Act 1984), it is illegal to mislead as to the nature, substance or quality of a food. Claims made about the health aspects of foods, should therefore be true. Under this legislation it is also illegal to claim that a food has medicinal properties and is able to prevent, treat or cure a disease. Sometimes "health" claims on food could imply that the benefits of consuming a food went beyond maintaining health and the status of disease-risk reduction claims, referring to risk factors for disease, such as cholesterol, rather than the disease itself was in need of clarification. Some EU member states, for example, banned such claims.

Consumer organizations were also critical of the way in which the burden of proof worked. In the UK, it was up to local authority trading standards officers, who were responsible for enforcing the legislation, to challenge a food company to provide scientific evidence to back up their claim. Given the many other responsibilities that trading standards departments had, this rarely happened. Where a claim is used in an advertisement, it is the responsibility of the Advertising Standards Authority (ASA) to police misleading claims, but this system also relies on the ASA responding to complaints of misleading advertising.

As the foods being produced and the claims associated with them could often be complex, the system relied – and still does to a large extent – on trading standards officers being able to develop sufficient expertise in often narrow and technical areas of nutritional science in order to stand a chance of challenging a manufacturer's claim. The David and Goliath nature of this was illustrated clearly by a case taken in 2000 by Shropshire trading standards against Nestlé over a claim made on its Shredded Wheat suggesting that eating the cereal reduced the risk of heart disease (Marketing Week, 2000). In this case, Shropshire won the case.

The burgeoning so-called functional foods market made it essential to clarify some of these issues and provide a clearer framework for the responsible use of claims that would help consumers as well as incentivise and enable manufacturers to promote the products that they had invested in developing. It is important that consumers can differentiate between products that offer genuine health benefits, from those that simply mislead with marketing hype.

9.7 MOVES TOWARDS SELF-REGULATION

In 1997, the Joint Health Claims Initiative (JHCI) was launched in the UK. This was a tripartite initiative, involving the food industry, trading standards officers and consumer organisations in order to develop a voluntary approach to more responsible use of health claims on foods. Founding members included the Food and Drink Federation (FDF), the British Retail Consortium (BRC) and the Proprietary Association of Great Britain (PAGB), Sustain and Which?.

The first stage in the process was the development of a Code of Practice that set out how claims should be used responsibly. For example, it was important that implied claims were included too. Companies which signed up to the Code agreed that health claims should assist consumers to make informed choices by ensuring that health claims were substantiated and checked for accuracy by independent experts prior to use. They also committed to comply with the spirit as well as the letter of the Code and agreed to the overriding principle that the likely consumer perception of the health claim is paramount.

This was followed by the establishment of a Code Administration Body, comprised of a Council, Secretariat and Expert Committee. Claims submitted by interested parties, such as trade bodies, were assessed based on the totality of the evidence by the expert committee. In the time that they were in operation, five generic claims were "approved" by the JHCI, relating to:

- reduced saturated fat and blood cholesterol
- wholegrain foods and heart health
- soya protein and blood cholesterol
- oats and blood cholesterol
- omega 3 polyunsaturated fatty acids and heart health.

The JHCI was therefore an important step forward. It had to work through an approach to substantiation of claims and drew on experience in other countries, such as Sweden and the United States which already had some form of vetting system in place for health claims. But as it was voluntary, there was no obligation on companies to submit claims for approval or even agree to comply with the Code.

The JHCI came into being at the same time that there was an increased call for EU legislation to tackle this complex area. Many of the signatories to the JHCI, therefore supported it on the understanding that it was an important step forward in the absence of regulatory measures.

9.8 REGULATORY ACTION

The 2000 European Union (EU) White Paper on Food Safety committed the European Commission to come forward with a proposal for the regulation of health and nutrition claims on food. In 2003 a proposal was published which included the important requirements that nutrition claims should be defined and that health claims should be independently substantiated.

One of the most controversial aspects of the proposals – that continues to be unresolved – was the issue of whether or not criteria should be established for the nutritional content of foods that carried health and nutrition claims on foods. In the United States for example, disqualifying criteria for claims were included in the Nutrition Education and Labeling Act 1990. These were intended to ensure that products of a poor nutritional quality could not suggest that they were healthier than they were. The argument was accepted in the EU and the concept was included in the final Regulation on Health and Nutrition Claims on foods which was adopted in 2006 (European Commission, 2006).

This Regulation covers claims used in the labeling, presentation and advertising of foods and makes it an offence for nutrition and health claims to be false, ambiguous or misleading; give rise to doubt about the safety and/or nutritional adequacy of other foods; encourage or condone excess consumption of a food; state, suggest or imply that a balanced and varied diet cannot provide appropriate quantities of nutrients in general; or refer to changes in bodily functions which could give rise to or exploit fear in the consumer, either textually or through pictorial, graphic or symbolic representations.

There are several important aspects that underpin the Regulation. First, a claim is defined as "any message or representation, which is not mandatory under Community or national legislation, including pictorial, graphic or symbolic representation, in any form, which states, suggests or implies that a food has particular characteristics."

Fundamental to the approach set out in the Regulation is independent scientific substantiation. Nutrition and health claims are both required to be based on and substantiated by generally accepted scientific evidence. But the regulation also makes reference to the need for health claims to be well understood by the average consumer.

9.9 DEFINITIONS FOR NUTRITION CLAIMS

Only the nutrition claims that are included in the annex to the Regulation are allowed to be used. The regulation came into effect from 31 July 2009 but some aspects have taken much longer to implement than anticipated.

The annex is also based largely on the Codex definitions on nutrition claims; for example low fat is no more than 3 g per 100 g; products making low sugar claims can contain no more than 5 g of sugar per 100 g. Comparative claims such as "light" or "lite" and "reduced" are also included. A claim, or any phrasing that suggests, a particular food as been reduced may only be made where the reduction is at least 30% compared to a similar product. Light or lite has the same meaning as reduced.

The list of permitted claims can be extended through the EU's commitology procedure involving member states and the Commission. This procedure is subject to scrutiny by the European Parliament. A particularly hot topic has been the issue of whether manufacturers should be able to promote the fact that they have lowered the levels of certain nutrients even where this is not significant enough to meet the criteria for a reduced claim. On the one hand, it has been argued that manufacturers need to be able to communicate to consumers that they have lowered levels in order to provide an incentive to invest in reformulation to lower fat or salt levels for example. But on the other, it is argued that it could be confusing to suggest that a product is healthier than similar products that have not been reformulated, even if they contained lower levels to start with. On this basis, the European Parliament achieved a sufficient majority to reject the addition of this type of claim to the annex (European Parliament, 2012).

9.10 APPROVAL OF HEALTH CLAIMS

Health claims are distinguished from reduction of disease risk claims, with different approaches set out for each. In both cases, it is the responsibility of the European Food Safety Authority (EFSA) to assess the scientific evidence and determine whether claims can be substantiated. It also has responsibility for advising on nutrition claims.

Certain health claims are specifically prohibited. These are claims that suggest health could be affected by not consuming the food, claims which make reference to the rate or amount of weight loss that could result from consumption of a particular product; and claims using recommendations from individual doctors, health professionals or other associations (other than those permitted under national legislation).

Claims that refer to the reduction of disease risk or children's development and health can only be made if they are authorised according to the procedure set out in the Regulation (Article 14). Companies have to submit an application through a member state. They then submit the claim, and evidence to substantiate it, to EFSA. EFSA has to give its opinion within a specified time, but it is up to the European Commission to put a proposal to Member States for authorization, based on this advice.

Examples of claims that have so far been approved under this process with specific conditions of use include:

- *Plant stanol esters*: plant stanol esters have been shown to lower/reduce blood cholesterol. High cholesterol is a risk factor in the development of coronary heart disease.
- *Oat beta-glucan*: oat beta-glucan has been shown to lower/reduce blood cholesterol. High cholesterol is a risk factor in the development of coronary heart disease.
- *Calcium and vitamin D*: calcium and vitamin D are needed for normal growth and development of bone in children.

9.11 CATEGORIZING HEALTH CLAIMS

The claims that have caused a lot more difficulty are those known as Article 13.1 claims. These are health claims which refer to the role of a nutrient or other substance in growth, development and the functions of the body; psychological and behavioural functions; or slimming or weight control, reduction in the sense of hunger or an increase in the sense of satiety or reduction of available energy in the diet. This type of claim is allowed if based on generally accepted scientific evidence and well understood by the average consumer.

Member States had to provide the European Commission with a list of claims that fell under this category, initially by January 2008. Under the Regulation, a list of approved and rejected claims would be adopted following consultation with EFSA.

While it was known that many claims were in use on foods, the scale of the number of claims that would be submitted under this part of the Regulation was not anticipated. From an initial list of around 44 000 submissions from Member States, EFSA considered over 4000 claims. Just around a fifth of claims were able to meet the criteria under the regulation and be substantiated based on generally accepted scientific evidence.

A first Community list of approved and rejected claims did not, therefore, come into effect until 14 December 2012 – and this list is still incomplete. Some claims are on hold, including many claims made about botanicals which have failed to be substantiated but could be permitted if used on traditional herbal medicines. The European Commission is therefore looking at how these should be controlled.

A further category of health claims is also distinguished in Article 13.5 of the Regulation. These are claims other than reduction of disease risk claims or those referring to children's development and health, which are based on generally accepted scientific evidence and/or which include a request for the protection of proprietary data. These also have to be assessed by EFSA and then authorized as for the other claims through the EU's commitology procedure. A claim for a water-soluble tomato concentrate has, for example, been authorised under this part of the regulation, allowing it to claim it helps maintain normal platelet aggregation, which contributes to healthy blood flow.

9.12 THE SUBSTANTIATION PROCESS

EFSA has published guidance (EFSA, 2011) to clarify how it approaches the assessment of the evidence submitted for the different types of claims that it has to look at. The assessments are carried out by a panel of experts on its Panel on Dietetic Products, Nutrition and Allergies (NDA). There are three basic criteria:

- whether the food/constituent is defined and characterized
- whether the claimed effect is defined and is beneficial to health
- whether a cause and effect relationship is established between the consumption of the food/constituent and the claimed effect for the target group under the proposed conditions of use.

If a cause and effect relationship is considered to be established, the Panel considers whether it is reasonable to consume the quantity of food or follow the pattern of consumption needed for the claimed effect. It also takes into account whether the wording reflects the scientific advice and complies with the criteria in the Regulation and whether the proposed conditions of use are appropriate.

Some claims which did not meet the criteria were given a second chance at substantiation. This included, for example, claims about probiotics where manufacturers had been unable to characterize the constituent.

Claims are not, however, authorised until they have been approved by the Member States through Standing Committee procedure, based on a proposal by the European Commission. This therefore provides an opportunity to take into account other considerations – although this process should not undermine the scientific substantiation process. For example, a product containing sodium claimed that the sodium could have a beneficial function but this was removed at committee stage as, although it was accurate, it contradicted another claim that stated that it was beneficial to health to reduce sodium intake.

9.13 TAKING STOCK

The process has put EU resources to the test, but has largely been a vindication of the need for much tighter regulation in this area. The quality of claims and supporting evidence has shown how many of the health claims used to sell products to consumers would just lead them to waste their money. Only around 20% of claims submitted have been scientifically substantiated so far. The process should, therefore, finally enable consumers to have confidence in the claims that they see on foods. It should also encourage responsible innovation by food companies.

There is, however, still some way to go until complete consumer confidence is achieved. The issue of botanicals still remains unresolved. These are on hold while borderline issues with traditional herbal medicines, which have a lower level of substantiation, are resolved. It needs to be ensured that such claims when made on food are not allowed a lower level of supporting evidence than those permitted for other substances. The claims that have been given longer to try and provide evidence to support them also need to reach a conclusion. Once this process is completed, however, there are still likely to be further claims put forward for endorsement.

The other important issue is that of nutrient profiles. How can it be ensured that health claims can only be made on genuinely healthy foods? Under Article 4 of the Regulation, the European Commission is required to establish:

> specific nutrient profiles, including exemptions which food or certain categories of food must comply with in order to bear nutrition or health claims and the conditions for the use of nutrition or health claims for foods or categories of foods with respect to the nutrient profiles.

The intention is to ensure that a product high in fat, sugar or salt, such as a doughnut for example, could not carry a health claim even if it had a beneficial ingredient or nutrient added to it. Nutrition claims will also be banned from foods that fail to meet the nutrient profile. But if the product is only high in one of the nutrients that are included in the nutrient profile criteria, a nutrition claim could still be made, but the high level of the "negative" nutrient would also have to be emphasized. For example, if a breakfast cereal was low in fat, but high in sugar, it could only claim to be "low in fat," if it also stated that it was "high in sugar."

The nutrient profiles have to be established, taking into account:

- The quantities of certain nutrients and other substances contained in the food, such as fat, saturated fatty acids, *trans*-fatty acids, sugar and salt/sodium.
- The role and importance of the food (and or categories of the food) and the contribution to the diet of the population in general or, as appropriate, of certain risk groups including children.

• The overall nutritional composition of the food and the presence of nutrients that have been scientifically recognized as having an effect on health.

The Commission should also take into account the advice of EFSA in setting these profiles. This was initially required to be completed by 19 January 2009 in the Regulation. But the profiles do not exist and there is currently no sign of any progress in developing them.

An EFSA Opinion on nutrient profiles was published in 2006, which set out possible approaches. An initial draft proposal was also published by the European Commission for discussion in 2007. But the issue has become very political. The question of which foods can be classed as healthy or not has implications for individual companies and, it has emerged, for national pride. The example of German rye bread is one particularly high profile example of this: parts of the German press took great offense to the implication that German rye bread could not carry a claim because the salt content was considered too high. A similar outcry followed in Italy over Nutella chocolate spread. No profiles exist and the signs are that if profiles are ever developed, they may be far removed from the over-riding objective of the nutrition and health claims regulation which is to protect consumers.

9.14   EVOLUTION OF PROFILES

This issue does, however, need to be resolved. Failure to do so undermines the whole basis of the Regulation. Even if claims are accurate, they could have a negative rather than positive effect on people's diets if they encourage them to eat more foods high in fat, sugar or salt.

Since the regulation was adopted back in 2006, nutrient profiling as a concept has developed significantly. Different models have been developed of varying levels of complexity. In the UK, for example, a threshold model for fat, saturated fat, sugar and salt underpins the traffic light front of pack nutrition labeling scheme. A model has also been developed for TV advertising restrictions to children which is based on a simple scoring system and takes account of energy, saturated fat, total sugar, sodium and fruit, vegetables and nuts, fiber and protein. This model has also been developed in Australia as a basis for restricting health claims on foods.

The WHO has piloted guidance on the use of nutrient profiling models for different purposes. The important principle that health and nutrition claims should only be made on foods that really are healthy, therefore needs to be maintained and implemented.

9.15   COMMUNICATION IN CONTEXT

The development and legitimisation of health claims on foods is based on the assumption that there are certain types of foods that are good for people – or at least not as bad for them as other foods – and therefore they should be identified so that people can eat more of them.

Many products have been developed with this in mind. If the deadlock over the development of nutrient profiles can be addressed, concerns about whether or not these claims are being made on foods that people should really be encouraged to eat might be resolved. The substantiation process means that claims used on products are true, assuming that the legislation is effectively enforced.

This does, however, still leave a wider issue of whether or not the types of products that are being produced and the claims about their health-benefits are really that relevant in the wider context of the diet-related diseases that need to be tackled. While they may appeal to people actively seeking healthier choices and possibly already more conscious of their food choices, they may not appeal to those who are less interested in eating healthy food and therefore most need to be choosing healthier products. Measures such as product reformulation "behind the scenes" to reduce saturated fat, salt and sugar levels in foods may have a greater impact on reducing levels of diet-related disease but are not always advertised if they affect people's perception of quality.

Health and nutrition claims are clearly not a panacea – and neither are the functional foods that are being developed to make best use of them. There remains a need for clearer, actionable advice and information that puts these claims in context. It also needs to be made clearer which are the 'bad' foods, as well as the "good" foods – which foods do people need to eat less regularly and which do they need to eat more of regardless of whether or not they carry a claim telling you to do so.

9.16 WIDER NUTRITION INFORMATION

The claims regulation makes it a requirement that nutrition information is provided if a claim is made on a food. But until recently, this was the only way that nutrition information was legally required. Many retailers and manufacturers in the UK have provided it for many years, but this has been on a voluntary basis and has not always been in the most user-friendly format.

Nutrition information has, however, evolved considerably over the last few years and the EU legislation was updated last year to include mandatory provision. There has also been a move away from mere presentation of the facts to some form of interpretation of the way that the information is provided.

There have broadly speaking been three main stages on the way to giving consumers meaningful information. First, guideline daily amounts (GDAs) were developed. These took population dietary goals that had been developed by the UK government's Committee on Nutritional Aspects of Food Policy (COMA), now replaced by the Scientific Advisory Committee on Nutrition (SACN), and turned them into estimated amounts that men and women should eat in a day. In the case of fat, for example, where COMA recommends that no more than 35% of food should come from fat, the GDA breaks this down into a more meaningful figure of 95 g fat a day for an average man and 70 g for an average woman.

The GDAs, although widely used, were not developed by the government, but by the UK food industry body, the Institute of Grocery Distribution:

Guideline daily amounts		
	Men	**Women**
Fat	95 g	70 g
Saturates	30 g	20 g
Sodium	2.5 g	2 g
Fiber	20 g	16 g

Source: Institute of Grocery Distribution.

Efforts were then made by the government (initially the Ministry of Agriculture, Fisheries and Food (MAFF) and then the Food Standards Agency) to develop this guidance into advice to consumers on how much fat, sugar or salt was 'a lot' and how much was 'a little' (Rayner *et al.*, 2004).

A lot and a little	
A lot	**A little**
These amounts or more:	These amounts or less:
10 g of sugars	2 g of sugars
20 g of fat	3 g of fat
5 g of saturates	1 g of saturates
3 g of fiber	0.5 g of fiber
0.5 g of sodium	0.1 g of sodium

Source: MAFF/ Food Standards Agency

Several retailers included the GDAs on the back of pack as a first step. This generally appeared in close proximity to the nutrition information panel in order to help consumers make sense of the figures as an indication of the contribution of consuming the product to meeting the GDA.

The third key development was that nutrition information started to appear on the front of pack, as well as on the back of pack. This recognized that many people shop in a hurry and so do not always have time to closely study the nutrition information panel, unless they are looking for specific information. Having more up-front information on the front of pack is more user-friendly as it can be seen "at a glance."

For a number of years Which? has been calling for a simple, consistent front of pack labeling scheme including traffic light color coding to indicate whether levels of key nutrients are high (red), medium (amber), or low (green), be used by supermarkets and manufacturers to make fat, sugar and salt levels clear. In 2004, the Health Select Committee (House of Commons, 2004) as part of its inquiry into obesity, recommended that the Government introduced a traffic light system for labeling foods, according to criteria devised by the Food Standards Agency, which would apply to all foods.

Two main forms of front of pack nutrition labeling have since evolved with several hybrids in-between: a multiple traffic light labeling scheme and a percentage guideline daily amount labeling scheme. Both of these approaches evolve from the GDAs. The traffic light labeling scheme is a more comprehensive way of expressing what is "a lot" and "a little" on the front of pack. It shows whether levels of fat, saturated fat, sugar and salt are high, medium or low using red, amber, and green color coding. The "green" criteria is based on what is considered to be a "low" claim as part of the health and nutrition claims regulation definition of nutrition claims. The amber and red cut-offs are set as proportions of the GDA. A specific benchmark figure was set for sugars in order to try and take account of added sugar, rather than naturally occurring sugars and because of concern that the GDA figure for sugar was too generous for this purpose.

Traffic light labeling criteria		
Green (low)	Amber (medium)	Red (high)
Sugar ≤5.0 g/100 g	>5.0 to ≤12.5 g/100 g	>12.5 g/100 g
Salt ≤0.3 g/100 g	>0.3 to ≤1.5 g/100 g	>1.5 g/100 g
Fat ≤3.0 g/100 g	>3.0 to ≤20.0 g/100 g	>20.0 g/100 g

Source: Food Standards Agency

The percentage GDA approach also interprets information based on the GDA, but does not go so far. It shows what proportion of the GDA for fat, saturated fat, sugar, salt, and calories is met by eating a recommended sized portion of the product.

Some retailers now include both %GDA information and traffic light color coding, which is based on the levels included in a 100 g of the product – although also takes account of products consumed in larger portions. Asda has also included "high," medium," and "low" in text, in addition to the traffic light color coding and %GDA.

While this move to interpretation has been positive, the continuing use of different schemes has created unnecessary confusion. Different retailers and manufacturers use different approaches. Some of those using %GDAs, such as Tesco and Aldi, also color code the nutrients (not the nutrient levels) which may confuse some people with the traffic light labeling scheme. It has been estimated that approximately one third of the market uses the traffic light labeling scheme.

In an effort to try and consolidate and simplify the proliferation on front of pack nutrition labeling schemes, the Food Standards Agency commissioned an independent evaluation of front of pack nutrition labeling schemes which reported in 2009 (Food Standards Agency, 2009). This was overseen by the government's chief social scientist and included a combination of research approaches aimed at testing the performance of the different schemes when used by consumers. The conclusion of the research was that two schemes performed best overall. Taking into account what the research showed consumers also preferred, it recommended that a scheme including %GDAs, traffic light color coding and "high," "medium," or "low" in text was the best approach. Two companies were already doing this: Asda and McCain. But it suggested that a small compromise across the board could easily achieve the objective of a common scheme. The results were consistent with research previously carried out by Which? (Which?, 2006). This had found that when consumers were presented with tests using mocked up labels on food products, traffic lights worked best for most people.

Unfortunately, such a change would have to be achieved on a voluntary basis. Food labeling is an EU competence and so the UK could not legislate on this issue. The nutrition labeling directive has been amalgamated with the wider food labeling directive as part of the EU's Food Information to Consumers regulations (EU, 2000). The only way to make all manufacturers and retailers use the best scheme, therefore, was to secure a change requiring this within the legislation.

9.17   LEGAL LABELING FIGHT

The Food Information to Consumers Regulations may turn out to be one of the most heavily lobbied pieces of legislation in recent years, with Members of the European Parliament claiming that they were surprised at the extent of the resources the food industry put into fighting any such move. Despite efforts by consumer and public health groups to secure an evidence-based approach, the final version of the regulations does not mandate traffic lights. It does, however, still allow for them.

The Regulations make it a legal requirement for nutrition information to be provided for the first time, regardless of whether or not a claim is made. It states that a mandatory nutrition declaration should include the energy value and the amounts of fat, saturates, carbohydrate, sugars, protein, and salt. This can be supplemented with information about monounsaturates, polyunsaturates, polyols, starch, fiber and vitamins and minerals. It allows for information about the energy value alone or the energy value, fat, saturates, sugar and salt to be repeated on the pack, for example on the front of pack. It allows these to be expressed as a percentage of the reference intakes (i.e. GDAs) per 100 g or per 100 ml.

Provision is however also made for additional forms of expression or presentation in Article 35 of the Regulations. This can be done using other forms of expression. The levels can be presented using graphical forms or symbols in addition to words or numbers when the following requirements are met:

- they are based on sound and scientifically valid consumer research and do not mislead the consumer
- their development is the result of consultation with a wide range of stakeholder groups
- they aim to facilitate consumer understanding of the contribution or importance of the food to the energy and nutrient content of a diet
- they are supported by scientifically valid evidence of understanding of such forms of expression or presentation by the average consumer
- in the case of forms of expression, they are based either on the harmonised reference intakes set out in the Regulation, or in their absence, on generally accepted scientific evidence on intakes for energy and nutrients
- they are objective and non-discriminatory, and
- their application does not create obstacles to the free movement of goods.

This therefore leaves a situation in the UK which enables traffic light labeling to be added to front of pack labeling, but does not require it. It is therefore necessary to incentivise retailers and manufacturers to move to a consistent scheme that includes the traffic light colours. In August 2012, Tesco, the largest UK retailer, announced that it would adopt traffic lights on front of pack. It was quickly followed by other retailers who had not been using the scheme, leading to a commitment by all major UK retailers to use both traffic light colors and %GDAs. The UK Government is now working on a national scheme which is expected to be in use during 2013. At the time of going to print, it is unclear whether the main food manufacturers will also commit to use this on a voluntary basis.

9.18 HEALTHY AND SUSTAINABLE?

Ensuring consumers have useful and reliable information about the nutritional content of their food has been a long struggle over many years. It is important that the same mistakes are not repeated as greater focus shifts to helping consumers make food choices that are sustainable in a wider sense, including taking account of the environmental impact of the food that they buy. This discussion is becoming more prominent. Consumers are becoming more interested in making more sustainable food choices, but will also be encouraged to do so as the government and food industry focus more on reducing the environmental impact of food production and consumption in the coming years.

"Sustainable food consumption" is covered in detail in Chapter 8. However, in a similar pattern to the evolution of health and nutrition claims and nutrition labeling, a plethora of claims and labeling schemes about different ethical and environmental aspects of food production have appeared on the market in recent years. These have been described as falling into two broad categories, practice based, such as organic schemes, and outcome based, such as the Carbon Trust carbon footprint label (Defra, 2010). A mixture of regulation and certification schemes underpin the schemes which are not necessarily obvious to the consumer.

Going forward this is an area where much can be learned from the area of nutrition labeling and health claims. It remains to be seen whether voluntary approaches to the regulation of "green" claims will succeed where such an approach failed for health claims. More research is still needed and clearer government advice needs to be provided around the most important elements of sustainable choices to communicate to consumers. But as with nutrition information, mere presentation of the facts is unlikely to be adequate and broader interpretation of the information that puts into the context of good or poor practice is also needed. Which? research (Which?, 2010) has indicated for example that the carbon footprint label, while an interesting first step in this area, means little to consumers without some form of reference to enable them to assess whether a particular figure is good or bad.

The following principles (Box 9.1) should therefore help guide the development of clearer information about sustainable food choices, building on the experience from informing consumers about healthier food choices:

This is an area where regulators have paid limited attention to date. But this is set to change. There is a greater focus nationally as the UK Government considers how best to deliver the recommendations in the Foresight report on the Future of Food and Farming (Government Office for Science, 2011). But also at EU level. A European Commission Communication on Sustainable Food is scheduled for 2014, but discussions around potential environmental labeling schemes for food are also likely to see more specific proposals emerging.

Box 9.1   Sustainable food information – lessons from nutrition labeling:

- The importance of using terms and labeling schemes consistently across different products and brands to avoid confusion
- Providing consumers with facts and figures alone is unlikely to be helpful enough – they need to be interpreted so that consumers know what is significant and how they relate to broader government advice
- Schemes have to be underpinned by scientific evidence and be verifiable
- Labeling schemes need to be linked up to broader government messages
- Food company buy-in is needed at an early stage, otherwise the competitive nature of food labeling will result in many different schemes, causing confusion and requiring a legislative solution.

9.19 CONCLUSION

There is now consumer-focused regulation in place to help ensure that nutrition and health claims made on food are reliable and help consumers make healthier choices, rather than undermine them.

This Regulation (EU, 2011) has been a long time coming and many aspects of its implementation still need to be resolved so that it can be ensured that it is effectively enforced. Consumers are increasingly interested in eating healthily – and need to be given rates of obesity and diet-related disease. It has, however, been difficult for people to work out which products carrying health and nutrition claims are genuine and which ones are not.

But controls over claims are just one aspect of the many measures needed to help consumers make healthier choices. A more responsible approach is therefore needed from food companies to the way that they develop, promote and label their products. Clearer and consistent messages are needed about how to make healthier choices and the overall nutritional content needs to be presented simply and consistently on front of pack so that it is as easy to identify the less healthy foods as the healthy ones.

The development of health and nutrition claims and evolution of nutrition information also hold important lessons for communication about wider sustainability aspects of foods so that consumers can more easily choose foods that are healthy and have a lower environmental impact.

REFERENCES

Codex Alimentarius Commission (1997). Guidelines for use of nutrition and health claims (CAC/GL 23-1997).

Defra (2010) Effective approaches to environmental labelling of food products, Research Project Final Report, University of Hertfordshire, Defra, July 2010.

Department of Health (2011a). Healthy Lives, Healthy People: A Call to Action on Obesity in England, Department of Health 13 October 2011.

Department of Health (2011b). National Diet and Nutrition Survey: Results from Years 1 and 2 of the Rolling Programme 2008/9-2009/10, 21 July 2011.

EFSA (2011) European Food Safety Authority Scientific Opinion: General Guidance for Stakeholders on the Evaluation of Article 13.1, 13.5 and 14 health claims, EFSA Panel on Dietetic Products, Nutrition and Allergies, 25 March 2011.

European Commission (2006) Regulation No 1924/2006 of the European Parliament and of the Council of 20 December 2006 on nutrition and health claims made on foods.

European Parliament (2012) MEPs veto "misleading" food labelling, European Parliament News Release 2.1.12.

Food Advisory Committee (1990). Food Advisory Committee report on its review of food labelling and advertising, HMSO, London.

Food Standards Agency (2009) Comprehension and use of UK nutrition signpost labelling schemes, May 2009. FSA, London.

Government Office for Science (2011) The Future of Food and Farming, Challenges and Choices for Global Sustainability. Final Project Report. Foresight January 2010. Government Office for Science, London.

House of Commons (2004) Obesity, Health Committee, Third Report of Session 2003-04, May 2004, House of Commons, London.

Marketing Week (2000). Nestle lawsuit sets labelling precedent, Marketing Week, 8 June 2000.

Mintel Oxygen (2011) Functional Food and Drink UK, September 2011, Mintel.

NACNE (1983). A discussion paper on proposals for nutritional guidelines for health education in Britain, National Advisory Committee on Nutrition Education, The Health Education Council, London.

Official Journal of the European Union (2000) Directive 2000/13/EC of the European Parliament and the Council of 20 March 2000 on the approximation of the laws of the Member States relating to the labelling, presentation and advertising of foodstuffs – as amended September 2003.

Official Journal of the European Union (2011) Regulation (EU) No 1169/2011 of the European Parliament and of the Council of 25 October 2011 on the provision of food information to consumers.

OECD (2012) Obesity Update 2012, Organisation for Economic Co-operation and Development (OECD), February 2012. OECD, Paris.

Rayner M, Scarborough P and Williams C (2004). The origin of Guideline Daily Amounts and the Food Standards Agency's advice on what counts as a lot and a little. *Public Health Nutrition* 7(5):693.

Which? (1997) Under wraps – what lies behind the label? Policy Paper, Consumers' Association/Which? November 1997. Which?, London.

Which? (2000) Functional food: health or hype? Policy Paper. Consumers' Association/Which?, London.

Which? (2006) Healthy Signs?, Policy Paper, Which?, London.

Which? (2009a) Hungry for Change? Policy Paper. Which?, London.

Which? (2009b) Which? food is best for you? Which? magazine, March 2009.

Which? (2010) Making sustainable food choices easier: a consumer focused approach to food labels, Policy Paper, September 2010, Which?, London.

Which? (2011) Just how healthy are health drinks? Which? magazine, October 2011.

World Health Organization (2004). Global Strategy on Diet, Physical Activity and Health, WHA 57.17, 2004. WHO, Geneva.

World Health Organization Europe (2006) European Charter on Counteracting Obesity, 16 November 2006. WHO, Geneva.

10 Worker exploitation in food production and service

Charlie Clutterbuck

10.1 INTRODUCTION

Food and farm workers have been exploited since the beginning of time as we know it. All civilizations were built on eight crop centers (Vavilov Centers, 2012). Aztecs in Mexico developed maize, Incas in Peru grew potatoes, while wheat originated in Mesopotamia (now Iraq). In each civilization, farm and food workers always seemed near the bottom of the pile.

Most of the major exporting crop regions are now on the other side of the world from which the crop originated. The most popular explanation for this is that the crops can escape their natural pests and predators (Purseglove 1968). However, this distribution of world crops can be ascribed to labor conditions as much as it can to natural or geographic conditions. Sugar from Africa to the Caribbean and the use of slaves is obvious. Other examples are tea (China origin) plantations in India/Sri Lanka, which are based on migrant Tamil Indian populations and coffee ranchers dependent on Spanish migrant workers. What grows where is very labor dependent. Yet its role is largely overlooked.

The role of land, labor and technology needs much greater analysis than can be produced here. According to Hyams "The agricultural idea, tools and techniques were brought to Atlantic Europe from the south-east by immigrant peoples. The crafts of tillage do not, in Europe, slowly develop before the eyes of the archaeologist, as in the Fayum or in Mesopotamia" (Hyams, 1952). Things have not changed – particularly the dependency on migrant labor for much farm and food production. The UK depends on East European workers for most of the horticultural crops, and the US fruit crop is dependent on Mexican workers. Land and labor needs much greater examination.

As the world's food system becomes more globalized, there will be increasing dependence on vast numbers of migrant workers, as the search for ever cheaper labor scours the earth. The pressure is to produce even cheaper food to fuel the world's economy. The drive for cheap food to boost the economy started with the UK's Repeal of the Corn Laws to allow industrialists to feed industrial workers more cheaply; and it goes on

Practical Ethics for Food Professionals: Ethics in Research, Education and the Workplace, First Edition.
Edited by J. Peter Clark and Christopher Ritson.
© 2013 John Wiley & Sons, Ltd. Published 2013 by John Wiley & Sons, Ltd.

today as enabling more women to go out to work in the economy. The retailers' "price wars" end up with somebody else paying the costs – and it is usually workers. Whether it is banana or coffee workers, they always bear the brunt. The rich want cheap food at any price.

This drives down wages in the food chain. Throughout the world, employees in both food and farming are always among the poorest paid. It is the same everywhere, in any country. In many cases the same people producing the food are starving. Of the billion who go hungry, it is estimated that 80% of these are rural poor. US farmers receive an average of 30% of the retail price of most foods (USDA, 2011) Similar figures would apply to family farms in UK. Farm labor costs are typically less than a third of farm revenue for fresh fruits and vegetables. It means that farmworkers wages could go up by 40% and households would pay only about £$15/year more for their fruit and vegetables (FoodFirst, 2011).

Food and farm workers are poorly paid when compared with manufacturing workers, let alone civil workers and bankers. The UK is a good example, investing hundreds of billions of pounds more in the finance sector (displayed for example by the building labeled the "Gherkin" in the City of London), rather than in growing cucumbers (a kind of gherkin), as displayed by closing the last Vegetable Research Station in 2009, for want of £2m (Clutterbuck, 2009). These reflect the values of society at present, and cheap food is considered more important than the costs to land and labor.

The days of *The Grapes of Wrath* may be long gone, but there are still many incidences of "unethical" work. Exploitation is endemic in the food chain. Yet the dominant players in the food world – the retailers, claim that they behave "ethically." All companies have teams of "ethical" specialists. Some may have an ethical training, but many will use their personal judgments to determine whether any particular practice is "ethical."

Nevertheless, however bad it has been, a new set of ethics is developing. This questions whether it is right and fair to expect other people to do what we are not prepared to do ourselves. Unemployment remains high in 2012, yet there are still twice as many migrant rural workers in the UK and the US, as there are permanent rural workers. Why do not local students and young unemployed workers do this work? Presumably because they regard the wages as too low for the hard work involved

The question that follows is how do we decide whether this work is exploitative, when some people are prepared to do this kind of work for the wages offered? That is what is being asked when companies set up systems through their supply chains to ensure that they are trading "ethically." This can mean many different things, but usually includes consideration about treatment of workers, especially in developing countries.

The main big retailers determine the employment practices which should be avoided (e.g. child labor), as having been pre-determined as "exploitative." The status of many other employment practices is less clear. Most operating in a world market expect their suppliers to follow the internationally agreed ILO Conventions. If the suppliers operate to the Conventions, that trading is usually called "ethical" (Ethical Growers, 2012). There are a lot of auditors "ticking boxes," but does anybody ask the food and farm workers: "Is that fair?"

This chapter sets out to help identify ways that an 'ethical practitioner' could make judgments about whether the employment is "exploitative" or whether it is 'ethical'.

10.2 FOOD EMPLOYMENT – IS IT "EXPLOITATIVE" OR "ETHICAL?"

Clearly many employees will tend to say that the employment is exploitative, while most companies will tend to say they are behaving ethically. What may be exploitative to one person, is another person's opportunity. Nor will two people have the same ethical code. Yet ethics does require a yes or no – a judgment. "Is your Employment 'Exploitative' or 'Ethical'?"

10.2.1 Ethics

What, then, is "ethical?" This question has a history which we can call upon. Many people in the past have tried to work out how we come to making that crucial ethical decision – is it right or wrong? The issue is covered in detail in Chapter 3, but a few issues are worth highlighting here.

Just like the legal system, there are precedents and both have to arrive at a guilty/not guilty (or ethical/unethical) decision. Both ethics and law have to deal with specific sets of circumstances, always different Ethics is the "big picture." Food/farm ethics takes in sustainability (primarily environment but also social concerns), welfare of animals and people, but also wider concerns – that go beyond what may be easily measurable.

To quote Mepham, food ethics is:

> concerned with the issues of right and wrong, and/or good and bad, as they impact on the production, processing, distribution, and utilization of food – encompassing the interests of the current and future generations of humans and sentient beings.

> Such ethical concerns seek to pay due concern to the diversity of political, cultural, spiritual and technological influences on the world's peoples, while prioritzing the principle of fairness.

10.2.2 Exploitation

What is exploitation? There are several distinct groups in the food chain including retailers, manufacturers, suppliers, growers, managers and workers.

The balance of power has changed in last 25 years, before which the food manufacturers dominated. In last 25 years, retailers have grown in power, as women on their way to join the workforce (because two wages are now needed) dash into the supermarkets to get ever cheaper food from off the shelf. Again cheap food feeds the workforce.

The retailers have become some of the biggest companies in the world. In the EU, forthcoming agricultural policy reform recognizes this power and seeks to encourage producer co-operatives as a way to combat retailers.

Retailers are always having price wars, the costs of which are borne all along the supply chain – to the farmers and workers at the end. The power position of farm workers, at the bottom, has remained the same. Industrial workers have generally fared better and financial workers even best. The forms of exploitation can take many forms, from not paying the proper wages, to long hours of work, or the state of accommodation. This chapter identifies a wide variety of injustices, that give a flavor of what is happening, without attempting to provide a comprehensive list.

10.2.3 Employment

There are two completely different industrial relations models running on the land, in parallel. These are plantations with hundreds of migrant workers often with quite sophisticated labor systems; and family farms employing less than five permanent workers with as few systems as possible. One set of standards is needed for one, and another set for the other, but they are getting mixed up. There are 150 000 permanent farmworkers (not farmers) on family farms in the UK compared with 300 000 migrant workers on plantations.

Plantation agriculture developed, on the model of the sugar and cotton slave plantations. To quote Courtney:

> the growing of new crops in new areas, particularly Asia .... they were generally large and located in sparsely populated tropical areas, in many cases inland, and in consequence of the low indigenous population densities of these areas, the recruiting of labor from elsewhere was necessary ... Production was specialized, concentrating usually on one crop which was destined for the European or North American market (Courtney,1965).

Investment was predominantly in colonial countries where major investments could be better protected. Employers have learnt the lessons and applied them in the developed world, again with a lot of help from migrant labor. Plantation agriculture is not now just in the tropics; it is in the US and UK, at least in the East and South. There, very large organizations now employ up to 4000 workers in UK – almost all migrant employees from Eastern Europe and Africa. This contrasts with the west side of Britain, which is predominantly pasture, and family farms. Conditions are quite different.

This parallel land universe applies across the world. Plantations came earlier to Africa and Asia than EU, as the employers could control labor better there. From sub-Saharan Africa where they import a lot of food – about half comes from small farmers/coops and $^1/_2$ from plantation farms with up to 20 000 workers. In US, 70% of agriculture is carried out by migrants, mainly from Mexico (Martin, 2010).

How do we inform and explore this dialectic between exploitation and ethics with regard to employment? In this chapter we use the ethical matrix on how to apply ethics to issues in food and farming. There are three principles to help determine what is ethical: wellbeing, autonomy and fairness.

10.3 ETHICAL MATRIX

The ethical matrix was developed by Ben Mepham and is described in Chapter 3. It is set out in generic form in the following table.

Respect for	Wellbeing	Autonomy	Fairness
Farmers	Satisfactory income and working conditions	Managerial freedom of action	Fairtrade standards
Consumers	Food safety Acceptability, quality of life	Democratic informed choice	Availability of affordable food
The biota	Conservation	Biodiversity	Sustainability
Crop	Flourishment	Adaptability	iIntrinsic value

People are free to develop or reinterpret it at will. It is a tool for ethical deliberation, and does not prescribe an outcome. It helps people arrive at and justify to others, their ethical reasons for reaching a judgment. The matrix is there to be adapted.

Thus, it can be used for labor conditions. There are two main groups of people on the land – those who control their labor (farmers), and those who labor for others (farm workers). Standards to look after the growers/farmers are usually called "fair," whereas those standards that look after the workers in the chain are called generally "ethical."

Respect for	Wellbeing	Autonomy	Fairness
Farmers	Satisfactory income and working conditions	Managerial freedom of action	Fairtrade standards
Food & farm workers	Health & safety	Trade union rights	Ethical codes

When discussing the impacts of certain technological innovations it is not unusual to see these listed as "economic, safety, environmental and ethical." The logic of that approach implies that it could be acceptable for us to countenance unethical economics, unethical safety and unethical environmental protection measures.

The matrix approach casts ethical issues in a much broader context, so that they become open to rational discussion. It draws on ethical theories which underpin widely accepted principles defining the right and the good. In practice, all the theories behind these are likely to contribute, to varying degrees, to people's attitudes on what should be done in specific circumstances. Each of us blends these theories (consciously or unconsciously) with intuitive responses, and subject to cultural influences.

It is important to challenge the view that ethics is simply a matter of opinion and therefore carries little weight compared with the objective reality of scientific knowledge. Science does not work in a societal vacuum. Surely we suffer wrongs if: violently attacked (a violation of our wellbeing), wantonly deprived of our liberty (an infringement of our autonomy) or convicted of a crime of which we were innocent (a miscarriage of justice). There similar practices in food and farming.

Let us therefore look at conditions for food and farm workers, according to the three main principles, so as to help ethical practitioners determine whether those conditions are "ethical."

10.4 WELLBEING

Respect for wellbeing corresponds to issues prominent in utilitarian theory, which characteristically employs a form of cost/benefit analysis to decide on what it is right to do. Most famously articulated in the eighteenth and nineteenth centuries by Jeremy Bentham and John Stuart Mill, it may be summarised as aiming for "The greatest good for the greatest number" (Mepham and Tomkins, 2003).

According the United Nation's World Health Organization constitution "Health is a state of complete physical, mental and social well-being and not merely the absence of disease or infirmity." (WHO, 1946) Food and farming is unhealthy work, despite the clean outdoor image.

10.4.1 Food

Work in the food sector hardly represents "the greatest good for the greatest number." Worldwide, the main source of information about disputes between employees and employers in food and farming is the International Union of Foodworkers (IUF). The IUF highlight a present dispute with the Tetley Group owned by large Indian company Tata, which also own Jaguar cars and UK Steel. The IUF has alleged that the famous Tetley teas are pushing nearly 1000 tea plantation workers and their families into starvation on the Nowera Nuddy Tea Estate in West Bengal, India. It also alleges that an extended lock out has deprived them of wages for all but 2 days in 3 months. The IUF claims that Tata wants the workers to renounce their elementary human rights, including the right to protest against extreme abuse and exploitation. The lockout allegedly started when workers protested at the abusive treatment of a 22-year-old tea garden worker who was denied maternity leave and forced to continue work as a tea plucker despite being 8 months pregnant (IUF, 2011a)

The wages any tea picker gets is minimal. In Sri Lanka at the invitation of a Plantation Workers Union the author took a photograph of a tea picker and asked the union organizer, whether the picker should be given a few rupees to say "thank you." He said "No that was far too much." Many of the pickers are Tamil, brought there by the British from India. A lot of the "Tiger Trouble" had been generated when many of the tea plantations were given back to local people in the mid-1950s during Bandaranaike's premiership. This was to reduce the country's dependency on one cash crop. However, the Tamils – whose presence and language was recognized under British rule – were excluded from access to that new land, despite then being the third or fourth generation in the country.

The IUF also campaigned in support of over 100 workers locked out at the meat processing plant ANZCO. The union claimed that this was a new tactic that "locks out the workers in an attempt to force a signature on an essentially non-negotiated agreement." Some of their meat goes from New Zealand to the British retailer Waitrose, who were pulled into the dispute. They recommended both sides enter "facilitation," which turned out to resolve the dispute (IUF, 2011b).

In a quite different dispute, the IUF supported a strike by student exchange workers at US chocolate maker Hershey that they claimed: "exposed the trail of outsourcing and exploitation." Over 300 foreign student workers sat in and then walked off the job at a distribution plant in Palmyra, PA. The students, from countries as diverse as China, Moldova, Nigeria, Turkey and Ukraine, had come to the US on a program established as a 2-month work and travel program for foreign students. They were paid 8 dollars per hour to perform what were formerly union jobs. The IUF claimed that charges for rent for housing and compulsory fees for company transportation and other deductions left many workers with less than $100 for a 40-hour workweek (IUF, 2011c).

10.4.2 Farm

The agricultural sector employs an estimated 1.3 billion workers worldwide, that is half of the world's labor force. In terms of fatalities, injuries and work-related ill-health, it is one of the three most hazardous sectors of activity (along with construction and mining). It is also the largest sector for female employment in many countries, especially in Africa and Asia, and accounts for approximately 70% of child labor worldwide. According to International Labour Organization (ILO) estimates, at least 170 000 agricultural workers are killed each year. This means that workers in agriculture run twice the risk of dying on the job compared with workers in other sectors (ILO, 2011a).

The UK reflects these statistics almost exactly (HSE, 2011). There are three main reasons for these high rates. First, there are many hazards, including just about every biological, chemical and physical hazard going. Second, cheap food requires everybody to cut corners. One year when fatalities were much lower than the average in the UK was 2008 – the year of high grain prices. Finally, poor union organization in the industry supports the theory that "healthy workplaces are union workplaces" (Hazards, 2011). Not having roving safety reps or any recognizable Health and Safety (H&S) structures is a contributory factor. Most of the UK fatalities are on family farms.

The Health & Safety Executive's campaign "Revitalising Health and Safety" (HSE, 2000) identified the two worst sectors – construction and agriculture. Since then the construction industry has halved its fatality rate due to big companies taking the lead. But there has been no improvement in UK agriculture. The last fall in farm fatalities was due to the introduction of tractor regulations in the early 1990s. There are no corporations in the sector, so it is left to the main employers' organization to take the lead. The National Farmers Union (NFU) held their first "Safety Summit" in 2010.

These statistics do not include the single biggest food-related disaster in Britain, which involved the deaths of 23 Chinese cockle pickers in the sea off Morecambe Bay in 2000. This incident led to the formation of the Gangmasters Licensing Authority (GLA). The then Transport and General Workers Union had been campaigning for years to bring controls on Gangmasters, but it was the trigger of this disaster that bought it to the statute book (BBC, 2005a). The GLA have done a good job at clearing the top tier of suppliers, but recognize that the subcontracted lower tiers are much more difficult to access. The GLA covers many more aspects of employment law – the main being that labor providers have a license to show they know what the rules are. They have revoked over 150 licenses of labor providers (GLA, 2011)

There was a combined inspection operation in October 2011 called Safe Haven with the GLA, HSE, Gas, and Fire and Rescue in the Spalding area of Lincolnshire – where the plantation farms predominate. They inspected about 30 pack-houses and related accommodation and served 8 Prohibition Notices, 44 Improvement Notices and took

2 prosecutions. Clearly, company auditing does not always work (Lincolnshire Fire and Rescue, 2011).

The health of farm workers (395 males and 210 females) was measured through the use of standard health instruments to see if working on organic farms was healthier than traditional farms, as required by organic standards. Farm workers' health was significantly poorer than published national norms for three different health instruments of measurement. There were no significant differences in the health status of farm workers whether working on 'conventional' or 'organic' farms – although the 'organic' workers were 'happier' measured on another scale (Cross *et al.*, 2008).

10.5 AUTONOMY

Respect for autonomy corresponds to the notion of rights advanced in the 18th century by Immanuel Kant, which appeals to our responsibilities and duties to "treat others as ends in themselves" – in essence, the Golden Rule: "Do as you would be done by." For Kant, ethics was about respecting others as individuals, not calculating costs and benefits (Mepham and Tomkins, 2003).

Most people would agree that if they are being badly treated, they have a right to organize against that oppression. The Golden Rule – do unto others as you would have them do unto you – is one of the oldest and best guides to good behavior and decision making. This most basic and useful ethical theory is sometimes called the Rule of Reciprocity. Most of us recognize that other people have the right to organize, as the best way to take on a more powerful group – in this case companies, whether large or small. We wish that because we wonder what we may do in the same position. That over 150 countries have ratified both the ILO "fundamental" Conventions on Freedom of Association (ILO, 1948) and Right to Organise (ILO, 1949) is testimony to the widespread acceptance of that Golden Rule.

10.5.1 Food

The IUF is often campaigning against Nestlé – the world's largest food manufacturer – on matters that go to the heart of the right to organize. This could offer a case study on how to determine whether employment is exploitative or ethical?

Nestlé state:

> More than five years ago, the world's largest food maker set out to standardize how it operates around the world. GLOBE, or the Global Business Excellence program, is aimed at getting far-flung operations to use a single system to predict demand (Nestlé, 2011).

They claim their GLOBE project depends on how well the market heads embrace what was a business initiative, not a technology initiative. They want to "find ways to streamline Nestlé's myriad and vast supply chains, for everything from paper to powders to chocolate to water; and to take the best administrative practices and spread them throughout the company's operations." The point is "to make Nestlé the first company to operate in hundreds of countries in the same manner as if it operated as one." Not even the British East India Company at the peak of its tea-trading power has done that in the history of global trade.

Have they managed it yet? According to the unions, Nestlé management pressurized workers at two sites – in Panjang, Indonesia and Kabirwala in Pakistan. At Panjang negotiations to make a Collective Agreement broke down, so workers went on strike – inside the plant, for about a month. According to the IUF, when they returned to work, over 50 union members' names were called and given "resignation" notices. At the newly expanded Kabirwala plant, according to the IUF, management attempted to undermine the union, interfering in elections and suspended the union president. Some years later, union concerns were more about the status of precarious/contracted workers – a common concern among food workers everywhere.

> Rather than meeting the union's demand to negotiate the employment status of precarious workers at this 'world class' facility, management has tried to mobilize local opinion against the union and its president and fomented a series of incidents and provocations involving false criminal charges (IUF, 2011d).

The dispute reached Geneva Council in 2012 when it was called upon:

> to insist that international labour standards be respected in Panjang and Kabirwala. According to the resolution, the Swiss government should be obliged to act in view of the fact that Nestlé is headquartered in Switzerland, that its products are frequently associated with the country's international image, and that Switzerland has ratified ILO Conventions on trade union rights which call on the government to take appropriate action to ensure that these rights are respected, among other compelling reasons (IUF, 2012a).

The UK Nestlé union representatives have worked out what Nestlé may mean by "operate in hundreds of countries in the same manner as if it operated as one." They say "Our members believe that when a conflict is over, and an agreement is signed, it is over. They are now asking themselves if it is Nestlé policy to continue a conflict after it is formally resolved, and whether this could happen to them as well" (IUF, 2012b).

10.5.2 Farm

Farm workers have always found it difficult to organize. From the sharecroppers forced to work on plantations in *The Grapes of Wrath*, there is a well-trodden path to today. The first farm workers' union was founded by Joseph Arch in middle England. To this day, the National Farmers Union – the employers' – the farmers', organization, does not recognize the farm workers' union in the legal meaning of "recognize" (i.e. negotiate on a range of issues). They may talk with farm workers' but that does not mean they have to then consult with them on other matters. The farmers' union opposes the right for the farmworkers' union to have "roving" Safety Representatives – where legally recognized health and safety representatives can go from farm to farm. Without these it is asking far too much for individual farms to organize for better health and safety.

The reasons for poor organization are clear. It is hard to discuss matters with fellow workers, as they are a long way away. Travel is difficult and farmers are not known for letting their workers have time off for such matters. It means the whole process of organization is more difficult than for their industrial allies. Regulations that encourage the creation of safety committees to discuss issues have no relevance whatsoever in the countryside.

Union organization has been more successful in the packing houses and the nearby food processing factories. The larger companies – the top tier, know what their responsibilities are. Companies turning a million chickens a week into precooked meals not only have good food safety systems, they have recognizable negotiating structures.

As UK plantations have grown up in the last 20 years employing migrant eastern Europeans, the existing unions find it difficult to recruit the migrant workers, who often see themselves only here for a short time.

10.6 FAIRNESS

Respect for justice corresponds to John Rawls' notion of justice as fairness.

> Justice is the first virtue of social institutions, as truth is of systems of thought. A theory, however elegant and economical, must be rejected or revised if it is untrue; likewise laws and institutions, no matter how efficient or well arranged, must be reformed or abolished if they are unjust (Rawls, 2010).

However, there is a problem in defining what fairness means: e.g. does it mean that goods should be distributed according to need, or ability, or effort (Mepham and Tomkins 2003)? Of the approximately 1.1 billion men and women working in agricultural production nearly half labor for wages. Millions of these workers earn the lowest wages in the rural sector, lower even than the amount required to subsist. This situation exists despite rising agricultural trade and labor productivity worldwide. Working conditions are sometimes appalling and child labor is pervasive (FAO, 2011). It is hard to see how anybody can call this "fair." But somehow most people prefer not to see this big picture.

10.6.1 Food

Fairness for food workers has never really been evident in any civilization. The phrase "eat humble pie" comes from the habit of feeding servants the "umbels" (insides) of deer while the landowners fill themselves on venison.

In the UK, we tip much less than in the US. UK tipping is usually just for restaurant and hotel staff, the hairdresser and Christmas "boxes" for the postman/milkman. However, tipping for food work is common throughout the world, although expected amounts vary across regions and countries, with Japan food service workers not expecting any (Travel & Leisure, 2010). There is a relation between the core salary and that raised in tips. The two lowest paid jobs (in the UK) in 2011 were waiter and bar staff, but they receive tips and tips are often expected. This custom represents a nod in recognition that food service workers are not properly paid. Failing to give an adequate tip when one is expected may be considered very miserly, a violation of etiquette, or unethical (Wikipedia, 2011). This contribution to try and make some amends extends only to the front staff. While many restaurants share the tips, that bit of 'guilt money' does not get passed back down the food chain.

Yet, the most obvious aspect of fairness for most people is how much they get paid. This has been taken up by the Fairtrade movement, described in Chapter 13. This started in Netherlands in the late 1980s to protect coffee farmers from the worst excesses of a free market in coffee – that replaced an international quota system that kept farmers decently paid. Fairtrade sets out to guarantee a decent living for farmers – for those who own their crop, as opposed to laborers of other people's crops. Fairtrade protects mainly farmers – although it has been extended to planation systems too. However, there are no "fair" standards for those laborers through the long food chains from the fields to our forks. Is that fair?

As well as equity, equality is an aspect of fairness. An Equal Pay Act came in to determine whether cooks (women) should be paid the same as chefs (men). In the UK, in the Equality Act 2010, it was recognized – having been established by previous case law, that a woman cook preparing lunches for directors and a male chef cooking breakfast, lunch and tea for employees were considered as "like work" – the basis of any equality claim.

There are massive inequalities between industrialized and developing countries, and they have widened over the past 15 years, but there are also considerable inequalities within developing countries. Recent food price rises and shortages have increased inequalities, making many more in the food chain –who do not benefit from the food price rises, vulnerable (Millstone, 2011).

10.6.2 Farm

For a typical household, a 40% increase in farm labor costs translates into a 3.6% increase in retail prices. If farm wages rose 40% (as happened in the US in the mid 1990s when the campaign to get rid of unauthorized workers was so successful that it created a labor shortage) and this increase was passed on to consumers, average spending on fresh fruits and vegetables would rise about $15 a year. For a typical seasonal farm worker, a 40% wage increase could raise earnings from $10 000 for 1000 hours of work to $14 000 – lifting the wage above the federal poverty line. (Martin, 2010) Is it fair that food and farm workers are paid so little for producing so much?

In the UK, the Agricultural Wages Board (AWB) was set up after the Second World War to regulate pay. It carried on when all the other Wages Boards were abolished by Thatcher's conservative government. The weak bargaining position of farmworker unions was recognized by the National Farmers Union (who represent the owners not the workers) who were happy for the AWB to continue.

Pay negotiations were carried out at national level with representatives from various sides, mediated by legal professionals and independents. A penny or two was added to the basic national minimum wage. This was too much for the plantation owners in the east of the UK. They lobbied and with NFU support, convinced the Coalition government to abolish the AWB as part of their drive for "austerity" (Labour, 2011).

However, Welsh farmers support the AWB – which reflects the dual nature of the workforce in the UK. Wales and the west of the UK consists primarily of family farms. Family farm workers (150 000) rely on a skills ladder to progress through working life – usually on the same farm. Recognizing this skills building is part (80%) of the AWB. The plantation owners are preoccupied with wages of migrant workers (300 000), so object to the few pence above the national minimum wage. Yet it is permanent workers who will suffer most. The idea of lots of family farmers negotiating with farmworkers who work alongside makes for difficult negotiating conditions (Clutterbuck, 2011). The plantation model has damaged the family farm model that has been so important for many years – and should be in the future too.

There is not widespread recognition of these two very different styles of farming; laws and standards for one are not necessarily appropriate for the other. This is leading to application of laws made for one group being applied to the other. One particular case involves six farmers in Devon working together to provide work for an apprentice being deemed to be "labor providers" and thus subject to the GLA (see above).

There is a constant drive to undermine farm and food workers' wages and conditions. Most people realize that this is not fair, but turn a blind eye. Politically it is very important to have cheap food. Ex-President Richard Nixon and his Agricultural Minister Earl Butz were very aware of this, leading to vast increase in corn growing. But few ask about the costs to workers. They are always taken for granted. Is that ethical?

10.7   ETHICS IN PRACTICE

How can we turn all this theory into something that an "ethical practitioner" in the field can monitor? Can somebody make decisions about what employment is ethical or exploitative; and then provide a justification for the judgment?

It should be possible to build on the set of principles explored in the previous paragraphs, turning these into policies at company level, which can then be turned into practices in the workplace. This is what managers do all the time. There are already good policies and standards, so it should be possible to identify good/bad practices for ethical matters.

A company policy should be based on checking the following:.

1. *International guidelines.* The UN's ILO has passed many conventions over the years (ILO, 2011b). These are established by agreement between business, unions and government bodies. Each Convention is then ratified by each country, who are then required to put the convention into their law. But if a country does not want to tie its hands, it does not ratify the Convention. For example. the UK denounced the Convention for Occupational Diseases. The US has ratified only 14 Conventions, compared to the UK's 68 The latest Convention is a Health and Safety Code, agreed November 2011 (ILO, 2011c).

   The Ethical Trading Initiative is an "alliance of companies, trade unions and voluntary organisations, working in partnership to improve the working lives of people across the globe who make or grow consumer goods – everything from tea to T-shirts" (ETI, 2012). It uses the ILO Conventions as minimum standards for its Ethical Base Code (ETI, 2011). This Base Code goes a bit beyond the Conventions and includes other guidance on how to treat workers, for example by extra encouragement to recognize and reward safety training with H&S vocational qualifications.

2. *Government laws.* By regulations, the government can prevent the worst aspects/excesses, as evidenced by the GLA and HSE in the UK. Authorities with legal powers can inspect employment for transgressions. But as some controls are seen as trade barriers, there are no world-wide laws. The ILO has no power – the only power is if the country signs up for the particular Code. Regulations mean inspection and possible prosecutions, resulting in publicity.

In the EU a new Agency Workers Directive was passed into force in October 2011 that ensured that "agency" workers – those employed by a third party 'labor provider' should be paid the same as permanent workers (AWD, 2011). Under the new law, around 1.4 million agency workers, many in the food and farm chains, are eligible for the same wages and benefits as permanent staff after they have held the same post for 12 weeks. Within a few weeks, Tesco, and then Morrison's, held discussions with their recruiters, and, adopted the so-called "Swedish derogation" model, that waives temporary staff's rights to the new benefits. A spokesman for Morrisons said that "they were using agencies that had adopted the get-out-clause, and stressed that it was a 'legitimate route' for businesses to take." (International Supermarket News, 2011).

3. *Company standards.* Most major food companies now have "ethical groups"; and ethical policies, many as part of their Corporate Social Responsibility (CSR) agenda. Much of the CSR agenda is concerned with the social aspects – in particular, their treatment of employees. They are concerned that their brand can suffer a lot with bad publicity. Retailers were accused of buying leeks 'picked by slaves." They got the bad publicity, even though their suppliers were found not guilty some months later (Hastings, 2010).

Retailers have shown a concern when various examples of exploitation appear, and have sorted some industrial relations issues out, When 400 Polish workers went on unofficial strike over conditions at a strawberry farm in Hereford, neither the GLA nor unions could do much. It was Tesco who stepped in behind the scenes and told the supplier what they should do (personal communication and BBC, 2005b). Thus disputes have effects way beyond the acreage involved.

These voluntary standards are audited by third party bodies. They are not however 'independent' as the auditors are paid (indirectly) by the retailers, who often determine which auditors a supplier should use. There is no transparency, in that nobody from outside the company can see what is going on. Most of the main UK retailers work with the ETI and the author helped produced their website to help suppliers conform to the ILO standards (Ethical Growers, 2012).

4. *Independent standards.* There are a range of standards that companies and others can work to. The World Bank standards are used on "all investment projects to minimize their impact on the environment and on affected communities" (World Bank IFC, 2006). This has been put into an online form with permission from IFC by EPAW Ltd (EP@W, 2006).

Social Accountability International "is one of the leading global organizations working to advance the human rights of workers around the world." Their standard, SA 8000.is again based on the ILO Convention (SA 8000, 2008).

Global G.A.P is probably the biggest world standard body for farming, dealing mainly with sustainability issues. Recently they have added a new module dealing with employment matters called GRASP "Partners involved in the food sector are challenged to find innovative and meaningful approaches to ensure that their agricultural products are produced in line with internationally agreed labor requirements" (Global G.A.P., 2011)

Other independent standards that contain social elements include Fairtrade (see Chapter 13) and the Rainforest Alliance standard which includes a non-binding Social and Environmental Management System (Rainforest Alliance, 2010). None of these standards are deemed a barrier to trade by the WTO, as they are voluntary.

5. *Industrial relations.* Trade unions can use the various standards to monitor conditions. Trade unions – while poorly organized in the food and farming sectors, should be the starting point for any practitioner wishing to make ethical decisions. Ethical practitioners dealing with employment matters need to be aware that there are always two sides in this – those that buy labor and those that sell it. And there is no reason why these two sides should agree.

An "Ethical Policy," produced by the directors, would include something like the following: "To provide direction to employees in making ethical choices, acting in a manner that demonstrates high ethical standard, and complying with the provisions of law" (Washington State Board, 2011)

It is the job of manager and their staff supervisors and technicians to translate the policies into procedures (both formal and informal) and practices which can be monitored. These could be said to be the "ethical practitioners." These are quite distinct from auditors, who are external personnel who are paid to check on given requirements. They tick boxes, while practitioners put ethics into practice.

10.8 CONCLUSION

Much more academic study is needed to understand the crucial role of labor in the food chain. Many discussions about the future of food ignore the importance of work in the food process. "Labor" is much more than a line on an accounts sheet. It is as if the markets and science work like magic – without a hand planting and picking the crop. What happens in the US and UK in 10 years' time, when a fair proportion of the aging farm population has died, with few skilled workers to replace them? Or two thirds of their farm workforces decide they do not want to leave home and come hundreds of miles to work in poor conditions, for little pay?

Who is predicting the sort of workforce we may need on the land in 10–20 years' time? With much debate about the food security and sustainability of farm and food production, few are asking these labor questions. Many people seem to think that the necessary knowledge and skills will appear out of the field in a few years' time. But they will not. Many developed countries have cut back on their education and skills development on the land and the associated research. The whole food and farm process needs to be a lot more attractive. This blind spot of 'Labor' also means that the solutions proffered for food security and sustainability are usually technical – rather than social.

Is food fair? Conditions need to improve throughout the chain, but this will not happen while we continue to expect to get our food on our plates as cheaply as possible. It is easy to turn a blind eye to conditions, and to give a tip. An ethical practitioner will want to change behaviors, but recognize they will not be able to change the way our society thinks at present (i.e. ignoring the costs of cheap food).

Box 10.1

To help would-be practitioners, the author is developing the debate whether we can determine whether employment is "ethical" or "exploitative," by setting up algorithms to determine "ethical metrics." If you are interested, go to https://sites.google.com/site/ethicalmetrics/, where you can join in debates arising from this chapter.

There is a role for ethical practitioners (Box 10.1) – people who are not trained as accountants turning up as auditors, but people trained and educated in ethics and behavior. Just as we have systems built in for biological and chemical aspects of food safety, we could develop systems that monitor exploitative and ethical behaviour. There is no reason to treat anybody badly for producing the most vital commodities in the world.

REFERENCES

Agency Workers Directive (2011) [online] Available: http://www.agency-workers-directive.co.uk/ (accessed 14 December 2012).

BBC (2005a) Cocklers died 'due to negligence' [Online] Available: http://news.bbc.co.uk/1/hi/england/lancashire/4259226.stm (accessed 14 December 2012).

BBC (2005b) About Herefordshire Migrant workers ([online] Available: http://www.bbc.co.uk/herefordandworcester/content/articles/2007/11/23/migrants_01_feature.shtml (accessed 14 December 2012).

Clutterbuck, C (2009) What's Going on at Wellesbourne? EP@W Ltd Publishing. [online] Available: www.sustainablefood.com/wellesbourne.html (accessed December 22 2012).

Clutterbuck, C (2011) Abolition of AWB YouTube [online] Available: http://www.youtube.com/watch?v=5VffM_PbOeY (accessed 14 December 2012).

Courtney PP (1965) Factors behind the distribution of plantations in the twentieth century in: *plantation agriculture*. Bell & Sons, London, pp 50–68.

Cross P, Edwards RT, Hounsome B and Edwards-Jones G (2008) Comparative assessment of migrant farm worker health in conventional and organic horticultural systems in the United Kingdom. Science of the Total Environment 391: 55–65 [Online] Available: http://www.sciencedirect.com/science/article/pii/S0048969707011564 (accessed 14 December 2012).

EP@W (2006) Corporate Social Responsibility Planner EP@W Ltd. [online] Available: http:www.epaw.co.uk/CSR (accessed 14 December 2012).

Ethical Growers (2012) Ethical Growers Handbook. Online developed into online version by author for Impactts Ltd. [Online] Available: www.ethicalgrowers.org.uk (accessed March 2012).

ETI (2011) Base Code [online] Available: http://www.ethicaltrade.org/sites/default/files/resources/ETI%20Base%20Code%20-%20English_0.pdf (accessed 14 December 2012).

ETI (2012) [online] Available: http://www.ethicaltrade.org/ (accessed 14 December 2012).

FAO (2011) Farm Wage Labour: poorest of the rural poor. [online] Available: http://www.fao.org/docrep/x0262e/x0262e19.htm (accessed 14 December 2012).

FoodFirst (2011) Do we really want to pay farmworkers a living wage? [online] Available: http://www.foodfirst.org/en/Paying+farmworkers (accessed 14 December 2012).

GLA (2011) List of revoked GLA Licenses [online] Available http://gla.defra.gov.uk/Our-Impact/Revocations/ (accessed 14 December 2012).

Global G.A.P (2011) GRASP Risk Assessment on Social practice (GRASP) [online] http://www.globalgap.org/export/sites/default/.content/.galleries/documents/110105_GG_GRASP_MODULE_V1_1_Jan11_en.pdf (accessed December 2012)

Hastings, B (2010) Waitrose and Tesco sold leeks picked by 'slaves', court told The Independent 08 Dec 2010. [online] Available: http://www.independent.co.uk/news/uk/crime/waitrose-and-tesco-sold-leeks-picked-by-slaves-court-told-2154058.html (accessed 14 December 2012).

Hazards (2011) Choose Union workplaces for cleaner safer work, Hazards, Sheffield. [online] Available: http://www.hazards.org/unioneffect/ (accessed 14 December 2012).

HSE (2000) Revitalising Health and Safety. [online] Available: http://www.hse.gov.uk/lau/revitalising.htm (accessed December 2012).

HSE (2011) Agricultural industry. [online] Available: http://www.hse.gov.uk/statistics/industry/agriculture/index.htm (accessed 14 December 2012).

Hyams, E (1952) Soil and civilisation. Thames and Hudson. [online] Available: http://books.google.co.uk/books?id=EhE6AAAAMAAJ&q=Atlantic+Europe&source=gbs_word_cloud_r&cad=5 (accessed 14 December 2012).

International Supermarket News (2011) Morrison cuts temporary cuts. [online] Available: http://www.internationalsupermarketnews.com/news/5412 (accessed 14 December 2012).

ILO (1948) Freedom of Association Convention 087. ILO, Geneva.

ILO (1949) Right to organise Convention 098. ILO, Geneva.

ILO (2011a) Agriculture: Hazardous work. [online] Available: http://www.ilo.org/safework/info/WCMS_110188/lang--en/index.htm (accessed 14 December 2012).

ILO (2011b) ILOLEX Database of International Labour Standards. [online] Available: http://www.ilo.org/dyn/normlex/en/f?p=1000:1:0::NO::: (Accessed December 2012).

ILO (2011c) Safety and Health in Agriculture. Code of practice. ILO, Geneva. [online] Available: http://www.ilo.org/global/publications/ilo-bookstore/order-online/books/WCMS_159457/lang–en/index.htm (accessed 14 December 2012).

IUF (2011a) Ethical? Tetley's Tata Tea Starving Indian Tea Workers into Submission. [online] Available: http://cms.iuf.org/?q=node/111 (accessed 14 December 2012).

IUF (2011b) New Zealand lamb producer CMP ANZCO locks out 111 workers. [online] Available: http://cms.iuf.org/?q=node/1220 (accessed 14 December 2012).

IUF (2011c) Strike by student exchange workers at US chocolate maker Hershey. [online] Available: http://cms.iuf.org/?q=node/1096 (accessed 14 December 2012).

IUF (2011d) Nespressure returns with mass dismissal of union members. [online] Available: http://cms.iuf.org/?q=node/1144 (accessed 14 December 2012).

IUF (2012a) Deputies call on Geneva government to Stop Nespressure! [online] Available: http://cms.iuf.org/?q=node/1515 (accessed 14 December 2012).

IUF (2012b) Nestlé UK shop stewards say Stop Nespressure! [online] Available: http://cms.iuf.org/?q=node/1514 (accessed 14 December 2012).

Labour (2011) Fairness in the Countryside; Back the Apple. [online]Available: http://www.campaignengineroom.org.uk/countryside (accessed 14 December 2012).

Lincolnshire Fire and Rescue (2011) Operation Safe Haven fire Rescue. [online] Available: http://microsites.lincolnshire.gov.uk/lfr/news-and-events/latest-news/operation-safe-haven-%E2%80%93-helping-protect-the-vulnerable/107663.article (accessed 14 December 2012).

Martin P (2010) Immigration and the US Labor market. In: Human Resource Economics and Public Policy (Ed CJ Whalen) W.E. Upjohn Institute, Kalamazoo, MI, pp. 49–77.

Mepham B & Tomkins S (2003) Ethics and animal farming: a guide for students and lecturers. Compassion in World Farming Trust, Hampshire [online] Available: http://www.ethicalmatrix.net/guides/teacher_guide.pdf (accessed 14 December 2012).

Metro (2011) Blackburn Players star in fowl chicken advert Metro 28 July 2011. [online] Available: http://www.metro.co.uk/sport/oddballs/870738-blackburn-players-dunn-and-roberts-star-in-fowl-venkys-chicken-tv-advert (access March 2012).

Millstone (2011) Inequality and Food: 10 tasks for the UK government STEPS [online] http://stepscentre-thecrossing.blogspot.com/2011/06/inequality-and-food-10-tasks-for-uk.html (accessed 14 December 2012).

Nestlé (2011) Nestlé Pieces Together Its Global Supply Chain. [online] Available: http://www.baselinemag.com/c/a/Projects-Processes/Nestleacute-Pieces-Together-Its-Global-Supply-Chain/ (accessed 14 December 2012).

Purseglove JW (1968) The origin and spread of tropical crops. In: Tropical Crops Dicotyledons 1, 1st edn, Longmans, London, pp. 9–18.

Rainforest Alliance (2010) Sustainable Agriculture Standard 1. Social and Environment Management System. [online] Available: http://sanstandards.org/userfiles/file/SAN%20Sustainable%20Agriculture%20Standard%20July%202010.pdf (accessed 14 December 2012).

Rawls J (1972) A Theory of Justice. Oxford University Press, Oxford.

SA 8000 (2008) Social Accountability 8000. [online] Available: http://www.sa-intl.org/_data/n_0001/resources/live/2008StdEnglishFinal.pdf (accessed 14 December 2012).

Travel & Leisure (2010) Worldwide Guide to restaurant tipping. [online] Available: http://www.travelandleisure.com/articles/worldwide-guide-to-restaurant-tipping (accessed 14 December 2012).

USDA (2011) Food Marketing System in the US Price Spreads Briefing Rooms. [online] Available: http://www.ers.usda.gov/Briefing/FoodMarketingSystem/pricespreads.htm (accessed 14 December 2012).

Vavilov Centers (2012) [online] Available: http://en.wikipedia.org/wiki/Vavilov_Center (accessed 14 December 2012).

Washington Sate Board (2011)Agency Ethics Policy. [online] Available: http://www.ethics.wa.gov/ADVISORIES/Board_Approved_Policies/Accountancy%20Ethics%20031612.pdf (accessed 14 December 2012).

WHO (1946) Preamble to the Constitution of the World Health Organization as adopted by the International Health Conference, New York, 19 June – 22 July 1946; signed on 22 July 1946 by the representatives of 61 States (Official Records of the World Health Organization, no. 2, p. 100.

Wiki (2011) Tipping. Wikipedia [online] Available: http://en.wikipedia.org/wiki/Tip_ (gratuity) (accessed 14 December 2012).

World Bank IFC (2006) Environmental and Social Standards. [online] Available: http://www.ifc.org/ifcext/sustainability.nsf/Content/EnvSocStandards (accessed 14 December 2012).

III Examples and case studies

11 Ethical practices in the workplace

J. Peter Clark

11.1 INTRODUCTION

This chapter is based in part on a presentation given at a symposium at the 2009 Institute of Food Technologists Annual Meeting. That symposium also included papers on ethical theory and education. Additional authors and topics have been assembled for this volume.

Here, we discuss various workplaces and roles that employees play, which influence the ethical challenges they face. Then we discuss some of those ethical challenges as illustrations, not meant to be conclusive. One feature of many ethical challenges is that each can be unique, making it difficult to have ready-made solutions. Rather, one must be grounded in ethical principles, often called virtues as in Chapter 1. We discuss some of those principles and how they might be acquired. The analogy one can draw with physical sciences is apt: if one knows the laws of thermodynamics, for instance, one is well prepared to solve many practical problems in heat transfer, separations and refrigeration.

11.1.1 Who am I to discuss these topics?

I am educated as a chemical engineer, with over 40 years of experience in the food industry as a researcher, educator, research executive, engineering executive and consultant. This means I have been a student, educated and directed students, led multi-discipline teams of scientists and engineers, interacted with senior executives, and succeeded in competitive entrepreneurial environments. As a youth, I was an Eagle Scout, which had a significant role in shaping my values, as has my religious faith.

It is not necessary to belong to an organization like the Scouts nor to practice a religion in order to develop ethical principles, but these surely are examples of ways that people do develop their life-long foundational principles. Formal courses in college on ethics and self-study of the voluminous literature on the topic are other ways people find helpful. It is clear, however, that the foundation for ethical behavior is laid early in life, by parents and other early influences. Later in life, when people make poor ethical decisions, it is not usually because they are malicious, but rather that they are ignorant of the basic principles of ethical behavior.

Practical Ethics for Food Professionals: Ethics in Research, Education and the Workplace, First Edition.
Edited by J. Peter Clark and Christopher Ritson.
© 2013 John Wiley & Sons, Ltd. Published 2013 by John Wiley & Sons, Ltd.

11.2 WORKPLACES AND ROLES

We consider four broad categories of workplaces and within each the various roles that people play. Ethical challenges and responsibilities are closely related to these roles, and can change dramatically as one's career advances.

11.2.1 Government

There are federal, state and local governments and within these elected, appointed and career employees. Ultimately, all government employees should see themselves as working for tax payers because taxes in some form are the source of their salaries. Employees in general have a fiduciary responsibility to the owner of their employer, and in the case of governments, that owner is the tax payer. (I avoid saying "citizen" because some tax payers are not citizens and some citizens do not pay taxes.)

The level of responsibility obviously varies with grade level, from entry to elected executives, but at each level there can arise conflicts of interest between the individual and the owners. A conflict of interest is the most general and common ethical challenge that government employees face. We will discuss these in more detail later, but usually the conflict involves money. Government employees at every level face opportunities to enrich themselves at the expense of tax payers. They usually know right from wrong, but often rationalize that the harm is negligible or that they deserve more than they are paid.

It is true that many government employees are poorly paid for the service they render or the dangers they face. This is especially true for elementary and secondary educators, police and firefighters. It is generally less so for a city, county, state or federal clerk working safely in an office. Traditionally, many government employees were compensated for relatively low salaries with generous pension benefits that became available after relatively short lengths of service, often permitting them to have second careers.

Teachers, firefighters and police often were motivated by a sense of altruism that helped them accept low financial compensation. They felt rewarded in other ways. Many things have changed, including inflation eroding the value of salaries and savings, budget cuts imposing longer work days for the same pay, and some loss of respect by the public for services that are taken for granted. The combination of effects can lead to police seeking bribes, firefighters shaking down business owners and teachers taking excessive sick days.

What about food professionals in government? Most food professionals in government are probably in research at government laboratories or in regulatory agencies, ranging from the Food and Drug Administration, Department of Agriculture, and Environmental Protection Agency on the federal level through state departments of public health or departments of agriculture to county and city public health agencies. In these roles, the food professionals develop the science underlying regulations covering food, drugs and the environment, and then assist in interpreting and enforcing the regulations.

Regulators, in particular, have potentially significant power over manufacturers and other participants in the food industry, such as restaurants. In addition to their fiduciary responsibility to their ultimate employer, the tax payer, regulators have a responsibility to the public at large to protect against harmful food additives, improperly prepared food and damage to the air and water we all need and use. It is common for perceived responsibilities for protection of the public to conflict with the real or perceived rights of individuals and companies. Even relatively low rank government regulators may face these conflicts and need to resolve them on an ethical basis. In addition, just as employees, they will encounter many of the same ethical challenges that employees in other roles face routinely.

11.2.2 Industry

The food industry world wide is one of the largest segments of the economy, encompassing agriculture, food processing food distribution, and food preparation. In the United States, over 50% of food consumed is done so away from home. Within industry, food professionals may be employees, managers, executives or owners. Each level has its own responsibilities and challenges.

Employees have a fundamental obligation of loyalty and fiduciary responsibility to their employer. Often, these obligations are specified in a contract, employee handbook or other document, but even without such a written statement, the obligations exist. Loyalty implies that an employee seeks to do good and not harm for his or her employer. This includes doing the assigned job diligently, protecting physical and intellectual property, not misusing company property, and not slandering or libeling the employer. The fiduciary responsibility applies to matters involving money, such as submitting accurate expense accounts, not stealing or embezzling, and spending company money wisely and only as authorized.

Managers have increased ethical responsibilities because they typically lead groups of employees and are themselves led by executives. Managers have responsibilities for more money with generally less direct guidance than employees receive. In addition to the expectations imposed on all employees, managers also must exhibit fairness in supervision, set good examples to their subordinates, and work especially diligently in the company's interests.

Executives receive even less direct supervision than typical mangers, are paid more generously, and have responsibility for larger amounts of money. Accordingly, they may face greater temptations to enrich themselves improperly, to harass or favor subordinates, or to undermine perceived rivals unethically. Competition among ambitious professionals is normal and healthy, but with advanced responsibility comes advanced power, which easily can be abused.

Finally, *owners* can have unique ethical challenges and responsibilities. Owners may be individuals or members of families that literally own a business or they may be board members who represent public shareholders, the true owners who delegate their power to a board of directors. In either case, owners hire the top level of executives and usually exemplify the culture and values of the firm. Culture and values determine the ethical atmosphere of the firm more than any written document or individual act or decision. Owners have the responsibility to demonstrate integrity, humility, generosity and all the other virtues that underlie correct ethical responses to the inevitable challenges that every individual faces.

11.2.3 Education

Another workplace in which food professionals all spend some time, and some may spend entire careers, is education. Roles include student, educator, researcher and administrator.

It is common for individuals to play more than one of these roles sequentially or simultaneously. Here we discuss some of the challenges each role may face.

Students are tempted to cheat on exams or assignments, to plagiarize in writing, or to cut corners in research. Cheating may involve unauthorized collaboration, stealing answers or tests, or using disallowed tools or sources. Modern electronics, such as smart phones, have provided new ways to cheat, with unfortunate consequences for students' acquisition of skills. A student's obligations are primarily to him- or herself – to learn facts, techniques and skills that prepare the student for a career. Grades are one means of measuring progress, but, of course, they have come to mean much more. Grades may determine admission to desired institutions, offer of a first job, and level of starting salary. The significance of grades creates the incentive for unethical behavior among students.

Additional temptations may include laziness or incompetence so that one lifts written material without attribution or fabricates laboratory results. (Plagiarism in publications is discussed in Chapter 6.) The primary victim of unethical behavior by students is the person committing the violation. Consequences may go further, if the work is published, for instance. More significantly, unethical behavior as a student may establish an unfortunate foundation for a career in which virtue is lacking.

Educators are role models and teach more by example than explicitly when it comes to ethics. Explicit education in ethics is, unfortunately, relatively rare in higher education. Instead, students learn by observation. Accordingly, educators have a strict responsibility to be fair in their treatment of students and colleagues, to be diligent in their preparation for class, and to be scrupulous in their research and writing. There are strong temptations in education, especially for those with some experience, to cut corners in preparation, to use old notes and tests, and, perhaps, to be self-serving in requiring texts in which the educator has a financial interest. Keeping current in one's field, improving as a lecturer, and using modern tools to communicate are obligations of integrity in education. Neglect in these areas sends the message that laziness is acceptable and so might be other ethical lapses.

Research is usually an obligation of many educators in colleges and universities. It is a major component of graduate education, as compared with classroom teaching, which is more important in undergraduate education. Research is also an important endeavor in its own right. Absolute integrity is the solid foundation of all research. There are many temptations to unethical behavior in research. Priority, or first to publish, is a matter of pride but also can be financially significant. Slavish commitment to a theory has often led to unethical behavior, such as selective use of data, falsification of data or misleading description of protocols.

Science traditionally relies on replication of results by others as proof of a theory and on peer review before publication. Both practices depend on full disclosure of methods and results. It is unethical to mislead or deceive in these areas.

As it so often is, money lies at the root of other temptations. Most research is supported by grants from government or industry, obtained by successful proposals. It is unethical to obtain duplicate funding for the same work, though it is common to propose projects that may differ slightly. It is unethical to use funds obtained for one purpose to pursue another goal, though it is common for directions to change as work proceeds. Honest and timely communication is ethical behavior.

Finally, education *administrators* are the executives of universities and colleges. They are responsible for sizable budgets and are usually well paid. They are chiefly responsible for raising funds for their institution. They have obligations to treat faculty, students, staff and alumni fairly, to be fiscally responsible, and to avoid conflicts of interest. University presidents may be tempted to join corporate and foundation boards because of the prestige and financial rewards, but these may come at the expense of time and attention required for the university. An example of self-serving behavior sends the wrong ethical message to students, faculty and staff.

Educational institutions are challenging to lead because many faculty may have tenure, making them more secure than other types of employees in other situations. The best universities are collegial, led and governed by consensus and cooperation. Such an atmosphere promotes ethical behavior better than the competitive and authoritarian atmosphere of some corporate environments. Sadly, there is a trend in education to imitate corporate practices, driven by the need for financial efficiency. Those food professionals who reach educational administrative positions (department heads, deans, provosts, presidents) should consider the ethical impact of their leadership style.

11.2.4 Consulting

Consulting is a distinct workplace with its own roles and responsibilities. There are consulting firms in which people may be employees, managers, executives and owners, just as described in industry. Many consultants are self-employed, thus lack colleagues, subordinates or supervisors. Consultants have clients, while industry has customers. The relationship is similar but also different. Finally, groups of consultants may be partners, a less hierarchical structure with its own obligations and challenges.

Consulting *employees* have the same obligations of loyalty, diligence and fiduciary responsibility as other employees, but the nature of consulting introduces some other obligations. In particular, consultants are often privy to confidential and proprietary information of clients. Without such information, consultants cannot be very helpful. However, with such exposure comes a strict obligation to protect such information from deliberate or accidental disclosure to anyone. Often, consultants sign non-disclosure agreements (NDA), but these should merely serve as reminders of the obligation. There is a temptation, especially among newer consultants, to brag about information they may have and even to share it, primarily to enhance their perceived prestige. This is a serious violation of their obligation to confidentiality and is unethical. Violating the obligation for personal financial gain is not only seriously unethical, it is often a criminal offense.

Self-employed consultants share the obligation of confidentiality of information. In addition, they have an ethical obligation to know their own strengths and weaknesses. In particular, the self-employed consultant should only accept those assignments for which he or she is qualified and which can be completed in the required time with the available resources. Depending on the area of practice, there may be legal obligations, such as holding licenses, that a consultant must satisfy. For example, in most states, a consulting engineer must be a registered professional engineer. Many professions have published codes of ethics, which often address issues of unfair competition and competence.

Self-employed consultants usually build a network of colleagues to whom they can refer clients for assignments beyond their abilities or resources, or who can supplement the consultant where needed. Some consultants work through other organizations to whom they incur ethical obligations to respect client relationships and to adhere to preferred practices. The single most valuable asset a self-employed consultant possesses is integrity – integrity in professional competence and integrity in business practices.

Clients of consultants have their own ethical obligations. Consultants are not employees, but they are engaged for specific reasons for a specific time. The client is obliged to be clear about the scope of the assignment, the compensation and the schedule. Clarity of communication is an ethical responsibility that can be shared between the client and consultant. Clients are tempted to blame a consultant for advice they do not like and to seek excuses to avoid paying what are often perceived to be high rates. No matter what a client does with advice from a consultant, the client is obliged to pay the consultant in full and on time. Ethical clients understand the skills and limitations of consultants, define assignments clearly, have realistic schedule and budget expectations, treat consultants with respect, and pay fees and expenses as agreed.

Finally, some consulting organizations are *partnerships*, which incur unique ethical obligations. Legally, partners are each liable for the debts and obligations of the organization. In addition, partners are obliged to pull their weight in providing services, generating sales, being responsible with the firm's resources, and assisting one another. The concept of partnerships is that the whole is greater than the sum of the individuals. This demands integrity, trust, and generosity of all. These are the same virtues that underlie ethical behavior anywhere else.

11.3 ETHICAL CHALLENGES

An ethical challenge generally requires judgment, rather than being obviously covered by a law or regulation. Ethical challenges that a food professional might face include: economic, interpersonal, legal or regulatory, and others, each of which we discuss next.

11.3.1 Economic ethical challenges

As previously mentioned, money is at the root of many ethical challenges. Here we discuss some specific examples.

Expense accounts

Expense accounts are a means of reimbursing employees, job applicants, guest lecturers and other travelers for their out of pocket expenses. Every institution or firm has its rules and procedures for submitting claims. Normally a relatively simple form, often based on a computerized spreadsheet, is used so that addition is automatic and various categories such as meals can be identified. The basic principle of expense accounts is that the traveler should not suffer out of pocket costs that he or she would not have if he or she were not traveling. A good example is a hotel room. Air fare, rental car, gas and parking are other typical entries.

Some organizations, such as some state governments, do not reimburse for alcoholic beverages, while most companies do. There may be other rules, such as upper limits on total meal costs. Sometimes a flat allowance is provided for meals and other incidental expenses, called a per diem (Latin for "per day"). Usually, the established per diem is adequate. The US government and its contractors use that system. Except for per diem, most organizations require receipts for all expenses over $25 and for certain costs no matter the amount (parking and gas are examples).

Some common ethical challenges with expense accounts are inflating claims, claiming the same costs from two sources where a trip has several purposes, and claiming inappropriate costs. Violations of expense account policies can have serious consequences, including termination, if detected. Even if not detected, most violations are ethical lapses that can lead to insensitivity to other ethical issues.

Inflating claims is straight forward dishonesty – claiming for costs that were not incurred. One reason for demanding receipts is to deter such practices. As a result, it is generally difficult to inflate a claim very much, but it is still wrong.

More problematic is the situation where a traveler has two or more purposes for a trip, such as a student traveling to several job interviews or a consultant visiting several clients. Most organizations ask the traveler to allocate costs in some fashion. It is unethical to claim more in total than was spent. The exact allocation is a matter of judgment and can be on any reasonable basis, such as time spent.

Finally, ethical judgment must be applied to what is appropriate for a given individual to spend. Some organizations have rules that strictly personal costs, such as pay-per-view movies, are not reimbursed. In general a good rule about meals, hotel rooms and transportation is that if one would spend his or her own money at a given level, it is probably appropriate to claim. Thus, an hourly paid clerk should not order a $200 bottle of wine for dinner and expect the employer to pay for it. Sometimes there is no choice about the cost of a hotel room at a meeting held in a resort, but other times good judgment is demanded.

In general, the guide for expense accounts is to spend as if it were one's own money and not to use the system to enrich oneself.

Bribes and kick backs

Taking or giving monetary (or other valuable) considerations as bribes or kick backs is generally illegal, but there are enough gray areas that ethical judgment needs to be applied. Usually, the purpose is to influence a decision, often of a government official. In many businesses, cultivation of relationships with customers by suppliers is normal and usually ethical, but the practice can be abused. Most governments forbid employees from accepting anything from anyone, with certain small exceptions made for elected officials. Businesses can vary widely in their policies even within the firm. For example, people in purchasing may have more severe restrictions than other employees.

A good general rule is that gifts worth less than $25 are acceptable both to give and receive, under the assumption that such an amount is insufficient to influence anyone. Likewise, it is common to pay for meals (or accept being a guest) when the meal is not excessively elaborate and business is discussed, or a clear business purpose is served. For example, it is acceptable to have a celebratory dinner upon completion of a project.

Clearly, one should not allow the occasional free lunch to unduly influence significant decisions, nor should one expect to be favored because of picking up the tab.

Social occasions are the lubricant of relationship building in business, but they can be abused.

Embezzlement

Embezzlement is stealing and as such is a crime. It can range from petty stealing of office supplies, misuse of company vehicles, and misuse of company resources, such as internet connections and phones, to major diversions of funds. Companies typically have written policies or well-understood practices that govern many situations, but ultimately they must trust those employees who are responsible for financial matters and have access to cash and valuable assets. Employees are well advised to develop good habits early in their careers of protecting and valuing company property. As they become responsible for more valuable amounts, they will have ingrained the ethical principles discussed elsewhere and will deserve the trust they receive.

On a practical level, behavior that might be tolerated or over looked, but that still amounts to stealing, might be used as an excuse for termination when it suits management. It is wise not to provide such an easy excuse. As has been mentioned elsewhere in this book, development of the ethical virtues requires practice, and in matters of company property, practices start with the most minor items.

Taxes

In the United States we pay federal, state and local taxes on income and real estate, varying by location and situation. Taxes are governed by laws and violations can be punished severely, but still there is a temptation to cheat in various ways. As part of a consistent ethical life, one must pay the taxes owed accurately. The laws can be complex and violations can be inadvertent, in which case, usually, there may be a small fine or penalty. Deliberate under reporting of income or exaggeration of off setting expenses is a crime.

Normally employed workers paid a salary typically have relatively simple tax situations, with few opportunities to cheat. However, self-employed individuals and entrepreneurs can have more opportunities, and since the tax system largely relies on honest behavior, may be tempted to reduce their tax burden unfairly. It is important to remember that taxes unpaid by one person are made up from the other taxpayers, so it is an injustice to one's neighbors.

There is no need to pay more than is rightfully owed, but it is unethical to pay less.

Underestimating costs or bids

Many purchase decisions are made on the basis of proposals received in response to a request (RFP), which provided the scope of materials, equipment or services desired. The potential suppliers provide their qualifications, schedule and costs as requested in the RFP. Low cost alone is not usually the only basis for a choice – reputation, past experience, quality, references, and service may all be considered, but price is always important. It is tempting to submit costs that are unrealistically low to get the order. Not only is this unethical, it is ultimately a poor business model and may be illegal. Many professional societies and state regulations for certain professions, such as engineering, explicitly state that deliberate under bidding is unethical and illegal.

A business may, in good conscience, choose to take some business at a loss under certain circumstances. They may want to gain an entry to a new market or customer; they may want to keep people or assets employed, even at a temporary loss; or they may believe they can find savings that will convert an apparent loss to break even or a profit. Clearly, such tactics cannot be used for long or the firm will close.

When a firm submits a deliberately low bid with the intention of later raising its price through changes or other means, it is being unethical, and cooperating with such an effort is wrong. Refusing to help with a deliberately misleading proposal or bid may jeopardize that person's position, but they are probably better off to work for an ethical firm.

Selling goods below cost is a similarly challenging situation. Clearly, it is untenable over the long haul, but it often occurs in a promotion, a market introduction or as a response to a competitive threat. It can be illegal under some circumstances, but where it is legal, it must be done transparently and temporarily.

Cooking the books – deceptive accounting

Most food professionals do not encounter the opportunity to perform or condone deceptive accounting, but some might, as executives or entrepreneurs. The intent is to mislead stockholders, potential investors or creditors concerning the financial performance of a firm or organization. It may also be directed at managers for the purpose of earning a bonus or other reward. A project manager might choose to mislead others about costs or schedules just to avoid criticism.

Ultimately, such deliberate deceptions almost always are discovered and the perpetrator disgraced, but the short-term purpose may have been achieved. Rarely is the benefit worth the ultimate cost. Not only is such behavior unethical as a form of lying, but it may be illegal, and usually is a violation of company policy. In the face of almost certain discovery, one might wonder why it is ever tempting to cook the books, but it happens all too often.

11.3.2 Interpersonal ethical challenges

Interpersonal ethical challenges refer to relationships among co-workers, between supervisor and subordinates, and generally among people in the work place. Some are governed by laws and regulations, others by company policies, and still others by ethics and common sense.

Scandal and gossip

Scandal and gossip refer to information about other people that may be true, but can be harmful when conveyed. False information is libel and may be illegal. Gossip is information that may not be harmful, but would normally be considered private. Conveying scandal or gossip is not usually illegal, but it is often unethical because it can do harm to another's reputation. Not only is being known for providing gossip and scandal unethical, it is unwise because it establishes that the one who does it is unreliable and untrustworthy with private or sensitive information. Much better is to establish a reputation as discreet and sensitive with confidential knowledge. One should be very careful and selective in conveying negative information or opinions about others. This means doing so only rarely, if at all, and only when necessary for a greater good, and with verified facts and observations.

A person's reputation is a precious and fragile asset that should be protected from malicious, careless or inadvertent damage.

Harassment

Harassment refers to unfair, and unethical, treatment of others based on their gender, race, religion or other irrelevant factor. It can take the form of intimidation by a supervisor, mistreatment by co-workers, or creating a hostile environment. Under some circumstances, it can be illegal, and prosecution has resulted in significant financial settlements by employers who were alleged to have tolerated such behavior.

The ethical person not only does not participate in harassment, he or she tries to counter it by friendly gestures, reporting when appropriate, and defending the victim, if possible. This may demand some courage at times, but ethical behavior often does. As a practical matter, harassment hurts group morale and thereby group efficiency, creates distrust where reliance on each other may be critical (working under hazardous conditions, for instance), and in extremes can lead to violence.

Favoritism

Favoritism refers to unfair, and unethical, treatment of others based on such factors as family, romantic relationships, or other irrelevant issues. It is the mirror image of harassment; like harassment, it hurts group morale, creates distrust, and may reduce effectiveness by promoting incompetence. Unlike harassment, it is not usually illegal. It is not always obvious what the ethical person can do to rectify favoritism, because the practitioners of favoritism are often in positions of power. As a minimum, one should not use irrelevant factors in promotions, rewards or assignments. Complaining about unfair treatment is rarely effective. Superb performance to overcome possible favoritism is one approach. Finding a more fair environment is another.

Silence in the face of wrong

Speaking up in the face of wrong behavior is ethical and courageous; silence is wrong. There are laws to protect whistle blowers in some cases, and many companies and organizations also have policies to do so, but it still can be difficult to report perceived misbehavior. How can one be sure he or she has all the facts? There always is the issue of perception against reality. The ethical person is patient and tries to determine all the facts before reaching a decision to make a report. If possible, he or she discusses his or her concerns with supervisors or other appropriate parties. Additional information may clarify and remove concerns, or it may become clear that there is a widespread conspiracy. Depending on the industry and circumstances, there are various avenues for reporting possible criminal behavior, including federal agencies, local authorities and local media. Rarely is it wise or appropriate to protest alone.

Whistle blowers have occasionally reaped substantial rewards, but this should not be the motivation. Rather, one is concerned about potential damage to consumers, the environment or co-workers. Issues may include unsafe food, illegal discharges of air or water emissions, or unsafe working conditions. Not only are these unethical because of the harm they may do, they are unwise business practices because the penalties and consequences of inevitable disclosure are usually much larger than any temporary benefit.

Lying

Lying is almost always unethical and wrong because it is deceiving some one. It is also unwise, because it is almost always discovered, after which one's integrity is severely and permanently damaged. The temptations to lie in business are great: overly optimistic progress reports, exaggeration of accomplishments, excuses for absences, diverting blame, and commitments that will not be fulfilled. Even the seemingly harmless, "I'll call you back" is a lie when it is not carried out, and the other party mentally files that person under "unreliable, does not fulfill promises." The aspiring food professional wants to be in another file – the one labeled "keeps his word, does what she says she will."

Exaggeration on résumé

A particular case of lying deserves separate mention because it is, unfortunately, quite common and very harmful. Putting incorrect or misleading information on a résumé is unethical and very unwise, because it is deceiving others for personal benefit (usually, getting a job) and is relatively easy to discover. Credentials, experience and accomplishments are all relatively easy to confirm, yet people persist in falsifying these. There is a law against claiming an unearned military honor, but otherwise falsifying a resume has few legal consequences unless it is submitted under oath, when it could become subject to laws against perjury. However, the personal consequences can be severe once discovered – usually loss of position. The ethical approach is to make the best of what is true, without false exaggeration. I once examined a résumé of a talented engineer who explained a gap in employment history as his time in prison for a serious indiscretion. I admired his candor and would have hired him, confident that he was now trustworthy.

11.3.3 Legal or regulatory ethical challenges

Laws or regulations themselves do not pose ethical challenges, as we mean them here, because they must be obeyed and have mechanisms for enforcement. No significant judgment is required. Rather, ethical challenges arise on the fringe of laws and regulations.

Covering up

Disguising, hiding or delaying discovery of a violation may or may not be illegal itself, but it is almost always unethical. In the food industry, some regulations are largely self-enforced, because federal and state agencies are only capable of occasional inspections. Even in meat processing, where USDA inspectors are continuously present, they cannot see or be aware of everything. Thus food companies and their employees have numerous opportunities to bury violations of regulations. It is unethical to do so, because the regulations exist to protect the public. It is also unwise, because such cover-ups are often (not always) discovered and revealed.

The ethical food professional tries to comply with the relevant regulations and copes with possible deviations as appropriate.

Accepting assignments when not qualified

Consultants, contractors and vendors may be tempted to accept orders or assignments for which they are not qualified in experience, capability or capacity. This is tempting because it is always difficult to decline new business, but it is unethical because it amounts to misleading the potential customer. It is also unwise, because the deception will usually become obvious and the supplier's reputation, at least with that customer, will suffer.

A practical solution is to find a compatible and capable partner with whom to share the opportunity. There are several benefits from this ethical approach: one becomes known as honest and self-aware; one becomes known as generous, which may be reciprocated; one has the chance to grow in capability; and one retains or develops a customer. "Half a loaf is better than none." Likewise, half a loaf is better than a whole one that goes sour.

Selling dangerous products

Selling products known to be dangerous is usually illegal, but there can be gray areas calling for ethical judgment. The US Government is relatively slow and cautious in declaring a product or substance dangerous, and thereby illegal to make and sell. Thus, companies may manufacture a product, in good faith, that only later becomes known to them to cause harm. In such cases, the ethical approach is to cease manufacture and to recall product, if possible. Companies are required to have recall procedures and to be capable of tracing products and ingredients. However, they usually wait until ordered to perform recalls because of the cost. When a danger becomes evident, it is unethical to delay.

Food products may be dangerous because of an ingredient, because of improper processing, because of improper labeling, or because of adulteration or contamination. Discovery of danger may occur from consumer reaction, internal records or accidental inspection. In any case, the company must respond quickly both for its own sake and to protect the public. Prevention of such events is the goal of most companies and employees.

Misleading the public or enforcers of regulations

Deliberately lying to the public or to federal investigators is illegal, but misleading them by withholding information or disguising information is at least unethical. The temptation exists because often relations with inspectors is unfriendly and company marketing and sales people may want to convey information to consumers that is not exactly false, but is not complete either. Food professionals may be pressured to cooperate in such efforts or instigate them themselves. It is unethical to do so, because it borders on lying. It is not necessary to volunteer information to investigators; they are trained to request what they need.

The public, however, has no such opportunity, so the obligation, ethically, to inform the public accurately is significant. It is acceptable to convey information to the public in ways that are understandable, but which may be different than those that might be used with scientists. It is unethical, however, to deliberately mislead.

11.3.4 Other ethical challenges

Undoubtedly there are many other situations that cannot be anticipated in detail, but that will challenge one's ethical principles. Some are addressed in other chapters. Situations change as a career progresses, so managers and executives face conflicts that a new hire can barely imagine. The foundation laid in good behavior from the beginning will serve well later.

Taking credit for other's achievements

There may not be a law against it, nor a commandment, but it is wrong to take credit where it is not deserved. This can happen in many ways: a team is commended, but one or two individuals did all the work; a subordinate does the research, but his or her supervisor pretends the results are theirs alone; a presenter exaggerates his or her contribution; someone being complimented does not correct a misunderstanding of what really happened. One governing principle is the obligation not to deceive. Another is the rule against taking something of value that does not belong. Credit for achievement is valuable in the workplace and can result in promotions, bonuses, and recognition. It is tempting to seek as much as possible. Humility is the governing virtue and is much more admirable in the long run. Genuine achievement is usually recognized and appreciated, and those who seek it undeservedly are usually recognized, but not appreciated.

Innuendo and misdirection

Short of outright lying, which is clearly wrong, words can be used to deceive in more subtle ways, by withholding information, not correcting misinterpretations, and even by silence. The ethical principle is the need for honesty and integrity. As previously mentioned, in regulatory situations, there is no obligation to volunteer information, but it is wrong to mislead or deceive.

Go along, get along

The workplace is a community and there is great pressure to conform with group customs and practices. Sometimes, these practices are wrong, but it is hard to challenge them or not conform. Maybe "everyone takes long coffee breaks" or "everyone kicks in for a cash gift to the boss." Maybe the supervisor is always added to author lists of research papers, even though he or she does not contribute to the work. It is hard and lonely to break from such a pattern, but such practices are unethical and help to create an atmosphere in which increasingly unethical practices become accepted. An individual may not be able, alone, to reform an unethical culture, but it is possible to go one's own way or, if necessary, to leave for a more ethical environment. Students and workers usually have protective rights against retaliation if they refuse to conform to unethical practices. However, it can be unpleasant to do so, because of the subtle ways in which a group can punish individuals. It takes the virtue of courage to practice an ethical life.

"Don't work so hard"

One of the more insidious influences that can be exerted, especially on the younger employee, is the idea that they should not work too hard lest they make others look bad. This runs counter to the young person's enthusiasm and idealism. Listening to such advice can imbue cynicism and quench the eagerness that probably made the young employee attractive enough to hire. A good answer might be, "I like what I am doing so much that I want to stay late." As work forces have shrunk, many people are working hard, so this idea may be less common than it once was, but it has not disappeared. The ethical principle is that our employer deserves our best effort, and we should live up to our own standards, not those imposed by a group.

What is original?

Especially in research, but often in other assignments, it is important to understand what is truly original and what is not. It may be a matter of law, as in intellectual property, or of scientific credibility, but it can also be a matter of credit in ordinary affairs. Ethical lapses can occur in claiming originality falsely, thus plagiarizing or stealing credit from some one else. Consequences can be severe, as in lost precedence for patents, financial damages owed to injured parties, or termination for deliberate deception. Often, the same idea has been conceived simultaneously but independently by multiple parties. This is not an ethical lapse.

11.4 PRINCIPLES FOR GUIDANCE

Many other chapters of this book discuss the philosophical foundations of ethics. Here we try to provide easily understood and practical principles to help someone faced with the ethical challenges just described, and others we have not imagined, navigate the minefields of such challenges. There is a wide range of consequences for bad choices and a wide range in the world of cultures and beliefs. The bias of this book is toward a Western Judeo-Christian culture as seen in Europe and the Americas. Within such a shared culture, most people know right from wrong and generally agree with each other in such matters. That does not mean that everyone adheres to such knowledge.

11.4.1 Some generally accepted principles

Lying, stealing and doing harm to another person are almost always wrong. These generally accepted principles are often enough to resolve many ethical challenges, once the challenge is understood as involving such consequences.

There are several widely accepted and admired codes of behavior. Examples include:

- The Golden Rule, often expressed as "Do unto others as you would have them do unto you." There are alternative formulations, but they all say the same thing.
- The Ten Commandments of the Old Testament of the Bible.
- The Scout Law (Boy and Girl).
- The virtues (Seebauer and Barry 2001 and Chapter 1 of this book).

Two other guiding principles that also could be formulated in various ways are:

- What would your mother think? Most of us revere our mothers and know that they tried to imbue us with their codes of good behavior. Did her efforts succeed?
- Can you look in the mirror? We know in our hearts what is right. Confusing as many ethical challenges are, still we usually know the right course of action.

11.4.2 The virtues

The virtues are discussed elsewhere, but to emphasize their value, we list them here again. They are reliable guides through the dilemmas of ethical challenges in a practical sense.

- Justice
- Prudence
- Temperance
- Fortitude

11.5 SUMMARY

We all face ethical challenges. Mostly, we make good choices intuitively. This means we have acquired an instinctive understanding of the virtues and codes of conduct. However, some choices are difficult and require courage to make. In such cases, it helps to have core beliefs and values, and it also helps to think about, read about and discuss these issues. That is one purpose of this volume.

REFERENCE

Seebauer, E.G. & R.L. Barry 2001 *Fundamentals of Ethics*, Oxford University Press, Oxford.

12 Ethical thinking and practice

Louis B. Clark

12.1 EDUCATIONAL PROGRAM ON ETHICS

Suppose you are out walking and pass a fountain with a pond in which a toddler is drowning. You consider wading in (the pond is less than a meter deep) to rescue the child, but realize that doing so would ruin your brand new expensive shoes and it would make you late for work. So you do not rescue the child.

When presented with this scenario, virtually all respondents are shocked, and disagree vigorously. Of course we would rescue the child. Compared to the loss of a pair of shoes or one work day's tardiness, the potential life of a child is of no comparison.

This scenario, adapted from one presented by the ethicist Peter Singer, of Princeton University, indicates that there may well be a universal ethical sense in humans. There have been significant studies in recent social psychology to further confirm this. From where does this ethical sense originate? Is it "human nature"? Universally taught in our and other cultures? Or does our ethical sense derive from a higher power? Is there, in fact, a consistent, identifiable universal or even widely prevalent ethical understanding? What would provide the basis for such an ethic?

In 1837, Thomas Jefferson offered the opinion that morality was, in fact, a part of our human nature, part of our "endowment", just as humans are endowed with "inalienable rights" in the Declaration of Independence:

> He who made us would have been a pitiful bungler, if he had made the rules of our moral conduct a matter of science. For one man of science, there are thousands who are not. What would have become of them? Man was destined for society. His morality, therefore, was to be formed to this object. He was endowed with a sense of right and wrong, merely relative to this. This sense is as much a part of his nature, as the sense of hearing, seeing, feeling; it is the true foundation of morality (Jefferson, 1787).

So, according to Jefferson, there seems to be a "natural" human understanding of, and inclination to adhere to, moral principles. Twenty-first century professional philosophers and social researchers as well as neuroscientists agree, in this respect with early 19th

Practical Ethics for Food Professionals: Ethics in Research, Education and the Workplace, First Edition.
Edited by J. Peter Clark and Christopher Ritson.
© 2013 John Wiley & Sons, Ltd. Published 2013 by John Wiley & Sons, Ltd.

century thinkers (and, of course, also with thinkers dating back to the beginning of Western civilization). Can we ground this basic understanding, this apparent universal "sense" rationally? Perhaps in a way parallel to the way chemistry is grounded in physics, physics in mathematics?

The philosophic discipline of Ethics concerns itself with exactly this attempt at grounding. Unconcerned with adjudicating *specific* actions or classes of actions as "right" or "wrong", Ethics does concern itself with *fundamental* justifications.

For example, I might say "my action was right, because, after all, the end justifies the means."

Philosophical ethics asks "Is 'the end justifies the means' an adequate *groundwork* for morality?"

"Ethics" can be understood in three ways:

- A *"way of life"*: "His Buddhist/vegan ethic dictated that he avoid meat and all animal products."
- A *moral code*: "The ethics of medicine start with 'first, do no harm'."
- Inquiry *about* ways of life and moral codes: what *justifies* living according to one way of life versus another, according to one moral code versus any other?

In this chapter we return to to the work of the philosophers discussed in Part I of the book as a way of underpinning the search for ethical practice in the workplace covered in the previous chapter. This is using the third sense of "ethics" listed above, and any ethic at all in the first two senses. We will follow this examination with an account of the results of teaching philosophical ethics using concrete examples for the various philosophers' attempts at justification.

Ethics as a branch of philosophy was first discussed by Greek philosophers over 2300 years ago as a discipline that went beyond a list of "dos" and "don'ts". For example, Socrates (c. 469–399 BCE) and Aristotle (384–322 BCE) described an ethical mode of living as one that balances extremes, that always seeks wisdom and that approves an action only after thoughtful consideration of all the consequences that action might produce. Socrates loved to turn conventional wisdom on its head, showing "obvious" choices to be, in fact, the wrong choices. For example, definitions such as "the virtue 'courage' is *standing firm in battle*" or "courage is endurance" are shown by Socrates to be incomplete or even self-contradictory. In the case of "standing firm in battle", this is a good *example* of courageous behavior but other courageous behavior does *not* consist in "standing firm in battle" therefore "standing firm in battle" is not synonymous with "courage." Likewise, "endurance" is not synonymous with "courage." So what *is* the definition, the correct understanding of "courage?" From where do we get our idea of "courage?" Or, for that matter how do we define "virtue?" We may all agree that, for example, rushing into a burning building to save a child is a good example of courageous behavior. How can we come to such universal agreement? How do we "know courage when we see it?"

Socrates (through Plato (424/423–348/347 BCE), Socrates' student and apologist, and Plato's "Platonic metaphysics") believed that knowledge came through a kind of *remembering*. What we humans remember, when we come to (really) know something, are the "forms" – the perfect, ideal, insubstantial, eternal abstractions of which all that we sense in this natural world are a pale imitation. According to Socrates and Plato, we know courage, we *knew* courage because we (our soul) was in touch with this eternal, non-material world of forms before we were born. Our coming to know concepts like "courage", "good", "truth", "beauty" and the like comes from *remembering* the perfect forms that correspond to the definition of these concepts. We cannot define "courage" as Socrates' interlocutors attempted to do with "standing fast in battle", but, on the other hand, we already apparently know what "courage" is. Just as with the example at the opening of this chapter, we know what the right thing to do is in the case of the drowning toddler. Witnessing courage when we see an instance of it in this material world reminds us of the true nature of "courage" in the Platonic Real World.

12.1.1 Deontology

So for Socrates and Platonic philosophy in general, goodness was absolute, changeless, it applied to all situations, all people equally. Pursuit of the *recollection* of the pure, eternal form "good" is Socrates' definition of the purpose of life. Laws that embodied goodness therefore could never be broken. What is right is right in all contexts. This ethical position is known as *deontology*. "Deontology" is derived from the Greek *deontos* – duty or obligation. Ethics, from this perspective, describes systems of moral codes that can never, in any circumstance whatsoever, be violated. Never tell a lie. Never harm another person. Never. It makes sense that the earliest justifications of moral codes were universal, all-encompassing justifications. Compare the Judeo-Christian Ten Commandments. No commandment is couched with language that admits of exceptions. Coming from another culture, the ancient Code of Hammurabi (the "eye for an eye, tooth for a tooth" code) similarly admitted of no exceptions. All these early examples look to other-worldly powers or planes of existence for justification of their moral codes: Platonic forms, the God of the Hebrews, Marduk, the god of Babylon (Hooker, 1996).

For thousands of years, ethics in the West justified moral codes using deontological arguments. Eventually, the Platonic, other-worldly justification for the deontological position came under attack in European thought. The spirit of the 18th century Enlightenment was to rely on reason, and science, in short this world, to advance and support knowledge. Writing towards the end of the 18th century, the German philosopher, Immanuel Kant wished to preserve the deontological ethical position, but to give it support that did not rely on Platonic non-material forms, or an appeal to theology. Kant was awakened from his "dogmatic slumber" by the writings of the Scottish philosopher, David Hume. Hume proposed that all knowledge, rather than a memory of some eternal, immaterial form was in fact initiated entirely here on this material earth, gathered only through the senses. This approach to knowledge (the philosophical discipline known as "epistemology") is called "empiricism". Hume was one of the originators of empiricism.

If all knowledge is brought in through the senses, what is the basis for "moral knowledge?" What collection of sensed information together supports moral claims? For Hume, the answer is "no data." Moral judgments were essentially the "deliverance of sentiment." That is, ethical statements were not universal, binding. Ethical statements were not even true or false! Hume's ethical position is known as "Sentimentalism": moral judgments are the expression of sentiment (approval or disapproval). Sentimentalism can be viewed as the polar opposite of Platonic deontology. To Hume and the philosophers of the Enlightenment, sentimentalism followed naturally from empiricism and the new scientific bend to thought; if statements could not be backed by observable data, then those statements were neither true nor false, they merely expressed something else. In the case of morality, moral statements expressed "sentiments" of approval or disapproval.

Kant's counterproposal to Hume, using reason alone, and without resort to immaterial, abstract Forms, to support the deontological position, was his "categorical imperative"(Kant, 1785). In brief the categorical imperative was summarized by Kant: "Act only according to that maxim whereby you can at the same time will that it should become a universal law" (Kant, 1785). Kant placed the active, rational human subject at the center of the cognitive and moral worlds. However, he, and those who followed him, believed that he had rescued ethical discussion from a descent into mere opinion ("you think that 'x is good' well that is merely your opinion") and buttressed it on a foundation of reason.

Kant's ethics are another form of deontology. Modern philosophical discussions of deontology as opposed to other ethical justifications start with the Kantian position. What is right is right, what is wrong is wrong, and these principles hold in *every* human situation. Ethical claims can be debated as true or false. They are *not* merely opinion, the expression of sentiment. Ethical claims are statements of fact. However, the basis for establishing this deontological position goes no further than the study of and appeal to rational humankind.

12.1.2 Teleology

It should come as no surprise that Kant in expressing his political philosophy has been characterized as being an "apologist for Prussian absolutism" (Levinger, 2000). "Argue as much as you like", Kant wrote, "but *obey*" (Levinger, 2000). Kant's moral philosophy, Deontology after all, was also absolutist. Following the publication in 1848 of the *Communist Manifesto*, philosophers who were liberal but perhaps not as radical as Karl Marx addressed some of the demands for a more universal justice and an alternative to absolutism that was expressed in Marx's political philosophy.

One of these liberal philosophers was John Stuart Mill (1806–1873). Son of a founder of modern economics (James Mill) and godfather to Bertrand Russell, that towering figure of 20th century liberalism, John Stuart Mill is rightly considered the patriarch of contemporary non-Marxist liberalism. Mill's more populist (as opposed to the absolutist, authoritarian positions we have considered until now) ethical views start with defense of the freedom of the individual. In his work *On Liberty* (Mill, 1859), Mill defends the position that the individual ought to be free to do whatever she chooses just so long as that action does not harm others. So here is a radical redefinition of "right." No action is inherently right: we have to examine the *outcomes* which the action causes before we can determine the goodness or badness of the action. In short "the end justifies the means." In Mill's words: "actions are right in proportion as they tend to promote happiness, wrong as they tend to promote the reverse of happiness" (Mill, 1863). Mill referred to this position as "utilitarianism".

Teleology is another word built from Greek roots: *telos* – end or goal. So contrasting the two poles of classic western philosophic thought we have, dating back to the Greeks, "duty or obligation" versus "end or goal."

A 20th century version of utilitarianism is known as "situation ethics": *nothing* is right or wrong in and of itself, only the *situation* in which the action is performed determines its rightness. This moral theory accommodates troublesome circumstances in which it seems preferable to do something normally considered wrong (for example, violating one of the Ten Commandments) but in *this situation* that action is justified. The end produced by performing the action is far better than the alternative result. The well-known hypotheticals: "would you tell a lie to save a life?" Or even, "would you sacrifice one life to save others?"

There is a thought experiment used to highlight problems surrounding utilitarianism: you are standing near a railroad switch. By throwing the switch you can choose which of two rail lines a train approaching will take. There is a driverless train, out of control approaching at a high rate of speed. If the switch is not thrown, six people standing in the tracks just a bit further down the line will die. If, however, you throw the switch, only one person on the alternative track will die. What do you do? According to utilitarianism, the greater good will be served if you throw the switch. But if you do nothing, you have, by definition, done nothing wrong. Have you? You have performed no action for which you can be tried in a court of law. But, if you act to save more lives, according to the main principle of utilitarianism, then you actually did commit an act for which you could be tried. You caused the death of that single person on the alternative track. As with many philosophical thought experiments, there is no simple, easy, one-size-fits-all answer to this dilemma. The deontologist does have an answer: you do nothing. It is always wrong to take a life. The deaths of multiple people hit by the runaway train would not be caused by you, so they do not factor into your decision. It is interesting to ponder: if there were no one on the alternative track, the deontologist would have to recommend that you do throw the switch in that case. You have a moral obligation to save life, if you can.

This thought experiment does illustrate that there are cases where there are very different actions recommended by the polar extremes of ethical justification: deontology and teleology.

12.1.3 A middle way

Aristotle, as mentioned in the beginning of this chapter, advocated with Socrates a mode of living that balanced extremes. Taking off from his teacher, Plato, Aristotle maintained the forms do exist and that the forms are responsible for the phenomena we perceive, but instead of phenomena "descending" from heavenly forms, all the particular substances that we perceive have the forms "instantiated" within them. So to study the universal, the forms, we need to *start* with the particular. Particular objects each share in the universal qualities that make up the *essences* of things.

Aristotle's philosophic method starts with natural science, with observation. Through induction, the discovery of universality across a range of observations, Aristotle claimed we can discover the essences which are instantiated within the objects observed. This natural scientific approach to philosophy is reflected in Aristotle's ethics, which take a more pragmatic turn. Aristotle focused, not on the universal laws of a moral theory, but on an ethical life as embodied by a good, virtuous person. Aristotle's approach to ethics is called "virtue theory". Virtue, according to Aristotle is characterized by the "mean." Vice is characterized by excess and defect.

Aristotle's virtue theory is perhaps best illustrated by an example (once again): "Courage."

The deficiency of the character traits necessary for courage is cowardice: fear, lack of confidence. An excess of the character traits that are the essence of courage is foolhardiness: poor judgment, too little fear (fear can be appropriate), overconfidence.

Courage, the mean between these extremes is the strength of character to continue in the face of our fears. This practical, specific down-to-earth definition of courage contrasts with Socrates' dialectic which countered each attempt at a specific definition, as discussed earlier, by showing what we can learn as we narrow in on the definition of courage. "Courage" and "virtue" are not abstractions with existence only in some perfect non-material sphere (Aristotle, 1999). Courage is embodied (like Aristotle's conception of the forms) within courageous humans.

Virtue theory presents a nice, orthogonal perspective on the tension between deontology and consequentialism or teleology. Aristotle cuts the Gordian knot, in a way. Rather than attempt to force-fit a single justification for all potential actions, Aristotle asks us to look at virtuous humans. How do they live their lives? Gravitate towards the mean, avoid extremes. This philosophy of life resonates thoroughly with Buddhist ethics as well. Buddhists learn from the example of the Buddha, Gautama Siddhartha: neither embrace the materialistic life of the Prince to which Buddha was born, nor attempt the extreme yogic asceticism which he attempted as an alternative life, but rather choose the Middle Way of moderation in all things.

Returning to the thought experiment that challenges utilitarianism, but to which deontology's response is intuitively less than satisfactory: Aristotle's virtue theory would support throwing the switch. Yes, it is wrong in most circumstances to act in such a way as to cause a death. But in *this* circumstance, that action is justified. Avoid extremes: do not justify every action because of its result, but do not fail to act, to do the right thing because of excessive adherence to Law.

12.2 RESULTS

When I teach ethics, I combine the study of the original philosophical texts of Plato, Kant, Mill and Aristotle, as well as contemporary Ethicians such as John Rawls and Peter Singer with concrete examples of ethical dilemmas and/or the expression of one of these ethical philosophies.

Ethical arguments are common. Most editorials in newspapers are expressing a value-based argument. That is, the author is making an argument that can be understood as a variation of the directive "You *should* do this." For example, when supporting one particular candidate in an election the newspaper is arguing "You *should* vote for this candidate." When studying ethics as a philosophic discipline it is helpful to read such arguments and interpret them from the perspectives of Socrates, Kant, Mill and Aristotle. What would the philosopher say? Is the argument in the form: "You should do this because it is right, it follows an (obvious?) moral principle?" Or . . . Is the argument in the form: "You should do this because it results in this (beneficial) outcome and that outcome justifies the prescribed action?" Perhaps, following Aristotle, it is of the form "Perhaps you might not consider doing this, but in this case it is justified?"

Students respond positively to these suggestions. I assign term papers with alternatives for subject matter:

- An ethical dilemma the student has faced or of which she is aware
- An editorial, or
- An ethical dilemma represented in literature or film.

Present the ethical argument. Choose two philosophers with contrasting ethical approaches. Describe how each philosopher would evaluate the argument.

Of all the philosophy courses I teach, the essays I receive in response to this assignment show, in general, a greater understanding of the philosophical positions and how the positions differ from one another as well as a more personal application of philosophy to matters of daily concern than essays produced for other courses.

Film is a source of paradigmatic examples of ethical justification. I show clips from selected films that vividly and concretely illustrate each of the three significant ethical philosophies. Students respond well to the break that viewing a film clip provides as well as with enthusiasm for the association between the film example and the philosopher, as a memory-enhancing device.

12.2.1   Deontological film example: *A Man For All Seasons*

In the film version of Robert Bolt's play about Thomas More, *A Man For All Seasons*, a nearly perfect elucidation of the deontological position is articulated by the central character. More is approached by Richard Rich, the man who, in fact eventually betrays him. Rich (John Hurt) asks More (Paul Scofield) for employment. More turns Rich down at which point Rich exits. More's son-in-law, William Roper (Corin Redgrave) urges More to pre-emptively arrest Rich. But Rich has not broken any law. Though, as Chancellor of England, More is empowered to arrest, he refuses to do so (even though it is clearly in his own best interest). More says in response to Roper's argument "go he should if he were the Devil himself until he broke the law!" (Bolt, 1966). Roper cannot believe his ears. He insists that he would " . . . cut down every law in England (to get at the Devil)" (Bolt, 1966), But More eloquently defends his position and the deontological stance:

> What would you do? Cut a great road through the law to get after the Devil? . . . And when the last law was down, and the Devil turned round on you – where would you hide, Roper, the laws all being flat? This country is planted thick with laws from coast to coast, Man's laws, not God's, and if you cut them down – and you're just the man to do it – do you really think you could stand upright in the winds that would blow then? Yes, I give the Devil benefit of law, for my own safety's sake! (Bolt, 1966)

This is not only a vigorous defense of the "no-exceptions" deontological position, but note how it also makes a pragmatic consequence-oriented argument. Not only should one, under no circumstances, Rich's perfidy, or catching the Devil himself, break the law, if one is tempted to do so, consider: practically speaking, you have eliminated the only barrier between Evil and oneself. What do you do when you have no laws to protect you (having "downed" them all)? So the deontological position is not only right, Right with a capital "R", it is the course of action "for my own safety's sake."

The remaining plot of the play illustrates More's deft use of the law to protect himself from Henry VIII's will. More is eventually done in by Richard Rich, but only because Rich lies. Thomas More, falsely accused by Rich, and now destined for execution drolly confronts him, knowing that More's own soul has not been jeopardized by anything he has done: "Why Richard, it profits a man nothing to give his soul for the whole world . . . but for Wales?" (Bolt, 1966).

This clip, in fact viewing the entire film, prompts layered, multidimensional discussion. Among the discussion topics:

* Is deontology, as More argues, in fact the most *pragmatic* ethical position?
* When faced with one's own potential execution, is not some flexibility in the implementation of law justified?
* Is there another alternative? Why shouldn't More simply escape from England?
* Consider "pre-emption". Consider recent historical instances of pre-emptive action. Can justification be made *ahead* of the "end" that justifies the means?
* We probably know the utilitarian view of More's dilemma. What would Aristotle do? (Consider that, as Chancellor, More is not just another enforcer of the Law. More is the most senior officer of all the Courts of England).

This film, the clip described and More's ultimate fate at the hands of Henry VIII, through the (obviously morally wrong) actions of Richard Rich bring home to students the case for (and against) deontology.

Other considerations which I bring to students attention on the subject of deontological ethics:

* Deontology justifies actions based on following a code. Cadets at West Point follow an honor code, which is expressed in deontological terms: "no excuse, sir," whenever a cadet is questioned about an action. No extenuating circumstances. How does this code apply practically in the defense of actions taken in recent military encounters? My Lai? Abu Ghraib?
* The Japanese military code of *bushido* has many parallels with West Point's code of honor and other military codes. However *bushido* can justifiably be seen as the foundation for the behavior of the Japanese military in the Second World War, in particular in the Japanese treatment of prisoners of war. How does one distinguish among codes of conduct?
* For centuries, the feudal order dictated obligation. In fact, England during More's time was still a feudal society. Feudalism requires fealty to one's lord (which More also acknowledges). From the deontological perspective, how does one make judgments when one code (feudalism) conflicts with another (in More's case, Catholicism)?
* It is noteworthy that the Catholic Church waited until 1935 to canonize Thomas More. What was happening in the world in 1935? The events leading up to the Spanish Civil War (1936–1939) got under way in 1934. The rebellion against the Spanish government was led by and made up of Catholics. More's canonization can be viewed as the Church giving Spanish Catholics permission to choose their loyalty

to their Church over their loyalty to their secular government (supported by the atheist Soviet Union).

12.2.2 Teleological film example: *The Ten Commandments*

It may seem counterintuitive on the face of it to use Cecil B. DeMille's 1956 film *The Ten Commandments* as an illustration of consequentialism. After all, the Ten Commandments, as stated earlier are among *the* paradigms of deontological dictates. But in the film, Moses, while still in the favor of the Pharaoh as his adopted son, makes an eloquent and visually striking case for the teleological ethical position.

The background to the film clip I choose to illustrate teleology, or consequentialism is that Moses has been accused by his step-brother, Ramses of coddling the Hebrew slaves under his direction. The rationale put forth by Ramses for Moses' behavior is that Moses is planning an insurrection. The pharaoh, Seti and Ramses then personally confront Moses with a series of accusations:

1. Moses raided the temple granaries.
2. He then gave the stolen grain to the slaves.
3. Moses also gave the slaves "one day in seven to rest."

As each accusation is articulated, Ramses (Yul Brynner) places a measuring weight upon a balance scale. When the third accusation is made, the third weight dropped on the scale, it tips heavily to the weighted side. This is the case for the prosecution. Pharaoh asks "Did you do this to gain their favor?" (Wilson *et al.*, 1956).

In answer, Moses (Charlton Heston) does not attempt to refute the charges as stated, nor does he attempt to justify each action in turn as inherently right (the second and third charges could perhaps be justified on the basis of a higher law of humane consideration). No, Moses simply picks up a large clay brick. He says "A city is built of brick, Pharaoh. The strong make many. The starving make few. The dead make none" (Wilson *et al.*, 1956). As Moses pronounces the last sentence, he drops the brick on the opposite side of the balance scale to that containing the weights Ramses has placed. The scale tips mightily to Moses' side. Moses continues "So much for accusations. Now judge the results" (Wilson *et al.*, 1956).

This scene is perfect for illustrating consequentialism. Moses skips any intermediate defense. He bases his case entirely on results ("Now judge the results"). The actions taken led (as argued by Moses) directly to Moses' successful effort to build a city in honor of the Pharaoh's Jubilee (a task, not insignificantly at which Ramses had previously failed). The scene ends with Moses drawing back a drape so that a classic Cecil B. DeMille scene of epic proportions of the Jubilee city can be viewed in all of its pre-CGI magnificence. This is the case for the defense.

As with the scene from *A Man For All Seasons*, more ethical principles and discussion are layered in this one scene than simply the ethical position (teleology) which it principally illustrates. Moses' treatment of the Hebrew slaves (who, we know are the "good guys" in this movie), though viewed with suspicion, and violating tenets of the Egyptian religion, was inherently *moral*, not immoral. Unlike the choice More faces, that is, to break the law for greater good, Moses chose to perform morally praiseworthy (at least to the viewers' eyes) acts in order to obtain an even *greater* good. Nevertheless the scene provides us with a number of relevant questions concerning consequentialism. For example:

- If Moses had *failed* to build the city, would he (by his own argument) then be *wrong*?
- How many actions which would need to be justified by success could Moses have performed in his pursuit of his goal of building a city. For example, could he kill a task master in order to motivate others?
- The basis for the accusations was a conspiracy theory: Ramses suggested that Moses was currying favor among the thousands of Hebrew slaves in order to lead an insurrection. What if Moses (who, in fact, *did* lead an insurrection) had two goals: build the city for Pharaoh and lead the Hebrews? On what side does the moral equation balance then?

Consequentialism, though rational and congruent with other modern modes of thought, looks weak and insubstantial when scrutinized. The open and shut cases such as lying to a Nazi who is looking for one's Jewish neighbor (so justifying the normally immoral act of lying for the greater good of saving a life) must be viewed side-by-side with more problematic cases of consequentialism. Consider:

- Justification of "white lies" (albeit lies nonetheless) as long as no one ever finds out that one lied (the negative consequence is never actually realized).
- In particular, President Clinton's defense in the Monica Lewinsky scandal was clearly backed by "situational ethics" (not a defense of the subject behavior in general but a defense *in this case*). In particular though not explicitly articulated, Clinton's stance seemed to be one of "if I had not been caught, the deeds, in effect never happened, and therefore no harm was done."
- The problem of relative wrong versus a measure of the consequence. How many measurably wrong, but not significant actions are allowed to attain what positive outcome?
- In their work, *Freakonomics* (Levitt and Dubner, 2006), the authors show a correlation between a drop in crime and the coming of age of the first generation born after Roe v. Wade. In other words, the potential criminals who would have been born but unwanted were aborted when abortion became legal. So society as a whole benefited. Is this a consequentialist argument for abortion?

12.2.3 Aristotelian value theory film example: *My Fair Lady*

The film clip I choose to illustrate the Aristotelian mean as the defensible standard for ethics is a humorous one. In the film version of *My Fair Lady* (Lerner *et al.*, 1964), Eliza Doolittle's father, Alfred Doolittle (Stanley Holloway), approaches Henry Higgins (Rex Harrison) and Colonel Pickering (Wilfred Hyde-White) with a proposition: since the pair have taken in Doolittle's daughter (presumably, by Mr. Doolittle for all the wrong reasons), and they are now benefitting from the effort Doolittle has made in raising Eliza, Mr. Doolittle is owed some form of compensation. His proposed amount? Five pounds.

At first, both Higgins and Pickering are offended and put off by the very idea and the insinuations made by Doolittle, but as Alfred makes his case, Higgins (but not Pickering) is taken with the clever and eloquent manner in which Doolittle has made his case. In fact, Higgins is so delighted by Mr. Doolittle, that he suggests that he exceed the offer. "This is irresistible. Let's give him ten" (Lerner *et al.*, 1964). But Doolittle demurs. "No . . . Ten pounds is a lot of money. Makes a man feel prudent like and then, goodbye to happiness. No you just gimme what I ask guvner, not a penny less not a penny more" (Lerner *et al.*, 1964). Again, as with all chosen clips a perfect expression of the ethical position, in this case, Aristotelian value theory: gravitate towards the mean. In Doolittle's world, too much money will lead to unhappiness (as Aristotle wrote, either excess or deficiency detract from happiness in the philosophical sense, that which we pursue to give life meaning). Doolittle has already made the case that too little or no compensation is not justified either. He deserves his "fiver". The scene ends with Higgins declaring of Doolittle as he exits: "There's a man for you. A philosophical genius of the first water" (Lerner *et al.*, 1964).

And he is. Without an Oxbridge education, Doolittle (a common dustman as the dramatis personae of the original play describes him) has developed a "moral" philosophy that guides his every action and which he perceives as steering him toward happiness. He embraces his own position in the world "I ain't pretending to be deserving. I'm undeserving. And I mean to go on being undeserving. I like it" (Lerner *et al.*, 1964). His rejection of the ten pounds is a defense of his position. He has no intention of becoming a prudent, or God forbid a "deserving" person. He just wants his five pounds.

This clip has the added advantage of being humorous. The "moral" position Doolittle is taking is one of essentially selling his own daughter – and for five pounds ("not a penny more"). But it does illustrate Aristotelian value theory: happiness is a way of life. That way of life is the pursuit of moderation. The mean. Not too much, but not too little either. Too much "courage" is foolhardy. Too little we recognize easily: that is cowardice. Courage is found in the mean; and so with so many of life's decisions. In the extremes of hedonism and self-denial lies unhappiness.

I teach Aristotelian value theory at the end of an introductory ethics course. Further considerations on Value Theory can start with re-consideration of cases already examined:

- How would Aristotle view Thomas More? Is he an extremist?
- How would Aristotle assess Moses' argument? Does Moses strike a proper balance? What could Moses have done to head off the accusations before they got to the point they did (was Moses too extreme in his single-minded dedication to building the city that he became oblivious to appearances?)
- In the film, Higgins later arranges for Mr. Doolittle to lecture to the "Moral Reform League" based on Higgins' perception (and remember, as an English gentleman, Higgins was classically educated, no doubt including immersion in Aristotelian ethics). This new position elevates Doolittle into the middle class. Far beyond the additional five pounds he declined. Consider the morality of Higgins' action from Doolittle's perspective.
- Consider today's political climate. Legislation is a form of compromise. When one side does not compromise, that side typically does not get its way, rather nothing is accomplished. How could politicians apply Aristotelian value theory? Is value theory another way to utilitarianism? (the greatest good for the greatest number).

12.2.4   Results: two contemporary ethical challenges

Contemporary challenge: abetting sexual abuse of children

In the early years of the 21st century there have been a number of news stories about situations that all shared one feature. In all these news stories, prominent, respected, apparently highly ethical men (in all cases those concerned were men, not women) protected, hid or abetted sexual predators. The men were Roman Catholic archbishops, a renowned college football coach and an orchestra conductor, teacher and self-described "management guru" (Radin, 2006; *Boston Globe* Editorial, 2011; Rezendes and Wen, 2012). The ethical transgression I wish to discuss here is not the crimes of the child abusers. It is the acts of omission and commission that allowed the abusers either to continue in their activity or which protected them from discovery by legal authorities and/or institutional authorities. Had the authorities been alerted to the behavior, they would not have permitted the offenders to have access to the children who subsequently became additional victims.

When a respected member of society is found out, discovered committing a crime or performing an action which seems transparently *wrong*, one standard reaction is "What were they thinking?". That is exactly the question I am posing here. I wish to illuminate potential answers to that question using the ethical principles described in this chapter.

Let us first consider the deontological perspective: actions are right or wrong. The context does not change actions' correctness. So in the case of protecting men who prey upon children, assuming that the protector acted with an ethical perspective, the act that carries ethical correctness *must* be the act of loyalty to one's subordinates and colleagues. Taken in isolation, defending or protecting a colleague when he is attacked does seem "right". But to do so and to defend this action from the deontological perspective, the actors must *ignore* the simultaneous act that is performed while protecting colleagues, in these cases, namely the act of exposing children to further harm by sexual predators.

From the perspective of utilitarianism, these men must have considered the "greater good" the preservation and furtherance of the institutions (the church, the university, the music school) above the good of the children put at risk. This trade-off may have been easier to make if the men also discounted the possibility of future wrong actions by the abusers whom they were protecting. In fact, in the case of the music school this discounting was explicitly articulated (Rezendes and Wen, 2012). So is the act of guaranteed consequences (removal or shunning of past abusers who play a role in the support of the three institutions) warranted when it protects against an act that has not yet been performed? And may never be performed? When we consider that the archbishop, the coach and the conductor are all charged with protecting the children who enter the three institutions and that the responsibility of protection is among the highest responsibilities of these men, then, yes, the *potential* future negative consequences inherent in the exposure of children to known abusers is a "greater good" that must be chosen over *any* lesser end or goal. In the case of the football coach and the conductor, who were both fired for their failure to protect the children, no subsequent instances of abuse were identified. The men were fired for failure to consider the *potential* consequences.

Finally, what would Aristotle do? When considering the balance, the mean between extremes of considered courses of action, where does the center lie? Once again, taking the long-term negative psychological consequences that result from abuse into account, any considerations of loyalty to colleagues and support for the institutions of church, university or conservatory pale by comparison.

So, though there is something to be said, under all three of the classic ethical paradigms for the actions taken by these respected leaders (as counter-intuitive as that may seem), in no rational ethical system could the justifications for the protective actions taken outweigh the mandate to protect children.

Contemporary challenge: supporting human rights violations through consumer behavior

A final contemporary ethical challenge follows. Consumers of Apple™ products such as the iPod, iPhone and iPad project an image of youthful awareness. Aware of contemporary trends, certainly in technology, but extending to other trends, including the socially conscious, it is easy to imagine these consumers going out of their way to purchase cage-free eggs, free-range chicken and other products which are produced in environments that treat animals more humanely than more conventional food-producing enterprises. Yet these same consumers are purchasing products from Apple™, a corporation accused of producing products in environments that treat their workers inhumanely (Cooper, 2012).

Let us take a quick look at the three ethical paradigms and what they have to say about these actions: deontology: if it is right to choose products based on how well the producers treat animals, it must be right also to choose products based on how producers treat humans. Utilitarianism: the smaller "good" of the benefits derived from the purchase of an Apple™ product must be contrasted with the greater good of the treatment of thousands of employees. Finally, what would Aristotle do? The desire for the latest new technology, as compelling and useful as it may be must be tempered, moderated by the broader perspective of the impact our expenditures have on our fellow humans. I am reminded of the economic boycott brought to bear against the Apartheid government of South Africa in the 1980s. Change in one person's, one nation's economic behavior can be used to modify unethical behavior in another nation.

These two contemporary examples highlight one further truth about the persistence of unethical behavior: such behavior (some would characterize it as "evil") persists in an environment of ignorance. "Ignorance" in at least two senses: (a) a lack of relevant knowledge and (b) the act of ignoring (unpleasant) information. We can return to Peter Singer's thought experiment of the toddler in the wading pool: the difference between that child for whom everyone will sacrifice their expensive shoes and the starving children thousands of miles away is that the one is right there in front of us. The other, equally deserving of the appropriate ethical action is out of sight. Out of mind.

12.2.5 Results: conclusion

Among the philosophical disciplines which include: logic, metaphysics, epistemology, philosophies of science, history and religion, ethics is most easily recognized by students as having practical implications for their everyday lives. Books that have been written in the recent past, such as *How Philosophy Can Save Your Life* (McCarty, 2009) and *The Consolations of Philosophy* (De Botton, 2000) present philosophy and philosophers' advice as self-help. In all cases, the philosophy presented is ethical philosophy. "Do this. Do not do that. And here's why."

Philosophy has its own voice when it comes to matters of morality. It is a voice that often differs from culture, common sense, commonly held opinion, instincts, religion and politics. Philosophical ethics gives us a perspective on our place in the world. We do not have to accept our culture, nor do we have to accept our "base" instincts. Reason, logic, and philosophical consideration allow us to rise above them. Once the main principles of Western ethical reasoning are thoroughly understood, students can readily see their application.

Newspapers, novels, films, politics, television, even video games are all seen through a new lens: actions need to be justified. What of the *justifications* themselves?

REFERENCES

Aristotle. 1999. *Nicomachean Ethics*, trans. T.H. Irwin, Introduction. Hackett Publishing Company, Indianapolis IN.

Bolt, Robert. 1966. *A Man For All Seasons*. Perfection Learning, Logan, IA.

Boston Globe Editorial. 2011. "*Penn State shows danger of putting sports beyond reproach*". *Boston Globe* Nov 10, 2011.

Cooper, Dan. 2012. "Inside Apple's Chinese 'sweatshop' factory where workers are paid just £1.12 per hour to produce iPhones and iPads for the West." *Daily Mail*, UK Feb 21, 2012. Online: http://www.dailymail.co.uk/news/article-2103798/Revealed-Inside-Apples-Chinese-sweatshop-factory-workers-paid-just-1-12-hour.html#ixzz1p21lzQyW (accessed 17 December 2012).

De Botton, Alain. 2000. *The Consolations of Philosophy*. Pantheon, New York.

Hooker, Richard (ed.). 1996. "*Mesopotamia: The Code of Hammurabi*". Translated by L.W. King. Washington State University, Pullman, WA.

Jefferson, Thomas. 1787. *Letter to Peter Carr*. quoted in Brooks, David. 2011. T*he Social Animal: The Hidden Sources of Love, Character, and Achievement*, Random House, London.

Kant, Immanuel. 1785. *Groundwork of the Metaphysics of Morals*, 2nd revised edition. Translated by Lewis White Beck. 1989. Macmillan, New York.

Lerner, Alan Jay Loewe, Frederick and Shaw, George Bernard. 1975. *My Fair Lady* (screenplay). Penguin, London.

Levinger, Matthew Bernard. 2000.; *Enlightened Nationalism: The Transformation of Prussian Political Culture*. Oxford University Press, New York.

Levitt, Steven D. and Dubner, Stephen J. 2005. *Freakonomics*. William Morrow, New York.

McCarty, Marietta. 2008. *How Philosophy Can Save Your Life. 10 Ideas That Matter Most*. Tarcher, New York.

Mill, John Stuart. 1859. "*On Liberty*" ed. Gertrude Himmelfarb. 1985, Penguin, London.

Mill, John Stuart. 1863. *Utilitarianism* Hackett, Indianapolis IN.

Radin, Charles A. 2006. "O'Malley Asks Forgiveness For Church Sins", *Boston Globe* May 26, 2006.

Rezendes, Michael and Wen, Patricia. 2012. "Video maker had sordid past, strong advocates". *Boston Globe* Jan. 14, 2012.

Wilson, Dorothy Clarke, Ingraham, J.H., Southon, A.E., MacKenzie, Aeneas, Lasky Jr., Jesse, Gariss, Jack, and Frank, Fredric M. (1956) *The Ten Commandments*. (screenplay) Paramount Pictures.

13 The fair trade movement

Richard Norman

13.1 HISTORY

The idea of injecting ethical standards into international trade has a long history. Perhaps the earliest example is the activity of anti-slavery campaigners in the mid-19th century to encourage the purchase only of "free cotton" grown without any use of slave labor. The modern fair trade movement has its origins in the aftermath of the Second World War, when organizations such as Oxfam in the UK and SERRV in the US began selling goods made by displaced workers and refugees in war-ravaged Europe to help reconstruction. At about the same time, the Overseas Needlepoint and Crafts Project, subsequently to be renamed "SELFHELP: Crafts of the World" and then "Ten Thousand Villages," was founded in the US, prompted by the level of poverty in Puerto Rico, to alleviate poverty by selling goods produced by craftspeople in developing countries. The initial focus on handicrafts widened out to the selling of food and drink items produced in developing countries, and other fair trade organizations and shops proliferated. In the UK, the Christian organization Traidcraft was founded in 1979 and soon began selling fairly traded tea and coffee, extending its range to other food items and becoming an important and influential fairly trading company.

Meanwhile there was a growing realization at the political level that the way to tackle the challenge of international development, and the huge disparities of wealth and poverty in the world, lay not through aid in the form of handouts but through helping developing countries to trade on terms which would enable them to build up their economies. Achieving this would require the creation of a "level playing field" no longer distorted by the massive government subsidies given to farmers and producers in the wealthiest countries, and by the protective tariffs and trade barriers erected by those countries, which make it impossible for producers in the emerging economies to compete effectively in the international market.

These two strands – consumer-led purchasing, which pays a fair price to people living in poverty, and the political campaign of the trade justice movement for the reform of global trading institutions – come together in the idea of fair trade certification and labeling. The intention here is to go beyond the selling of individual fairly traded items

Practical Ethics for Food Professionals: Ethics in Research, Education and the Workplace, First Edition.
Edited by J. Peter Clark and Christopher Ritson.
© 2013 John Wiley & Sons, Ltd. Published 2013 by John Wiley & Sons, Ltd.

to a system of labeling that can be used more generally as a guarantee that the labeled items have been fairly traded and the original producers have received a fair price. Such a system can begin to create fairer institutions and structures of international trade, but without having to wait for the governments of the economically prosperous countries to reform their restrictive and unjust trading practices.

The first fair trade certification mark, the Max Havelaar label, was created in the Netherlands in 1988, taking its name from a 19th century Dutch novel which exposed the poverty and starvation produced by colonial policies. Over the next ten years the idea spread quickly, with fair trade labels being established in Belgium, France, Switzerland, Germany, Austria and Luxemburg. In 1992 the Fairtrade Foundation in the UK was set up, and its first products were launched in 1994, beginning with chocolate and soon followed by coffee and tea. (For a personal view of the history of Fairtrade in the UK, see Lamb, 2008.) In 1997 TransFair Canada was established, followed the next year by Transfair USA; 1997 also saw the creation of the Fairtrade Labelling Organizations International to coordinate the use of the FAIRTRADE Mark – the certification label initially employed by Max Havelaar and adopted by many (though not all) of the national bodies – and to guarantee consistent standards. "Fairtrade" as a single word is now used to refer to the labeling and trading system using the FAIRTRADE Mark as a registered certification mark (Figure 13.1).

The FAIRTRADE Mark certifies that products carrying the Mark meet the following standards:

- The price paid to the producers does not fall below a guaranteed minimum. This level is independently agreed with producers for each type of product. If prices on world commodity markets rise higher, producers benefit from the higher prices, but the Mark guarantees that the price will not fall below the guaranteed minimum, ensuring that producers can make a living for themselves and their families.
- Producers are able to enter into long-term contracts which enable them to plan for the future.
- The Mark guarantees that in addition to the guaranteed fair price, a Fairtrade premium is paid to the community, which decides democratically how it is to be spent, for example on improving local health facilities or schools, developing the local infrastructure such as better roads, or investing in new production techniques.
- Products are produced with respect for the environment. Farmers must avoid using the worst pesticides and the most damaging artificial fertilizers. Though the Mark does not guarantee fully organic farming methods, producers are encouraged and helped to move towards organic production. There is also increasing emphasis on helping farmers to cope with the effects of climate change through mitigation and adaptation, for instance by diversifying into new crops.
- Producers receive technical support and training in business skills, so that they can become effective business players rather than being at the mercy of an economic system over which they have no control.

There is some variation in the standards depending on whether production is by small independent farmers or on large commercial plantations. In the latter case, Fairtrade standards will include guarantees of working conditions, health and safety standards, and trade union rights. Where the producers are small farmers, they are

The FAIRTRADE Certification Mark was adopted by the Fairtrade Labelling Organizations International in 2002 to replace the different national labels and is now used in most countries including the UK.
Source: Fairtrade International.

Fair Trade USA's new Fair Trade Certified Label introduced at the beginning of 2012.
Reproduced by permission of Fair Trade USA.

The Max Havelaar label, the world's first Fairtrade Certification Mark, was officially launched by Stichting Max Havelaar in the Netherlands on 15 November 1988.
Reproduced by permission of Stichting Max Havelaar.

Transfair USA's original certification label, introduced in 1998.
Reproduced by permission of Fair Trade USA.

Figure 13.1   Fairtrade certification labels. *(For color details please see color plate section.)*

helped to organize themselves into cooperatives, which both promotes their economic empowerment and makes possible the paying of the community premium. (For a more detailed presentation of the international standards for Fairtrade certification, see: http://www.fairtrade.net/standards.html.)

More will be said below about the implementation of these standards, but what emerges from this brief history is the interaction between three overlapping strands within the fair trade movement.

1. *Specialist fair trade business and retailers:* These are the pioneering institutions which are wholly dedicated to the activity of fair trade as a response to poverty. They include innumerable small shops and traders, often focusing on the sale of fairly

traded craft items and clothes, for which it is difficult to develop generic certification standards, since the items are produced individually or in small quantities. Under this heading can also be included companies such as the coffee firm Cafédirect, the Divine Chocolate company, and Tropical Wholefoods, many of whose products carry the FAIRTRADE Mark but who put fair trade at the heart of their own business, and play an important part in introducing new products for which certification standards may subsequently be developed. Another organization to be mentioned in this context is Traidcraft, the Christian organization which sells Fairtrade food and drink products and has played a pioneering role in introducing new product categories, and which also makes available craft and clothing items which are fairly traded but do not carry Fairtrade certification. Many of these fair trade organizations, shops and companies are affiliated to IFAT, the International Fair Trade Association.

2. *The trade justice movement*: Included under this heading are both official intergovernmental bodies such as the United Nations Conference on Trade and Development (UNCTAD), set up in 1964 to promote international development through trade, and non-governmental organizations such as Oxfam and Christian Aid which campaign for more effective and more rapid implementation of the trade justice agenda.

3. *The Fairtrade labeling and certification system*: This brings the previous two strands together, building on the pioneering activities of the dedicated fair trade organizations but using certification to bring fair trade into the mainstream. Fairtrade-certified products are available not just through specialist fair trade shops and companies but on the shelves in supermarkets and on the high street where people do their regular shopping. The Fairtrade system uses consumer purchasing power to develop an alternative model of international trade which can prefigure the reforms sought by the trade justice movement, and have a significant global impact on poverty.

Most of the rest of this chapter will focus especially on the growth and functioning of the Fairtrade labeling and certification system, before returning to its relation with the wider fair trade movement.

13.2 THE GROWTH OF THE FAIRTRADE SYSTEM

The Fairtrade system in this third sense has seen a dramatic expansion since 1988. Sales of Fairtrade-certified products have grown substantially every year, with the year-on-year increase in sales in the UK often as high as 40%. Fairtrade labeling is widely recognized across the world and trusted as an independent guarantee. (See the most recent figures in GlobeScan, 2011.) And corresponding to the growth in the recognition of Fairtrade labeling and the growth in sales of Fairtrade products has been the growth in the number of farmers and workers benefitting from Fairtrade, growth in the value of sales to the producers, and growth in the returns from the community premium (Kilpatrick, 2011). There are three main reasons for this rapid expansion in Fairtrade sales and support for the Fairtrade system (Figure 13.2).

1. There has been a continuous growth in the range of products for which certification standards have been developed. Beginning with coffee, tea and chocolate, the range

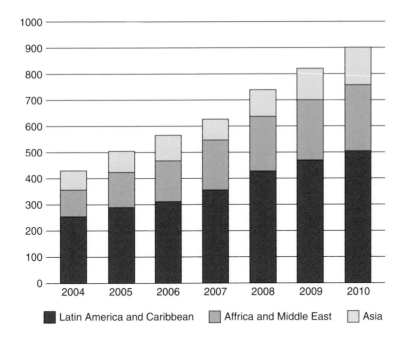

Number of Producer Groups benefitting from Fairtrade

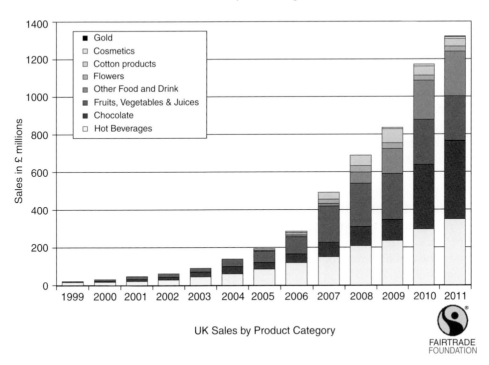

UK Sales by Product Category

Figure 13.2 Growth of Fairtrade sales to 2011. Source: Fairtrade Foundation, March 2012. (*For color details please see color plate section.*)

now extends to products such as sugar, tropical fruit (such as bananas, oranges, pineapples and mangos), fruit juices, nuts, dried fruit, wine, rice, and olive oil. The FAIRTRADE Mark can also be carried on non-food items such as flowers, cotton, beauty products, and gold. Another important development has been the expansion of composite products with Fairtrade ingredients, such as snack bars (made with Fairtrade chocolate and Fairtrade sugar) and breakfast cereals (with Fairtrade dried fruit and nuts).

2. Major companies, both manufacturers and retailers, have entered the system. In the UK, all the major supermarket chains stock substantial quantities of Fairtrade products, and often sell own-brand Fairtrade items such as tea and coffee. The Cooperative, with its long history of commitment to ethical retailing, has been in the lead, and the decisions by Sainsbury's and by Waitrose to make all their bananas Fairtrade were further ground-breaking steps. Café chains including Starbucks and Pret A Manger have made comparable moves. These commitments have been matched by major manufacturing companies, particularly chocolate manufacturers. Cadbury's, Nestlé and Mars have all in recent years committed to certification of one or more of their leading chocolate brands, leading to a scaling up of their sourcing of Fairtrade cocoa and sugar. Their reasons for making these commitments have been in part a recognition of the need to nurture their supply chains and safeguard the future of their suppliers, particularly in the case of those products for which they are dependent on small farmers. Their other main motivation has been consumer demand, and this leads to the third explanation but the rapid expansion of Fairtrade.

3. Outlets sell Fairtrade because consumers will buy it, and the consumer pressure has in turn been generated by campaigners who believe in the importance of Fairtrade and dedicate time and energy to raising awareness and generating support for it. An inspirational form of campaigning has been the Fairtrade Towns movement. This began in 2000 when the small UK market town of Garstang in Lancashire declared itself "the world's first Fairtrade Town." The Fairtrade Foundation responded by creating criteria for formal recognition as a Fairtrade Town, which could encompass anything from a village to a big city or a county, followed by comparable criteria for Fairtrade churches and other places of worship, universities and colleges, and schools. There are now over 500 Fairtrade Towns in the UK, and the idea has been picked up internationally, the figure of 1000 Fairtrade Towns worldwide being reached on 4 June 2011 with the declaration of Fairtrade status by towns in Australia, Japan, Europe and the United States. The energy generated by this movement stems from people's sense that, by working in their own local community to raise awareness, they can make a direct contribution to the struggle for global justice. This is perhaps, above all, the secret of Fairtrade's success – that it enables ordinary people to feel that they can make a difference, can by their everyday actions play their part in tackling what might otherwise seem the overwhelming challenge of world poverty.

13.3 FAIRTRADE AND FAIRNESS

The Fairtrade system has popular appeal and has garnered widespread support because it resonates with people's idea of fairness. It does not depend on any subtle or complex ethical arguments to justify it, but it is worth considering how it draws on certain widely accepted intuitions about the concept of fairness. These widely accepted intuitions could be regarded as components of the "common morality" as defined in Chapter 3. Drawing on this common morality and on philosophical theories of ethics, Ben Mepham develops in that chapter an "ethical matrix" employing three principles: one of which is a *principle of fairness*, alongside principles of *wellbeing* and *autonomy*. The principle of fairness is Mepham's version of the "principle of justice" to be found in an earlier list of principles, and it could be said that talk of practices as "fair" or "unfair" is the language in which principles of justice are applied in everyday life. This is not the place to go into the details of philosophical theories of justice, but I am going to look at how two intuitive ideas of fairness correspond to two dominant approaches in theories of justice, and how they apply to the case of fair trade.

13.3.1 Fairness as equality

One very common idea of fairness links it with equality – the idea that when goods are being distributed within a group, everyone should receive an equal share and no one should get more than anyone else. Think of the young child's typical protest: "It's not fair, he's got more than me!" When extended to a whole society, let alone to the whole international community, the idea of equality becomes more controversial. Few would defend simple strict equality – that everyone should be equally well off, with no differences of wealth or privilege – as either a feasible or a desirable ideal of social justice. Nevertheless a broader egalitarian conception of fairness would be widely accepted, that massive inequalities between people are morally unacceptable. It is not right that some should live in luxury while others starve.

That broadly egalitarian ideal can be seen as a plausible ethical underpinning for the Fairtrade system. It addresses the unacceptable inequalities, as described in Chapter 10, between the farmers and workers in developing countries who produce much of our food and the consumers in the more prosperous parts of the world who buy and enjoy it. The former typically struggle to survive, living at the margins of poverty and often lacking basic educational or medical provision. Consumers in the economically advanced countries of Europe and North America may not be particularly wealthy by the standards of their own society but most of them enjoy a standard of living which the producers can only dream of.

The intuition that such inequalities are unacceptable is reinforced by the recognition of our interdependence. This is encapsulated in some famous words from Martin Luther King's 1967 "Christmas Sermon on Peace" often quoted by supporters of fair trade.

> We are all caught in an inescapable network of mutuality . . . Did you ever stop to think that you can't leave for your job in the morning without being dependent on most of the world? . . . you go into the kitchen to drink your coffee for the morning, and that's poured into your cup by a South American. And maybe you want tea: that's poured into your cup by a Chinese. Or maybe you're desirous of having cocoa for breakfast, and that's poured into your cup by a West African . . . And before you finish eating breakfast in the morning, you've depended on more than half the world (King, 1991).

If the inequality between the world's rich and the world's poor were simply a disparity between two unrelated groups, it might be seen as less ethically problematic. But when the well-being of the world's rich is recognized to be *dependent* on the labor of the world's poor, then the big difference in the benefits reaped by them from this global network of interdependence is revealed as an inequality which runs counter to our sense of fairness.

A focus on the position of the least well-off is a feature of various theories of social justice as a complex version of equality. The most influential such theory is the one developed in great detail by the American philosopher John Rawls. Rawls' notion of *justice as fairness* is discussed briefly by Ben Mepham in Chapter 3, and this is not the place to go into the complexities of the theory, but at its heart is a simple principle which Rawls calls the "difference principle":

> Social and economic inequalities are to be arranged so that they are both (a) to the greatest benefit of the least advantaged..., and (b) attached to offices and positions open to all under conditions of fair equality of opportunity (Rawls, 1971, p. 302).

In the global community as a whole, the farmers of the global south may not be the most disadvantaged group. But if we focus specifically on the system of international trade in food commodities, then it seems plausible to maintain that those who benefit least from this system are the primary producers, and that the terms of trade can be made more equitable in ways which leave this disadvantaged group significantly better off.

Another concept which features prominently in theories of justice as complex equality is the idea of *basic needs*. The idea here is that although it may not be appropriate to require strict equality, a system of cooperation which enables some to prosper while leaving others unable to meet their basic needs represents an unacceptable degree of inequality. People cooperate for mutual benefit, but if the system of cooperation then leaves some participants struggling to satisfy their basic needs for things such as food, clothing, shelter, and health, they cannot be said to benefit sufficiently from it. Hence the conditions of Fairtrade certification, that the producers should be guaranteed at least a minimum price sufficient for them to be able to make a living from selling their produce, and that the Fairtrade premium should enable the producer community to meet basic needs for such things as health care and education, seem to match this intuitively plausible requirement.

13.3.2 Fairness as desert

A second influential conception of fairness is that people should get what they deserve, and that it is unfair if they receive less than they deserve. This idea may come into conflict with the ideal of equality, and indeed may partly explain why many think that fairness is different from strict equality. Some people, it is often thought, deserve more than others, so to reduce everyone to the same level is to fail to reward those who deserve more.

A fundamental difficulty with this idea is the problem of how to measure what people deserve. Are we looking for some notion of *moral* merit – that the greater rewards should go to those who are in some sense better people? It is difficult to see how any such idea could be applied in practice. There are, first, different and incompatible views of what constitutes moral goodness. Then, whatever we think it consists in, how on earth are we supposed to measure it? And even if we had some idea of what we were looking for, the secrets of people's hearts must prevent us from ever being able to apportion reward to moral worth.

More plausible might be the idea that greater rewards should go to those with greater *talents*, the more skilled or the cleverer or the more able. Again, however, there seem to be insuperable problems with measuring these things. And even if we could, why should we reward them? People's differing talents and abilities are to a large extent advantages which they are born with, the outcome of nature's lottery.

With this in mind, the most plausible version of the idea of "desert" might seem to be the idea of reward for *effort* – that those who work harder should receive the greater reward. Though this may seem to be a more acceptable interpretation of desert, there are again questions about how we could possibly measure it. Most defenders of the idea would, I think, say that it is more than simply a matter of rewarding people for the amount of *time* that they work. Some kinds of work, it may be said, are more valuable than others, or more demanding – which might seem to take us back to the idea of rewards for talents and abilities.

So whatever its intuitive appeal, the idea of people getting what they deserve turns out to be fraught with difficulty, and the difficulties may be insuperable. Nevertheless there seems to be something in the broad intuition that when people cooperate for mutual benefit, the benefits which they receive should match their contribution. And however we measure "contribution." the Fairtrade system once again seems to fit this intuitive idea of fairness. The food which comes to us from developing countries requires contributions at various different stages – from the farmers and plantation workers who are the primary producers, as well as the transport workers, the traders and middlemen, the company managers and administrators, the processors and packers, and the retailers. And in this system of global exchange and cooperation, the primary producers, on whom the whole system ultimately depends and whose lives are totally invested in it, are the ones who are most in danger of losing out by comparison with the other stages in the supply chain. The Fairtrade system, then, seeks to redress this imbalance, and to ensure that the farmers and plantation workers are fairly rewarded for their contribution.

13.4 CRITICISMS

The Fairtrade movement does seem to be rooted in widely-held ideas of fairness, but how does it work out in practice? Criticisms of Fairtrade have come from two opposite directions: (i) that it is inappropriate to try to introduce certified standards for fairness into international trade, and (ii) that the Fairtrade system is insufficiently fair.

13.4.1 Free trade rather than fair trade?

Some critics would argue that ideas of justice and fairness have no place in international trade, or in commercial exchange and economic life generally. Others would argue that fairness is best achieved by allowing economic markets to function freely, regulated only by the rule of law and the protection of private property. Both versions of this line of criticism typically appeal to the notion of "free trade" – that in any economic exchange both parties should be left free to make their own agreement on the terms of the exchange, and that any attempt to interfere with this process from the outside is both unethical and impractical. From an ethical perspective the argument might invoke the *principle of autonomy* which is another component of Ben Mepham's ethical matrix in Chapter 3. Producers and buyers, manufacturers and consumers, are, it might be argued, the best judges of their own interests and should be trusted to make such judgments for themselves, exercising their own autonomy.

A further argument might also invoke the *principle of wellbeing*. The claim might be that unrestricted competitive trade and exchange is likely to be more efficient and therefore more productive of overall well-being than any attempt to manipulate the process in the name of fairness. More specifically, it might be argued that the Fairtrade system subsidizes inefficient farmers who, if left exposed to the full rigor of the competitive market, would either have to improve their production techniques to become more efficient, or diversify into more profitable crops and markets, or would go out of business and leave the field clear for their more efficient competitors. (For examples of scepticism, from a "free trade" perspective, about the ability of Fairtrade certification to lift producers out of poverty in the long term, see Booth and Whetstone, 2007, and Mohan, 2010.)

There are a number of things which can be said in reply to this line of criticism. In the first place, the Fairtrade system seeks to make use of the free market rather than to replace it. It employs precisely that mechanism of consumer choice which the advocates of free markets are so keen on. The fact is that, in the exercise of their choice, many consumers with an ethical conscience will choose to buy products which give the primary producers a better deal. The Fairtrade system does not involve any coercive mechanisms which force producers and buyers to exchange on terms which they do not want. It has been suggested that the movement to promote Fairtrade Towns, Fairtrade Churches, Fairtrade Schools, Fairtrade Universities and the like is akin to a form of coercion, that by this route "Fairtrade is actively seeking a monopoly on those products it certifies" (Mohan, 2010 p. 98). But this is misleading. Commitments of that kind by local councils, churches, schools and universities are not coercive, they do not force people to consume only Fairtrade products and they do not prevent others from making different consumer choices if they so wish.

Similarly, the system does not eliminate the element of competition and the incentives which competition creates for greater efficiency. The Fairtrade price which producers receive is only a baseline. If the prices of the relevant commodity such as tea or coffee or cocoa rise on the world markets, then Fairtrade producers can still sell at that higher price if they can get it, so they still have the incentive to pursue quality in their product with the aim of getting as good a price for it as possible.

More fundamentally, however, the very idea of "free trade" needs to be examined more closely. There is no such thing as "free trade" in the sense of trade which is totally unregulated. Markets would be impossible altogether without external regulation and guarantees such as that the terms agreed to by seller and buyer will be adhered to, and to ensure that the exchange is not distorted by coercion or intimidation or fraud or any other improper pressure. Similarly, regulation of markets is often needed to prevent the operation of cartels or monopolies which would rig the market. The point is that how much autonomous choice the buyer or the seller are able to exercise in agreeing on a price will depend on the starting-point from which they make the agreement. It will depend on whether they start out from a strong or a weak bargaining position, and that will depend on factors outside their control and on the character of the economic institutions within which they operate.

We might compare the inequities and imbalances of the present international trading system with the relations between employers and employees in the newly industrialized countries in the 19th century. This was the period when campaigners and reformers were pushing for government regulation of mines and factories to protect workers and ensure that they were given decent and safe working conditions and not required to work for exhaustingly long hours without a break, to the detriment of their health. Many industrialists opposed such legislation on the grounds that it would interfere with the "freedom of contract" between employer and employee. If the workers chose to work on the terms and for the hours which the factory owners offered, that was their free choice and no one else, particularly governments, should interfere. In practice, of course, what "freedom of contract" all too often meant was the freedom of workers either to contract to work for long hours for a pittance in dangerous and unhealthy conditions, or to accept unemployment and starvation.

Similarly, in the current world trading system, farmers and producers in the global south find themselves at a great disadvantage in the market, for reasons which are not of their own making. The terms of trade are, for instance, heavily distorted by the government subsidies given to farmers in the most developed countries. As a result of such subsidies, imported American rice can often be sold in countries such as India, for example, more cheaply than locally grown rice, not because the American rice farmers are more efficient but because they are cushioned by government support. The dependence of third world farmers on cash crops is likewise a situation which they have inherited rather than created. Often it is a historical legacy of colonialism. It might be suggested that if they are so dependent on a fragile and volatile food commodity market, they have the choice to move into other crops or some other form of economic activity. But diversifying, or investing in some other form of business, is difficult unless it is done from a position of relative economic security and with capital to invest. The more precarious one's economic position, the more difficult it is to take risks. One of the strengths of the Fairtrade system is that it can give producers support, in the form of pre-financing or of advice and training, which makes it easier for them to diversify or re-invest.

More fundamentally still, we need to raise deeper questions about the meaning of the word "free" in the phrases "free market" and "free trade." As a concept which is intended to carry ethical weight, the word "free" draws on the recognition that human beings need to feel that they have some control over their own lives, that they can take responsibility for themselves and their families and shape their own future. This is the ethical force of the value of *autonomy*. Autonomy in economic life is not, however, something which can be achieved simply by the operation of unrestricted market pressures. In order to exercise autonomy, economic actors need to be *empowered*. What the Fairtrade system aims to do is precisely to help primary producers achieve a greater degree of control over their own lives, by equipping them with techniques and skills, and putting them in a position where they can take risks, invest, and introduce new and more efficient production methods or new crops.

13.4.2 Not fair enough?

I turn now to possible criticisms from the opposite direction – not that the aim of fairness is misplaced, but that Fairtrade is not fair enough. It may be criticized for setting its sights too low and not being sufficiently radical or transformative. It still leaves farmers and plantation workers with a disproportionately small share of the price which the products eventually fetch when they reach the retail shelves. The manufacturers and the retailers – the big powerful companies – still remain the principal beneficiaries of the system. Indeed, the suspicion has sometimes been voiced that retailers such as supermarkets actually exploit the Fairtrade system by putting an extra mark-up on Fairtrade products in order to take advantage of the commitment of ethically motivated consumers.

To the suggestion that Fairtrade does not do enough for the producers may be added the accusation that it does nothing at all to address the worst inequalities of the global economy. Small farmers who own their land may experience relative poverty, it may be said, but they are not living in absolute poverty. They are still better off than, for instance, the urban poor living in the slums and shanty-towns of the big cities in the global south, people who work in sweat-shops or are unemployed and with no social safety-net, with no recourse other than resorting to crime or scouring the rubbish tips for items to sell.

Moreover, the Fairtrade system helps only a relatively small proportion even of those engaged in food production and exchange. Indeed, it has been suggested that it is actually unfair to those farmers and workers who are not part of the Fairtrade system, and leaves them worse off as a result.

Finally, there have been specific accusations that Fairtrade does not do what it claims to do. There have been critical news reports involving interviews with farmers or plantation workers who say that they have seen no benefits from being supposedly part of the Fairtrade system.

The short answer to all these criticisms is that Fairtrade does not claim to solve all the problems. It is not a comprehensive solution to the problems of global poverty and injustice, but it does attempt to address particular aspects of that larger picture. Those who are helped by it will not of course find their lives totally transformed, but it is the beginning of a process, setting producers on a path where they have at least some measure of control over their economic destiny. The community premium, in particular, by facilitating the introduction of new machinery or the improvement of the local infrastructure, or better education and health facilities, can provide a powerful impetus to further economic development.

There is no reason to think that the farmers who are not part of the Fairtrade system find themselves at a greater disadvantage as a consequence of the system. At the worst, it leaves them where they were before, still able to sell their produce at the going market rate. There is also some evidence that Fairtrade actually improves the position of those outside the system, by raising overall standards in the sector so that buyers need to make better offers in order to obtain quality produce. Granted, the Fairtrade system does not of course provide any direct benefit to the urban poor. What its success can do, nevertheless, is to help stem the drift to the towns, by enabling small farmers to remain viable, to make a living and to stay in business.

The accusation that manufacturers and retailers benefit disproportionately from Fairtrade is neither proven nor relevant. There is no evidence that retailers increase their mark-up on Fairtrade products, and the major UK retailers are adamant that they do not do so, but in any case Fairtrade certification carries with it no claim either way about the level of profit made by the manufacturing and retailing companies. The system simply does not address this, it has no power to do so and indeed it would be illegal to try to limit or control the prices charged by companies and their level of profit. Fairtrade certification focuses on the situation of the producers, and the benefits which they receive.

If the farmers who are Fairtrade certified, and the workers on Fairtrade-certified plantations, do not benefit as much as they or we might hope, this is often because they cannot sell enough of their produce on the Fairtrade market. Gaining Fairtrade certification does not guarantee how much of a demand there will be. Often Fairtrade producers find that they can sell only a small proportion of their produce on Fairtrade terms, and this can be a major source of frustration for them. To take one particular case, a radio story in the UK included interviews with workers on two tea plantations in India who said that they had seen no evidence of any substantial improvement in their conditions of work. It subsequently turned out that although both tea plantations were Fairtrade certified, one of them had been able to sell none of its tea on the Fairtrade market for some years, and the other had sold only about 1% of its tea on Fairtrade terms. The lesson to be learned from such cases is not that the Fairtrade system itself has failed, but that its success depends on the level of demand from consumers. The more that shoppers can buy Fairtrade products, and the more that supporters of the system can persuade them to do so, the more impact the system can have.

There have undeniably been cases where the system has not worked smoothly, cases where the conditions required for Fairtrade certification have not been met. But of course the whole point of a certification system is to be able to identify such failings and to redress them. The Fairtrade Labelling Organizations International puts in place a structure of auditing and checking through annual inspections. If the conditions are not being properly met, if producers are not receiving the benefits which are due to them, or are themselves failing to abide by the standards, they are notified and are given time and support to enable them to come back into conformity with the requirements.

13.5 TENSIONS

The previous two sections have set out Fairtrade's appeal to popular ideas of fairness, and have responded to criticisms of it. Readers can decide for themselves how strong or weak the criticisms are. I think myself that the criticisms can be successfully answered. I also think, however, that there are problems which the Fairtrade movement faces, and in particular that there are tensions within it. How the movement develops in the future will depend importantly on how these tensions are addressed.

13.5.1 Niche or mainstream?

If fair trade is to have a substantial impact on global poverty, then sales have to increase. The purpose of the Fairtrade certification system has always been to bring fair trade into the mainstream, so that consumers can support it not merely with an occasional visit to a specialist fair trade shop or the occasional purchase from a specialist fair trade company, but with every visit they make to the supermarket and every weekly shopping trip to stock up on groceries.

The growth in sales has been impressive, especially in the UK, where sales of Fairtrade products have seen year-on-year increases often as high as 40%. As we have seen, this success owes much to the support from big companies – the retail supermarket chains, and manufacturers including chocolate brands such as Cadbury's, Nestlé and Mars.

These developments have been controversial, especially in the eyes of many people who are Fairtrade's natural supporters. They have been seen as an unacceptable compromise with, or even a capitulation to, big business and its dubious ethical standards. To take one prominent example, Nestlé has in the past been the target of a campaign protesting against its aggressive marketing of powdered baby-milk in developing countries. This is not the place to discuss the criticisms of Nestlé and the firm's response, but the fact is that some supporters of this campaign were indignant at what they described as "awarding Nestlé the FAIRTRADE Mark."

Some clarification is needed here. The FAIRTARDE Mark is given not to *companies*, but to *products*. It is not a statement about the overall ethical standards of the company which makes the product. It certifies only that the supply chain can be reliably traced back to primary producers for whom the Fairtrade conditions such as the guaranteed minimum price and the provision of the community premium are met, and that traders who deal with Fairtrade products meet trade standards such as providing pre-financing for producers who require it, and committing themselves to mutually beneficial long-term trading relationships. If a product does meet those conditions, then it would be perverse, indeed it would be unethical, to withhold certification. An essential feature of the Fairtrade system is that it applies objective standards, and whether they are met has to be a matter for impartial judgment which should not be influenced by views about other aspects of a company's business practices.

The commitments made by companies such as Cadbury's, Mars and Nestlé have had and will continue to have a substantial effect in improving the lives of cocoa farmers in countries such as Ghana and Côte d'Ivoire. However, they also risk eroding the support for Fairtrade from campaigners who have been vital to its growth. They perhaps risk also damaging the fair trade vision as an alternative model for international trade and business. And they risk marginalizing the dedicated fair trade companies which have pioneered that alternative model – companies such as Divine Chocolate and Tropical Wholefoods, which have put the empowerment of the growers at the heart of their business activities, and have played a vital role in bringing new products into the Fairtrade system. The danger is that every new commitment to Fairtrade certification on the part of big multinational companies may mean tougher competition for the dedicated fair trade firms, and if they were to go out of business the Fairtrade movement would itself be likely to be damaged.

This tension is in part also a tension between Fairtrade and the wider fair trade movement. The latter includes also specialist fair trade shops and mail order firms selling clothes and craft items for which it is difficult to develop generic certification standards. These enterprises too embody an alternative business model, often built on personal contacts with craftspeople in the global south working alone or in small workshops. It may be difficult or impossible for them to have a large-scale impact on world trade and the global economy, but they embody an alternative ethical vision which it is important to retain and foster.

13.5.2 Small farmers or large plantations?

A closely connected tension within the movement is that between different kinds of producer organizations. For many products, such as cocoa and coffee, Fairtrade has worked primarily with independent small farmers. They have been encouraged to form cooperatives to facilitate certification, and this has played an important part in the process of empowering the farmers and putting them in a position where they can exercise effective control over their economic future instead of being entirely dependent on the price which the local trader or middleman may offer.

Encouragement for cooperatives of small farmers is not, however, a model which can be universally applied. Some products are grown on large plantations employing hired labor, rather than on small farms worked by independent producers. Tea, for instance, though sometimes grown by small farmers and cooperatives, is more often grown on large plantations, in contrast to, say, coffee or cocoa, where smallholders form a larger proportion of the producers. There are also important geographical and historical differences in the scope for establishing cooperatives of small producers; the cooperative model has been developed more strongly in Latin America than in other parts of the world.

Certification standards for large plantations are therefore different from those for small producer organizations, focusing more on the working conditions of employees, their right to belong to trade unions, the banning of the use of child labor, and so on. There are perhaps narrower limits to what Fairtrade can offer them, compared with independent farmers, and the potential for their economic empowerment is more restricted.

The movement of Fairtrade sales into the mainstream, and the aim for substantial growth in sales, brings with it also pressure to source from large plantations and to extend this to products for which the Fairtrade focus has previously been on cooperatives of small farms. Fairtrade bananas are an example of a mixed model, sourced partly from the small farms of Caribbean banana growers but also from big commercial plantations in mainland central America and in west Africa. There is pressure now to do the same for other products, such as coffee.

Fair Trade USA has recently decided, as from the beginning of 2012, to operate independently from the Fairtrade Labelling Organizations International in order to move more rapidly in that direction. It is reviewing its certification standards with the aim of applying them to "other producer setups that have historically been excluded from the benefits of Fair Trade" (Fair Trade USA, 2012). In response, the Fairtrade Labelling Organizations International, and the Producer Networks which work with it, have expressed concern that the extension of fair trade standards to large coffee plantations and estates, though an admirable goal in principle, may work against the interests of existing Fairtrade coffee farmers. They point out that most large coffee plantations have a transient workforce, and that this makes it difficult to ensure that the benefits of fair trade reach the workers. They emphasize that:

> ... small-scale growers represent the majority of the world's coffee production and are at the heart of the Fairtrade system. One of the strengths of the small producer organization is the promotion of independent, democratic decision-making, something that is difficult to promote among a transient workforce. Opening the Fairtrade system to plantations with large coffee volumes could also threaten small producer organizations that cannot operate on the same scale ... (Fairtrade International, 2012)

The particular case of coffee exemplifies the tension within the fair trade movement, between deepening the benefits of fair trade, and extending them more widely at the risk of diluting them. It is a dilemma which the movement will continue to face and to grapple with.

13.5.3 Fairtrade or local produce?

The third tension stems from the growing awareness of the massive threat posed by climate change and global warming. One response to this has been the view that one ought to 'buy local' – to purchase food items in particular from local farmers and producers in order to reduce the CO_2 emissions produced by transporting goods over large distances. It has been suggested that we ought to measure the environmental impact of our purchases in "food miles," the distance which items have traveled from producer to consumer. For some ethically aware consumers who would be natural supporters of Fairtrade, this has prompted the question: should we really be buying, and encouraging other people to buy, Fairtrade products which have been imported from developing countries and which may have traveled halfway round the world?

The opposition between "Fairtrade" and "buying local" is simplistic. The concept of "food miles" is a crude and misleading measure of environmental impact. At least as important as the distance traveled by products is the mode of transport. The vast majority of Fairtrade products are carried in bulk by shipping, and the CO_2 emissions produced in consequence are proportionally very low. Moreover, it is important to look not just at emissions created by transporting a product, but at the product's total life-cycle (Fairtrade Foundation, 2009, p. 21). A study of food products in the United States showed that 89% of total emissions were associated with production, and only 11% with transport. Because of the production techniques involved, for instance, sugar made from sugar beet in European countries may well have a higher carbon footprint than imported cane sugar. And it has been calculated that over 90% of the CO_2 emissions involved in making a cup of tea are accounted for by the energy source needed for boiling the kettle.

Since most Fairtrade products are grown in tropical or semitropical zones, the alternative to buying them is not usually choosing to buy the same product from a local source, but only choosing not to consume items such as tea and coffee at all. A decision not to buy them would amount to imposing on farmers in the global south the costs of a problem which they are not responsible for creating, and to which they are already the most vulnerable compared with the countries of the industrialized north. Changing climatic conditions are already affecting farmers in developing countries whose crops are vulnerable to different temperature or rainfall patterns, or to the greater frequency of tropical storms and hurricanes, of which climate change is a likely cause.

In this respect, indeed, buying Fairtrade can be a positive response to the impact of climate change. The Fairtrade system provides support for farmers who are vulnerable in this way, helping them with measures of mitigation or adaptation. It can offer technical expertise and provide a cushion to facilitate diversification into new crops, for instance, which may be more suited to the changing weather conditions (Fairtrade Foundation, 2009).

To a considerable extent, then, the supposed opposition between Fairtrade and buying local produce is a false choice. It does, however, point to new challenges and new possibilities for the Fairtrade movement. In particular, there is scope for encouraging more regional trade between and within developing countries themselves, opportunities for Fairtrade farmers to increase their sales by selling to local and regional markets. The potentialities of Fairtrade could be significantly boosted by opportunities for local processing of primary commodities, so that west African cocoa for example, instead of being shipped to European or North American countries to be processed and go on sale as chocolate bars, could be processed and sold in the countries of origin. One of the most interesting recent developments in the Fairtrade movement has been that Fairtrade products have gone on sale in producer countries such as South Africa and India. It represents one of the most exciting potentialities for extending the benefits which Fairtrade can bring.

REFERENCES

Booth, Philip, and Whetstone, Linda (2007) *Half a Cheer for Fair Trade*. Institute of Economic Affairs, London. [Available online: http://www.iea.org.uk/publications/research/half-a-cheer-for-fair-trade-web-publication, accessed 17 December 2012].

Fairtrade Foundation (2009) Egalité, Fraternité, Sustainabilité: Why the Climate Revolution Must Be a Fair Revolution. [Available online at: http://www.fairtrade.org.uk/includes/documents/cm_docs/2009/c/climate_report_final.pdf, accessed 19 December 2012].

Fairtrade International (2012) Q&A on Fairtrade International and Fair Trade USA [Available online at: http://www.fairtrade.net/897.html, accessed 17 December 2012].

Fair Trade USA (2012) Standards Setting: Review and Development Process [Available online: http://fairtradeusa.org/sites/all/files/wysiwyg/filemanager/standards/FTUSA_Standards_Process_Overview.pdf, accessed 17 December 2012].

Globescan (2011) Shopping choices can make a positive difference to farmers and workers in developing countries: global poll [Available online at: http://www.globescan.com/news_archives/flo_consumer/, accessed 17 December 2012].

King, Martin Luther (1991) A Christmas Sermon on Peace. In: *A Testament of Hope: the Essential Writings and Speeches of Martin Luther King Jr.* (ed. James M. Washington), pp. 253–8. HarperCollins, New York. Also available online, for example at: http://portland.indymedia.org/en/2003/12/276406.shtml, accessed 17 December 2012.

Kilpatrick, Kate (2011) *Monitoring the Scope and Benefits of Fairtrade* [Available online: http://www.fairtrade.net/fileadmin/user_upload/content/2009/resources/Monitoring_the_scope_and_benefits_of_Fairtrade_2011.pdf, accessed 17 December 2012].

Lamb, Harriet (2008) *Fighting the Banana Wars and Other Fairtrade Battles*. Random House, London.

Mohan, Sushil (2010) *Fair Trade Without the Froth*. Institute of Economic Affairs, London. [Available online: http://www.iea.org.uk/publications/research/fair-trade-without-the-froth, accessed 17 December 2012].

Rawls, John (1971) *A Theory of Justice*. Oxford University Press, Oxford and New York.

The FAIRTRADE Certification Mark
was adopted by the Fairtrade Labelling
Organizations International in 2002 to
replace the different national labels and
is now used in most countries
including the UK.
Source: Fairtrade International.

Fair Trade USA's new Fair Trade
Certified Label introduced at the
beginning of 2012.
Reproduced by permission of
Fair Trade USA.

The Max Havelaar label, the world's
first Fairtrade Certification Mark, was
officially launched by Stichting Max
Havelaar in the Netherlands on
15 November 1988.
Reproduced by permission of
Stichting Max Havelaar.

Transfair USA's original certification
label, introduced in 1998.
Reproduced by permission of
Fair Trade USA.

Plate 13.1   Fairtrade certification labels.

Practical Ethics for Food Professionals: Ethics in Research, Education and the Workplace, First Edition.
Edited by J. Peter Clark and Christopher Ritson.
© 2013 John Wiley & Sons, Ltd. Published 2013 by John Wiley & Sons, Ltd.

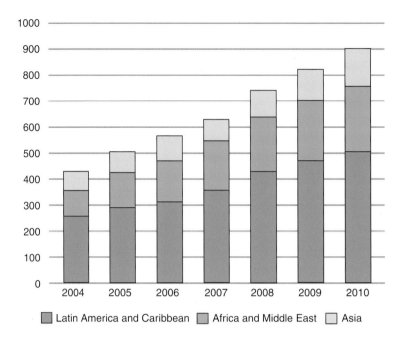

Number of Producer Groups benefitting from Fairtrade

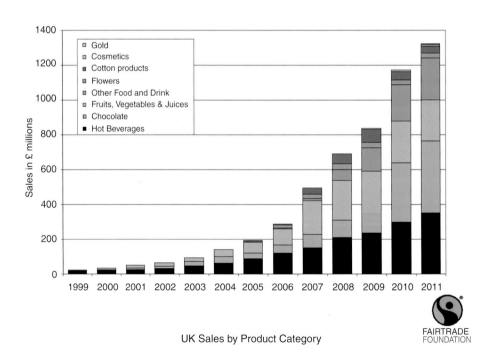

UK Sales by Product Category

FAIRTRADE
FOUNDATION

Plate 13.2 Growth of Fairtrade sales to 2011. Source: Fairtrade Foundation, March 2012.

14 A serious case: the Peanut Corporation of America

Mark F. Clark

14.1 INTRODUCTION

Shirley Mae Almer, Doris Flatgard and Clifford Tousignant had all made it into their 70s and 80s. The Brainerd, Minnesota, nursing homes where they were staying were in separate buildings operated by the same nonprofit company, Good Samaritan Homes. The three had a weakness for peanut butter. Those things were all they had in common until the holiday season of 2008. Within three weeks of one another they would be dead from salmonella poisoning (Erikson, 2009), felled by their favorite comfort food, Almer on December 21, 2008, Flatgard on January 4, 2009, and Tousignant on January 12 2009 (Sem, 2010).

Even before the third tragic death, the focus of a crack Minnesota Department of Health investigative team had narrowed sharply its intensive search for the cause of a multi-state outbreak of illnesses caused by salmonella which had curiously stricken disproportionately in Minnesota. The team discovered first that two, then that all three had eaten peanut butter from institutional sized tubs supplied to the nursing homes by King Nut Co. of Solon, Ohio. The peanut butter carrying the King Nut label was made in the Blakely, Georgia, plant of Peanut Corporation of America (PCA) (Anderson, 2010).

At the time it was tested for *Salmonella*, one of the tubs had been standing open in the kitchen for an indeterminate length of time. The fact that the container was open gave company officials an opening to slough off culpability by speculating that it could have been or might have been contaminated by a knife that wasn't cleaned after being used to cut raw chicken and before being plunged into the peanut butter (Pritzker, 2011). That rather unlikely suggestion would come to characterize the relatively small company's strange responses during ensuing inquiries by state and federal public health and law enforcement officials aimed at ending the gruesome epidemic that was affecting more people across the country every day and was soon to set the record as the worst food poisoning episode in US history (*Business Insurance Quotes*, 2011). Formal denials of responsibility would be contradicted by investigators' later discovery of internal PCA memos (Harris, 2009a; Gray, 2009) that pointed to a culture of, at best, carelessness with food safety and, at its appalling worst, as evidence increasingly showed, patterns

Practical Ethics for Food Professionals: Ethics in Research, Education and the Workplace, First Edition.
Edited by J. Peter Clark and Christopher Ritson.
© 2013 John Wiley & Sons, Ltd. Published 2013 by John Wiley & Sons, Ltd.

of substituting for health standards compliance a whatever-it-takes approach to meeting production and distribution demands (Rubenstein, 2009).

At times, documents would demonstrate, the firm knowingly allowed tainted product into the public food supply (Waters, 2009; Subcommittee on Oversight and Investigations, 2009). No matter how damning the internal communications, though, PCA insisted (Powell, 2010) – and still does (Parnell, 2012) – that no contaminated container of peanuts or peanut products was ever knowingly shipped. Following an unrelenting torrent of devastating publicity about the company, and having been subpoenaed before Congressional hearings about the scandal, top company officials, including CEO Stewart Parnell, settled into Fifth Amendment silence as their safest strategy (Subcommittee on Oversight and Investigations, 2009; Milbank, 2009).

The company filed for Chapter 7 bankruptcy two days after the hearing and was liquidated later in 2009 (Martin, 2009; Layton and Miroff, 2009). At a March 11 bankruptcy court hearing in his and the company's home city, Lynchburg, Virginia, Parnell invoked the Fifth Amendment again. His daughter, Grey Adams, who said she was the company's unpaid bookkeeper, did answer questions from Bankruptcy Trustee Roy Creasy and lawyers for the victims, including a statement that "to my knowledge," none of the company's shipped product had been tainted by salmonellae (Reed, 2009).

Shirley Mae Almer had been diagnosed with lung cancer in 2007 and survived it and the surgery that successfully treated it. On returning from a Florida vacation after the surgery she collapsed from the effects of a previously undetected brain tumor, was successfully treated with radiation for that second cancer and was just getting over a urinary tract infection, convinced that this time it was really going to happen – she would be home for Christmas – when the violent symptoms of salmonella poisoning set in and snatched away her repeatedly postponed dream of returning to her family. The homecoming, happily anticipated as well by her five children and four grandchildren, had been planned for December 22. She died on December 21 at the age of 72. Her daughter Ginger Lorentz, according to the *Minneapolis Star-Tribune*, called her mother's death, "pointless – with all the battles she overcame – to have a piece of peanut butter toast take her." On a visit the week before she died, Lorentz had made her mother's favorite snack, peanut butter toast, and fed it to her (American Institute of Baking, 2009; Frankin, 2010; Shaffer, 2009).

Doris Flatgard was 87. She'd been married to her husband for 65 years. She died from eating peanut butter on toast for breakfast, which is how she started every day (American Institute of Baking, 2009; Frankin, 2010; *Lakefield Standard*, 2009).

Clifford Tousignant, 78 when he died, was eager to celebrate his 80th birthday in a year and a half. His son Marshall lived close enough and visited frequently enough to be satisfied that his Korean War veteran and three-time Purple Heart recipient father was now well adjusted to nursing home life following almost eight decades of independence and initial reluctance to be institutionalized. By December 2009 he had lived in the nursing home just over a year, enjoying the food, including almost daily consumption of his favorite, a peanut butter sandwich. In late December he contracted a severe and extremely uncomfortable case of diarrhea that changed everything. A stool sample revealed the cause to be salmonella poisoning, and he was admitted to a nearby hospital to recover. Shortly after discharge he took a turn for the worse and was plagued by acute pain along with more diarrhea, vomiting and, soon, lethargy and near total unresponsiveness. He was returned to the hospital for further tests, which indicated the salmonella infection had migrated from the gastrointestinal tract to the bloodstream. On the morning of his death he was being prepared for a blood platelet transfusion in an attempt to remedy a drastic drop in his platelet count, which had contributed to an extreme level of unresponsiveness. He died before the procedure could be attempted. He left six children, 15 grandchildren and 14 great-grandchildren (American Institute of Baking, 2009; Frankin, 2010; Richardson, 2009).

Six others would die besides the Brainerd seniors. The nine deaths, along with 750 documented nonlethal illnesses, were all from consumption of peanut-containing foods processed at the Blakely plant and packaged in assorted ways ranging from jumbo tubs of peanut butter to convenience store snack crackers and peanut butter swirl ice cream (Severson, 2009a; US FDA, 2009).

14.2 JUST ANOTHER FOOD POISONING CASE

Early on it was viewed as merely one more food poisoning case posing the usual challenges of tracking down the source and removing the offending items from the nation's food supply. Rather than a case of criminal misbehavior it seemed simply an unlikely string of events. How could the most widespread and deadly outbreak of salmonella poisoning in US history have emerged from a place devoid of the usual suspects for causing salmonella contamination, namely, raw poultry, raw meat and raw eggs (Pritzker, 2011)?

Definitive or, at least, official answers to how the peanut products became tainted and how the shipments of poisoned peanut butter went so far and for so long before being recalled will presumably accompany completion of a federal investigation that was launched by the Food and Drug Administration (FDA) and Justice Department, including FBI work, and was still under way 3 years later with no criminal indictments issued nor even an estimate of when, if ever any would be (Gentry, 2011; Pritzker, 2011; Dumond, 2011; Sustainable Business Forum, 2011; Parnell, 2012).

We do know some of the answers already, though.

From Congressional and FDA investigative reports and from internal company documents now in the public domain it is evident that production line environments at all three Peanut Corporation of America plants were far below commonsense standards of basic cleanliness (Larkin and Woellert, 2009). Former workers and investigators have described filthy surfaces, standing water, and openings in walls, roofs and doorways large enough for vermin and birds to enter and for bird feathers, insects and other outside contaminants to get into the building, of heating and ventilation systems that actually sucked outside impurities into the atmosphere where processing was taking place, and of the presence of insects and disease carrying pests that might be randomly sighted anywhere in the plant almost any time of any day (Alonzo-Zaldivar and Blackledge, 2009; Aslum, 2009; Brumback, 2009; Gentry, 2009a; Harris, 2009b; Hill, 2009; Subcommittee on Health, 2009; Subcommittee on Oversight and Investigations, 2009).

We also know of officially journaled food safety citations as early as 1990 against an earlier version of PCA, which had been founded by Parnell's father in 1957, and against the later incarnation dating to 2001 (Alonzo-Zaldivar and Blackledge, 2009).

Parnell's lawyer, Bill Gust, now maintains that the PCA embroiled in the 2009 recall was a different entity from the one the Parnells launched in the 1950s, that the new one was not even headquartered in Lynchburg, and that Parnell's role in it was merely as a part owner and adviser (Gust, 2012). Memoranda leading up to the recall and the company's demise suggest, however, that if Parnell was not running the entire operation, he was doing a very effective job of pretending to do so, sending communications to employees at all its facilities with directives and statements of company policy (Asworth and Lattal, 2009; Committee on Agriculture, Nutrition and Forestry, 2010; Dumond, 2011; Gentry, 2009a; Glanton 2009; Gray, 2009, *Chicago Tribune*, 2009; Waters, 2009).

Even without concluding whether or not the company engaged in clear-cut unethical or illegal behavior such as shipping food products known to have tested positive for life-threatening pathogens, there are numerous horror-filled accounts that cannot be ignored from former employees. They describe appalling grime, lack of cleanliness and vermin infestation in food processing spaces that one would have assumed to be as free of germs and pests as a hospital operating room (Alonzo-Zaldivar and Blackledge, 2009; Atlanta Journal-Constitution, 2009; Cole, 2009; Falcone, 2009; Gentry, 2009b; Glanton, 2009; Gray *et al.*, 2009; Moss *et al.*, 2009; Raban, 2009; *Chicago Tribune*, 2009).

David James, a shipping department worker at the Blakely PCA plant, told the Chicago Tribune of finding baby mice nested in a pile of peanuts awaiting processing and packaging. He called the plant's atmosphere "filthy and nasty all around the place" (Glanton, 2009).

James Griffin, a cook at the plant, said he never ate any of the peanut butter produced there and would not let his children eat it (Glanton, 2009).

Blakely plant janitor Terry Jones recalled peanut oil left where it spilled and being absorbed into the floor. He told the Tribune that the roof leaked whenever it rained (Glanton, 2009).

James, Griffin, Jones and other former workers at the plant asserted that sanitation and food safety problems were mostly ignored by the company and tended to only when an inspection loomed. They recounted daily sightings of roaches and rats and described having to skirt around puddles of water that had been left to accumulate after rain or after floor moppings. Rat traps he set would catch three to four rodents a day, Jones said (Glanton, 2009).

James alleged that he saw company workers replace outdated expiration stickers on buckets of peanut paste with new ones. He said some bags of nuts had holes in them obviously created by rodents. When mold was discovered in peanut batches he would be told to sort through and retrieve the ones he thought were free of mold and pack them into new bags (Glanton, 2009).

14.3 FDA SANCTIONS

The law firm that sued for damages on behalf of the victims cited as indications of the company's culpability these sanitation violations documented at one time or another by the FDA:

1. Shipping product after it tested positive for at least two *Salmonella* subtypes.
2. Failure to clean and sanitize the peanut paste production line after *Salmonella* was isolated from the peanut paste.

3. Failure to confirm the effectiveness of the "kill step" heating process designed to kill bacteria during roasting of the peanuts.
4. Failure to store finished product safely by segregating it from raw, unroasted peanuts.
5. Failure to keep the roof from leaking.
6. Failure to use production line equipment that was capable of being cleaned properly.
7. Failure to direct air flow away from instead of into the production area from outside, possibly contaminated sources.
8. Failure to provide designated hand washing sinks for employees.
9. Failure to clean utensils and food production equipment thoroughly and properly.
10. Failure to prevent insect and pest contamination (Pritzker, 2009).

Fred Pritzker, a lawyer with the firm that sued on behalf of many of the victims said there was every indication that many or all the violations had been long-standing, "for months if not years." The remains of the company's liquidated assets after bankruptcy and proceeds of insurance policies were assembled into a $12 million settlement for the victims (Pritzker, 2009).

Every year more than a million food poisoning cases are traced to *Salmonella* (Layton, 2011). The earliest signs of the epidemic that would eventually lead investigators to pinpoint the Blakely plant of PCA emerged in late summer 2008, according to Dr. Stephen Sundlof of the FDA (Sundlof and Rogers, 2009). Local doctors reported cases as they arose to local health departments, which in turn reported them to the Centers for Disease Control and Prevention (CDC). CDC recognized in November 2008 that multiple cases were caused by the same salmonella strain, typhimurium (Sundlof and Rogers, 2009). Thirty cases were reported from almost as many states, says the CDC's Dr. Robert Trauxe. CDC interviewed the patients, thinking at first that the culprit might be chicken but soon determining from what the interviewees said about what they'd eaten that this was unlikely. They also talked to people in the same locales and households who had not gotten sick to find out what differences there were in their diets (Sundlof and Rogers, 2009).

With the discovery of the cluster of cases, including fatalities, in the Brainerd nursing homes, all of which involved eating King Nut peanut butter, followed by the positive test results of samples from those containers for *Salmonella typhimurium* on January 9, the FDA initiated the recall of products shipped from the Blakely plant. The authorities soon concluded that the firm had identified *Salmonella* and not taken appropriate steps to clean and maintain equipment safely, according to Michael Rogers of the FDA (Sundlof and Rogers, 2009). At the same time, an investigation in Connecticut found a closed container of peanut butter from PCA that was contaminated by the same *Salmonella typhimurium* (Sundlof and Rogers, 2009).

PCA made peanut paste, mostly used for cookies, crackers, candy and ice cream, as well as bulk peanut butter. It sold to 361 customers who in turn made 3918 products containing peanut butter or paste. All were recalled in what the FDA listed as the largest recall in history. The FDA was careful to announce to the public that the company did not make products used by the big peanut butter makers, and that name brand jars of peanut butter on supermarket shelves were not part of the recall (Sundlof and Rogers, 2009).

The ailment caused by salmonella poisoning, salmonellosis, was not only showing up in people. PCA's output was used by some pet food manufacturers, and dogs contracted salmonellosis, the first one being confirmed by Oregon officials on February 7 (US FDA, 2009).

14.4 *SALMONELLA* CAN BE COMMON

Like any item of produce grown outdoors and exposed to the droppings of birds and animals, peanuts can be contaminated by *Salmonella* in the field and arrive in a dangerous condition at a processing facility. While there is always a possibility that produce that is sold raw – lettuce, broccoli, various sprouts – can enter the food supply as health risks to unsuspecting consumers, that is generally less a concern when it comes to peanuts because most peanuts are roasted before being sold – whether as peanuts or after being processed (Shute, 2009). Internal and government-operated inspections at packaging plants are supposed to catch contaminated produce before shipping. With the extra measure of safety for peanuts, potentially deadly bacteria will normally be killed in the roasting stage, and one of the key safety measures is monitoring of roasting oven temperatures. Temperatures of 75° C (167° F) for 10 minutes or 55° C (131° F) for 90 minutes or other appropriate combinations of temperature and time in between are necessary to kill the organism. These levels are routinely achieved in roasting. The FDA asserted that records at the Georgia plant to verify roasting temperature compliance were incomplete, which raises the possibility that the fatal contamination occurred before the peanuts were further processed, or in other words, because of inadequate roasting temperatures. That is unlikely, however. According to the company's records, inadequate though they may be, temperatures were consistently monitored and kill levels achieved at all PCA facilities where raw peanuts were processed. It appears most likely that it was during post-roasting preparations that contamination occurred (American Peanut Council, 2009; Borrell, 2009).

Salmonella bacteria survive for long periods outside their hosts' bodies. Even tiny particles of excrement containing *Salmonella* can allow the organism to thrive if provided hospitable surroundings (Borrell, 2009; Gentry, 2009c). What is welcoming for *Salmonella*, therefore, should be totally absent in a food processing plant.

The ideal conditions for processing peanuts include a bone-dry environment, scrupulously maintained cleanliness, the absence of vermin, other animals and insects, and good personal sanitation among workers. A very important element is dry surfaces – floors, processing containers, counters, any place where the peanuts can come into contact with a preparation surface whether by design or by accident. Standing water, even what is left from mopping the floor to keep it clean, if not wiped dry, can become nearly perfect, extraordinarily friendly breeding grounds for bacteria, including *Salmonella*. The building used for peanut processing should be air-tight, with no gaps in door jambs, windows, walls or roofs that could let in bird feathers, insects, and airborne bacteria, all of which are also very hazardous for peanut processing (CDC, 2009, 2010).

The conditions in the three PCA plants were polar opposite of the ideal environment for safely turning peanuts into peanut butter.

The Plainfield, Texas, plant had been operating for 4 years below the radar of state and federal food safety regulation. No one had applied for a license to operate as a processer of peanuts when the company opened the plant in 2005. Food safety inspectors routinely drove past the well-identified plant on their way to other inspection destinations, but because it did not appear on their roster of operating food processors, they never conducted inspections there. Its output consisted of oil-roasted and dry-roasted peanuts, peanut meal and granulated peanuts. Authorities realized it was processing peanuts as a part of the PCA network only after the national salmonella outbreak was traced to PCA's Georgia plant. Once Texas health department officials realized they had a PCA plant in their jurisdiction, they conducted an inspection and found what might be described as food safety protectors' worst nightmare: dead rodents, rodent droppings and bird feathers in crawl spaces above production areas, and a ventilating system that was pulling these and other contaminants directly into the air surrounding the production line. Company officials had already halted production after an Illinois private laboratory hired by the company reported that granulated peanuts and peanut meal from Plainfield contained *Salmonella*. On February 11, 2009, 3 days after cessation of operations, the state ordered a recall of all products ever shipped from the plant (Harris, 2009c).

The company was also quietly running a third plant in the Tidewater area of Virginia, where the main operation was blanching peanuts to remove their red skins prior to shipping them to the Georgia plant for roasting and further processing. The FDA inspected the Suffolk, Virginia, plant after the salmonella outbreak and found no violation, but the plant had been cited in 2008 for sanitation violations including mouse droppings in a warehouse, a live bird inside the processing plant and mold on 43 1-ton containers of peanuts and on the peanuts themselves. In other inspections conducted by the Virginia Department of Agriculture and Consumer Services, reports noted gaps in doorways that could allow pests inside and flaking paint. It was also noted that the peanuts were handled "like a commodity and not a food," according to Michael Doyle, director of the University of Georgia's Center for Food Safety. Following the outbreak, the US Department of Agriculture suspended the company and the Suffolk plant from doing business with the federal government (Lindsey, 2009).

14.5   THE BLAKELY PLANT

The center ring of this circus was the Blakely plant, which was the source of the nationwide *Salmonella* outbreak. It was cited multiple times in 2006 and 2007 for unsanitary surfaces, grease and dirt accumulation, and in 2008 for failure to meet cleanliness standards expected of a food production facility (Gray *et al.*, 2009). The Georgia state Agriculture Department, which had been contracted by the FDA to conduct food safety inspections, listed rusty surfaces that could drop particles into food, gaps in doors wide enough for rodents to enter, unlabeled spray bottles and other problems (Gray *et al.*, 2009). A 2007 report described dirty preparation surfaces and empty, ready-to-be-filled peanut butter containers being stored uncovered. Two reports in 2008 documented sanitation violations (Gray *et al.*, 2009). Another listed the presence of roaches, mold and dirty surfaces (Gray *et al.*, 2009).

Poor sanitation practices dated back to when the plant was acquired by PCA. Nine months after the acquisition in 2001, an FDA inspection found possible insecticide contamination and dead insects near peanuts (Alonzo-Zaldivar and Blackledge, 2009).

A low point in the Georgia plant's food safety compliance history was the assertion by the FDA that it had discovered a practice by the company of shopping around for favorable *Salmonella* tests after receiving positive test results for the pathogen. But that was not the worst of it, according to the agency. It amended its report to allege that the Georgia plant had shipped chopped peanuts twice, on July 18 and July 24, 2007, after salmonella was confirmed by private lab tests and before receiving contrary information, and that it had at other times shipped products following separate confirmation of *Salmonella* (Atlanta Journal-Constitution, 2009; Harris, 2009a, b); Hauter, 2009; Hill, 2009. Jalonick, 2009; Maugh and Engel, 2009; Powell, 2010; Scott-Thomas, 2009).

A buyer for a snack company, David Brooks, said it was common knowledge among people in the peanut business in the states where PCA operated that the company's sanitation practices were substandard. He called PCA "a time bomb waiting to go off," according to the *Washington Post* (Layton and Miroff, 2009).

Brooks was not relying on hearsay. The Plainview plant was the second location for that operation. During the 1980s, before PCA bought the facility and moved it to Plainview, he'd inspected the operation in Gorman, Texas, three separate times in trying to reach a decision whether to buy its peanuts and peanut butter. He said he found unsanitary conditions, leaky roofs, and birds in the buildings. He gave the plant failing grades each time. The change of ownership, he said, had resulted in no improvement (Layton and Miroff, 2009).

After the director of the Georgia Bureau of Investigation concluded that convictions of violations of applicable state laws would result in only minor penalties and requested federal intervention, the Federal Bureau of Investigation announced its participation in a federal criminal investigation (*USA Today*, 2009).

Parnell wrote a number of memos that suggested he was more anxious to ship peanuts no matter what their condition than to assure public protection from harmful agents in PCA products. Pritzker states that there were more than 12 known tests in 2007 and 2008 showing salmonella contamination in products (Pritzker, 2011). Yet Pritzker also states that Parnell wrote to company employees on January 12, 2009 that "we have never found any salmonella at all … anywhere in our products or in our plants" (Pritzker, 2011).

In one memo after learning *Salmonella* had been found, he wrote, "I go thru this about once a week. I will hold my breath … again." He stated in another that the positive tests were "costing us huge $$$$$" (Pritzker, 2011).

After discovery of the national outbreak, he pleaded in writing with federal regulators to permit him to get back to shipping product from Blakely because he and the company "desperately at least need to turn the raw peanuts on our floor into money" (Pritzker, 2011).

The *Chicago Tribune* quoted Caroline Smith DeWaal, director of food safety for the Center for Science in the Public Interest, as saying the Georgia PCA plant's practice was to keep sending samples that showed the presence of *Salmonella* to other labs until a negative result came back so the suspect product could be sent to customers for packaging and sale to the public (Glanton, 2009).

This plant was running tests for their own information but ignoring all the positive test results," said DeWaal. "They ignored anything they did not like" (Glanton, 2009).

Parnell and the company maintained that the repeat tests were merely attempts to find the most competent testers, not the most lenient (Martin, 2009).

PCA liked to cite the findings of outside private firms that would perform "third-party audits" as testimony of its safe, clean operation and of the integrity of the food safety testing it employed (Martin, 2009).

One such firm that was brought in was the American Institute of Baking, which awarded PCA a "superior" rating as a food processor, a sort of *Good Housekeeping* seal of approval for food safety. But knowledgeable retail officials were appalled that the baking institute's blessing would be taken seriously for a peanut butter factory (American Institute of Baking, 2009; Powell, 2010).

"The American Institute of Baking is bakery experts," said Costco's senior food safety official, R. Craig Wilson. "You stick them in a peanut butter plant or in a beef plant, they are stuffed" (Powell, 2010).

Costco was one of several national brand name retailers, food manufacturers and restaurant companies that rejected the third party auditors' favorable findings at PCA and conducted their own inspections for food safety, after which they went elsewhere to purchase peanut butter and peanut paste (Powell, 2010).

Seattle food safety consultant Mansour Samadpour dismissed the validity of these supposedly neutral examiners, saying: "The contributions of third-party audits to food safety is the same as the contributions of mail order diploma mills to education" (Powell, 2010).

Douglas Powell, professor of diagnostic medicine and pathobiology at Kansas State University, and an advocate for food safety systems reform, said in his online blog that audits alone do not enhance food safety. Shortly after the PCA fiasco he called third-party food safety audits a "Ponzi scheme that is ... starting to collapse" (Powell, 2010).

Powell noted that the baking group auditors were not aware at the time they were called in by PCA that peanuts are susceptible to salmonella poisoning and did not even test for it before awarding the "superior" rating. The audit was being conducted for Kellogg, which used PCA peanut paste for snack products. PCA paid for that audit, and Kellogg's use of it was later cited in legal filings as a reason it should be held liable for using PCA products in making its own branded foods, some of which led to consumer illnesses (Powell, 2010).

After the outbreak, the auditor who had done the inspection wrote, "I never thought that this (salmonella) bacteria would survive in the peanut butter type environment. What the heck is going on?" (Powell, 2010).

The company's proclivity for ignoring or suppressing health and safety risks in its shipped products extended beyond the salmonella question. There was, according to accounts of incidents documented in the investigation, a pattern of salvaging damaged goods and turning them into sources of income whenever there was a way to do it out of sight of the food safety regulators. On April 11, 2008, for example, a Canadian distributer refused a shipment of PCA chopped peanuts because metal fragments were discovered among the peanuts (Theimer and Alonzo-Zaldivar, 2009). Five months later, the FDA declined to allow PCA to re-import the metal-adulterated peanuts back across the border because besides the metal pieces, the peanuts also had mixed in with them a "filthy, putrid or decomposed substance" (Associated Press, 2009). It is not clear what the company had in mind. It would seem that even converting the chopped peanuts into animal feed would be cruel and dangerous. If the intent was to turn them into fertilizer, they would be some expensive soil nutrients given the expense of transcontinental freight.

14.6 TRACING A FOOD POISONING OUTBREAK

Tracing the source of a food-borne illness is challenging under the best circumstances to say nothing of when consumer protection laws are weak or there's a permissive environment for business operation and hostility to government regulation, as was the case in the period when PCA was shipping batches of poisonous peanut products, some of which had tested positive for the bacteria that would ultimately sicken hundreds and kill nine.

Because the PCA case became so deadly and was so widespread, it drew enormous public attention. Swift resolution of the mystery elicited a huge collective sigh of relief. It is useful to re-examine how and why the origin of the contamination was discovered as quickly as it was. Health authorities across the country were perplexed by the sudden surge of *Salmonella typhimurium* infections. The search was on for a pattern to connect the cases. A Minnesota team of epidemiologists and health officials narrowed it down to a shipment of King Nut peanut butter that was produced at Blakely (Anderson, 2010; Greenhalgh, 2010).

According to Food Safety News's Ross Anderson, Minnesota public health detectives achieve consistently superior performance because of outstanding individuals, sufficient funding, availability of forensic resources and a centralized state system. The man in charge of the super sleuths is Health Department Food Borne Illness Unit Supervisor Dr. Kirk Smith. Smith notes that the health and agriculture departments work extraordinarily well together and that the health department has a close working relationship with nearby University of Minnesota, where he and many other staff professionals serve as adjunct faculty members. When, for example, a stool sample arrives from the other end of the state, it can be dispatched to the adjacent campus. Within hours instead of the weeks it takes elsewhere, a diagnosis, DNA analysis and other results can be produced and comparison theories can be posited and built upon in real time (Anderson, 2010).

Smith cites as perhaps most responsible for cracking the PCA case the employment of a team of University of Minnesota students who are routinely pulled together on short notice to focus on finding the source of a new outbreak of illness. The student detectives are affectionately known as "Team Diarrhea." They get on the phones and conduct thorough interviews with patients who have exhibited symptoms. Well before a particular foodborne culprit is suspected, before even the point where the case is confirmed to be food-related, the interviewers concentrate on building a picture in as much detail as possible about the individuals' recent activities. This always includes where they dined and what they ate in the previous several weeks (Anderson, 2010).

The student teams "give us the people power we need to do rapid investigations," Smith says (Anderson, 2010).

He also credits a Minnesota law that requires direct reporting to his unit whenever doctors and hospitals diagnose a food borne illness. Elsewhere, when such reports go to local health departments for initial investigation, there is a tendency to search for a local source of contamination, which can mean failure to consider broader commercial causes or at the very least, long delays before centralized analysis can be undertaken (Anderson, 2010).

Salmonella is a genus of bacteria, many species of which are found in animals and humans. Although it thrives in a living body, it can survive for weeks outside a host, especially in conditions such as moist surfaces. Humans generally acquire the most harmful species of *Salmonella* through ingestion of contaminated foods. These are most commonly insufficiently cooked meat products from animals that have been infected and from eggs. But various species have also been transmitted through consumption of vegetables that become contaminated in the field through, for example, contact with feces of infected insects, birds and animals. When a processing plant's sanitation is deficient, the bacterial contamination can be transferred to a product that arrived from the field or orchard clean (American Peanut Council, 2009).

The very young and very old and those with compromised immune systems are most vulnerable to serious consequences of salmonella poisoning (CDC, 2010).

Acid in the digestive system will kill much of the salmonellae that enter the body of a healthy person eating, say, a piece of cooked chicken that was inadvertently mixed on the kitchen counter with raw chicken liquid that was not cleaned up immediately or spreading tainted peanut butter on a piece of toast. The bacteria that survive enter the bloodstream and can produce virulent, even deadly, poisons, causing vomiting and diarrhea, severe loss of body fluids and electrolytes, which can lead to septic shock and toxemia and, ultimately, as in the case of nine people who ate the contaminated PCA products, death (CDC, 2010).

14.6.1 Peanuts are nutritious and popular

Peanuts are consumed as shelled and in-the-shell whole foods, peanut butter (either in the jar or as ingredients in other foods such as peanut butter crackers, cookies, candy and ice cream), oil and less evident ingredients in other foods. The rich dietary value of the peanut makes it a favorite of professional nutritionists as well as parents who want to ensure good lunchtime eating habits while their children are out of sight. Thus millions of peanut butter and jelly sandwiches appear in school lunch bags and on school cafeteria menus. The legume contains protein, niacin, vitamin E, calcium, copper potassium, iron, magnesium and, in the peanut skin, resveratrol, an ingredient also found in red wine that some research studies have linked to anti-aging results. That and other nutritional claims for resveratrol are largely unproven and still controversial, but it has passionate adherents (Mayo Clinic Staff, 2012). The fat in peanuts is the kind now considered beneficial for good health and, according to studies by a University of Toronto team, can help people with diabetes control blood sugar and people at risk of heart disease control harmful cholesterol (Jenkins *et al.*, 2007).

Cuisines featuring peanut ingredients range from Thai and Indonesian in the Far East to Peruvian and other South American dishes dating to the discovery of the plant in the Western Hemisphere (Smith, 2002). Americans consume about $900 million worth of peanut butter a year (Severson, 2009b).

It has become a key piece in a worldwide hunger alleviation strategy. In Africa, peanut butter is the main ingredient of a product created in France known as Plumpy'nut, that has been instrumental in fighting hunger on the continent. Plumpy'nut also contains powdered milk, sugar and vitamin-enriched vegetable oil. Provided in foil pouches, it has an extra-long shelf life without refrigeration. Fed to starving children twice daily, it has resulted in healthful weight gain and, when fed on a disciplined, monitored regimen, remarkable reduction in malnutrition (Severson, 2009b).

The popularity of peanuts, peanut butter and peanut butter snacks helps explain why the PCA recall reached such record proportions.

That episode would not be the first widespread salmonella case involving peanuts, nor would it be the last, but it was easily the most spectacular and far-reaching as well as the one that most directly raised questions of ethics in the conduct of food processing officials.

14.7 OTHER INCIDENTS

An outbreak of salmonella-caused illness in the summer of 2006 that eventually spread to 47 states and sickened at least 625 people was traced to a ConAgra peanut butter plant in Sylvester, Georgia. It triggered recall of certain jars of Peter Pan and Great Value (the Wal-Mart store brand) peanut butter that had been produced as early as 2004. It was thought to be the first peanut butter related salmonella outbreak in US history. New York, Pennsylvania, Virginia, Tennessee and Missouri had the largest number of reported infections. Of all those sickened, about one in five required hospitalization. None died. ConAgra later attributed the origin of the problem to a leaky roof and malfunctioning sprinklers that allowed water accumulation in the Sylvester plant, which in turn became a breeding ground for the bacteria to grow and taint its products. Vigorous cleaning of the plant following discovery of the contamination failed to eradicate the *Salmonella* entirely, and some peanut butter was packaged that had come into contact with the poisonous bacteria. Tainted containers continued to leave the plant until a wave of salmonella infections was traced to ConAgra by Centers for Disease Control and Prevention testing. ConAgra voluntarily recalled all Peter Pan peanut butter in February 2007 and did not reintroduce the brand until the following August. Wal-Mart stopped shipment of its store brand coming from Sylvester but continued producing and shipping peanut butter from plants elsewhere that also supplied the Great Value brand. ConAgra, based in Nebraska, later estimated that its Peter Pan and Great Value recall cost $50 to $60 million. By August 2007, when it had thoroughly fixed the problems that allowed the contamination, ConAgra issued a "100 percent guarantee of its brands," and Peter Pan peanut butter once again appeared on grocery shelves (Funk, 2007).

The next nationwide outbreak, which would eventually develop into the infamously deadly case of salmonella poisoning and triggered a total national recall of items made in

Blakely began in 2008. It was traced to PCA in early 2009, and the strain of *Salmonella* was identified as *typhimurium*.

As though heretofore rare health emergencies involving peanut butter were settling into two-year cycles of fright and recovery, Skippy, one of the pioneering peanut butter brands, in March 2011 voluntarily recalled jars of a single size (16.3 oz.) manufactured in six production runs of its smooth and chunky reduced fat peanut butter because of possible salmonella contamination. Unilever, Skippy brand's corporate owner, said the recall of shipments to 16 Eastern and Midwestern states was prompted by discovery of contamination during routine sample testing (US FDA, 2011).

The name-brand peanut butter recalls of 2007 and 2011 differed fundamentally from the 2009 PCA recall. In the earlier and later cases, discovery that some shipped food was tainted impelled the manufacturers immediately to issue a call for the offending products to be removed from retail shelves and for customers who had bought from the affected lots to throw their peanut butter jars away and claim reimbursement (Funk, 2007; Miller, 2011). The 2009 PCA case involved accusations of searching for backup test labs to contradict negative results, shipping bad food after tests revealed contamination, production in dirty, disease-prone plants, failure to follow prescribed food safety measures, and, in one case, an attempt to reimport to the United States a shipment of peanut pieces riddled with metal fragments and decomposed biological matter that had been rejected by Canadian authorities. The 2009 recall ranks first among the top ten food recalls in US history, according to "Business Insurance Quotes," a service financed by major business liability insurers (Business Insurance Quotes, 2011). In addition to being the top recall in volume, the PCA case blows away, at least to date, any competition for outrageous behavior. The case epitomized food safety consultant Robert A. LaBudde's characterization of some food processors' sense of priorities:

> The only thing that matters is productivity. You only get in trouble if someone in the media traces it back to you, and that's rare, like a meteor strike (Moss and Martin, 2009).

In the meteor strike that was the PCA case, recalled products ranged from convenience packages of what a prudent grocery shopper might consider an ideal healthful snack like Trader Joe's brand "celery with peanut butter" and "sliced green apples with all natural peanut butter" to national brands and store brands of lunch bag treats like "Keebler cheese and peanut butter sandwich crackers." There were energy bars, candy, ice cream, and many varieties of pet foods. The FDA's complete list of 3918 items ran to 227 pages, each typically packed with 20 to 30 product descriptions including brand name, lot number, sell-by date and universal product code so that retailers could remove them from their shelves (US FDA, 2009).

14.8 PCA CEO MAINTAINS INNOCENCE

Stewart Parnell has been a peanut man all his life, continuing to advise any company that will hire him. He rejects the notion that he knowingly exposed people to tainted food, asserting in a brief interview that the excerpts from e-mails that have been used as evidence of wrongdoing were wrenched out of context. In March 2012 he had still not been charged with a crime nor apprised of any criminal investigation but said he would like the investigation to be concluded, whatever its outcome, so that he can get on with his life (Gentry, 2009d, 2011; Parnell 2012).

> They (Federal authorities) have called everybody and gone over everything. As of a year ago the suits had all been concluded, people paid, waivers signed. It has been that long, and still there is no sign whether or not they are planning to bring any action against me (Parnell, 2012).

As this book was being finalized, the US federal government announced indictments on fraud and conspiracy charges of Parnell and three others connected to the Peanut Corporation of America salmonella poisoning case (Tavernese, 2013).

Before he drew laser-like attention as the heavy at the center of the peanut butter recall firestorm, Parnell had been a prominent peanut industry figure and was among the business elite of Lynchburg. He was appointed by the George W. Bush Administration to advise the US Department of Agriculture as a member of its Peanut Standards Board, which determines standards for quality and proper handling of peanuts (Blacklisted News, 2009). He was removed from that position by Obama Agriculture Secretary Tom Vilsack when his company's alleged misdeeds were exposed (Akre, 2009). Beyond urging that anyone who is examining the case "read the entire memos," he will not discuss for the record the actions of PCA leading up to the epidemic of illnesses and deaths (Parnell, 2012). He has been quoted in numerous news accounts denying that any product was ever shipped after being found to be poisonous (Powell, 2010). He also made that claim writing to company employees during the crisis (Glanton, 2009).

Parnell comes across on the telephone as a man long since overwhelmed and to some extent bewildered by the sequence of events that left his business in ruins and his reputation shattered. He says media mischaracterizations of his behavior while head of PCA have taken away his good name (Parnell, 2012). He would like to have a chance to repair the damage and restore some of the esteem that once belonged to his family. He is frustrated with delay of prosecution of the case and wishes the FDA and Justice Department would move ahead, either to bring charges or to clear him and his colleagues (Gentry, 2011; Dumond, 2011).

His lawyer, Bill Gust, says media coverage of the case has been fraught with inaccuracies from details of the corporate structure of PCA to the serious allegations that the company behaved in the most unethical and irresponsible manner possible in the years and months before the outbreak. He insists that the company was not, as portrayed in much of the media coverage, a Virginia-based company with plants in Georgia and Texas but rather a firm with no direct ties to the original PCA that Parnell and his father and brother founded in 1957 in Virginia. That firm was sold to another processor in Virginia and the name abandoned. Subsequently Parnell went to work for a peanut processing firm in Blakely and resurrected the discarded name for it (Gust, 2012).

In February 2009, having been in office just 12 days, President Barack Obama confronted the spreading consumer-protection nightmare that was already building to what would become the largest food recall in the nation's history (Reinberg, 2009).

In a Today Show interview aired on February 2, 2009, President Obama said the Food and Drug Administration's oversight of food safety "has not been able to catch some of these things as quickly as I expected them to," and described how close to home the tainted peanut problem came, recounting his fears for daughter Sasha, who he said consumed peanut butter sandwiches "probably three times a week."

"I don't want to have to worry about whether she's going to get sick as a consequence of having her lunch," he said, adding that he would be taking steps to be sure the public could better rely in the future on protection of the food supply (Layton, 2009).

Although a toughened food protection law backed by the White House would quickly be introduced and start moving through Congress, it would be nearly two years before the principal vehicle for food safety reform reached his desk. Obama signed the Food Safety Modernization Act on January 4, 2011. Its main goal was to put the FDA in a better position to prevent contamination of food rather than continue its traditional focus on responding to contamination cases (US FDA, 2012).

In January 2013 the FDA issued for public comment two new food safety regulations designed to ensure that food producers will better protect the public from food-borne illnesses (US FDA, 2013).

Leading up to that reform were contentious Congressional hearings and extensive media attention to the ineffective, deficient and loophole-ridden system of guarding against contamination in some agricultural processing facilities. Much of the focus was on the PCA recall (Committee on Agriculture, Nutrition and Forestry, 2010; Subcommittee on Health, 2011; Subcommittee on Oversight and Investigations, 2009).

Even that disposition, though, may pale next to the promise of more reliable food safety protections written into law by the Food Safety Modernization Act, whose genesis, as much as anything else, was the peanut butter *Salmonella* outbreak of 2009.

REFERENCES

Akre, Jane, 2009, "Parnell Off Federal Peanut Butter Board," Injury Board National News Desk, February 6, 2009.

Alonzo-Zaldivar, Ricardo and Brett J. Blackledge, 2009, "Insecticide Found in Peanut Butter Plant During 2001 Tour," Associated Press story published in *The News and Advance*, Lynchburg, VA, February 6, 2009.

American Institute of Baking, 2009, Chronology of Events Related to Peanut Products Recall Involving PCA, http://www.aibonline.org, April 2, 2009.

American Peanut Council, 2009, Statement on APC Voluntary Code of Good Manufacturing Practices including specific procedures for peanut processors, December 10, 2009.

Anderson, Ross, 2010, "How Minnesota Cracked the Peanut Butter Case," *Food Safety News* Blog, http://www.foodsafetynews.com, December 13, 2010 (accessed 18 December 2012).

Ashworth, Cindy and Darnell Lattal, 2009, "Nuts! When Leadership Abandons Ethics in the Name of Profit," PMeZine (performance management online magazine), http://aubreydaniels.com/pmezine/leadership-abandons-ethics-name-of-profit (accessed 8 January 2013).

Aslum, Andrenne, 2009, "Eat, Drink and Be Safe," The Cavalier Daily, University of Virginia, Charlottesville, VA, http://www.cavalierdaily.com (accessed 18 February 2009).

Associated Press, 2009, quoted in AboutLawsuits.com, February 2, 2009.

Atlanta Journal-Constitution, 2009, "Food Risks Draw Little Urgency," http://www.ajc.com/feeds/content/metro/stories/2009/02/08/peanut_0208.html?cxtype=rss&cxsvc=7&cxcat=13.

Blacklisted News, 2009, "Troubled Peanut Firm's Chief Also an Industry Quality Adviser," reprinted from the *Atlanta Journal-Constitution*, January 31, 2009.

Borrell, Brendan, 2009, "How Does Salmonella Get Into Peanut Butter? And Can You Kill It Once It's There?" *Scientific American*, http://www.scientificamerican.com/article.cfm?id=salmonella-poisoning-peanut-butter, January 13, 2009 (accessed 18 December 2012).

Brumback, Kate, 2009, "Peanut Company in Salmonella Probe Shuts 2 Plant," Associated Press story published in *The News and Advance*, Lynchburg, VA, February 10, 2009.

Business Insurance Quotes, 2011, "10 Biggest Recalls in US History," http://www.businessinsurance.org/10-biggest-food-recalls-in-u-s-history/ (accessed 8 January 2013).

Centers for Disease Control and Prevention, 2009, Podcast: "Salmonella Serotype Typhimurium Outbreak in Peanut Butter and Peanut Containing Products," http://www2c.cdc.gov/podcasts/player.asp?f=10684, January 21, 2009 (accessed 18 December 2012).

Centers for Disease Control and Prevention, 2010, "What is Salmonellosis?" definition document by CDC's National Center for Emerging and Zoonotic Infectious Diseases, Division of Foodborne, Waterborne and Environmental Diseases, http://www.cdc.gov/salmonella/general, September 27, 2010 (accessed 18 December 2012).

Chicago Tribune Staff Writer, 2009, "FDA: Peanut Butter Maker Knowingly Shipped Salmonella Tainted Product," *Chicago Tribune*, February 6, 2009.

Cole, Rebecca, 2009, "Salmonella Alerts Ignored," *Los Angeles Times*, http://articles.latimes.com/2009/feb/12/nation/na-peanut-salmonella12, February 12, 2009 (accessed 18 December 2012).

Committee on Agriculture, Nutrition, and Forestry, US Senate, 2010, "Examination of Federal Food Safety Oversight in the Wake of Peanut Products Recall," February 5, 2009, Hearing Transcript, S. Hrg. 111-231, Government Printing Office.

Dumond, Chris, 2011, "Former Peanut Corp. Head Fights to Restrict Release of Records," *The News & Advance*, Lynchburg, VA, September 28, 2011.

Erickson, Matt, 2009, "All Three State Salmonella Victims Lived in Brainerd," *Brainerd* (MN) *Dispatch*, http://brainerddispatch.com/stories/012809/new_20090128050.shtml (accessed 8 January 2013).

Falcone, Michael, 2009, "Plant Supplies Banned from Federal Business," *New York Times*, http://topics.nytimes.com/topics/reference/timestopics/organizations/p/peanut_corporation_of_america/index.html, February 6, 2009 (accessed 18 December 2012).

Frankin, Sen. Al (D-MN), 2010, US Senate floor speech, September 15, 2010.

Funk, Josh, 2007, "Peanut Butter Recalled Over Salmonella," Associated Press story published in the *Washington Post*, February 17, 2007.

Gentry, Bryan, 2009a, "FDA: Peanut Firm Found Salmonella Multiple Times in Two Years," The News & Advance, Lynchburg, VA, January 27, 2009.

Gentry, Bryan, 2009b, "FDA Cited Peanut Corp. in 1990 for Toxin Problem," *The News & Advance*, Lynchburg, VA, February 4, 2009.

Gentry, Bryan, 2009c, "Peanut Corp's Suffolk Plant Showed Problems in Inspections," *The News & Advance*, Lynchburg, VA, February 12, 2009.

Gentry, Bryan, 2009d, "Friends Praise Peanut Corp. Chief Parnell, *The News & Advance, Lynchburg*, VA, February 19, 2009.

Gentry, Bryan, 2011, "Two Years after Recall, Peanut Exec Looks for Way Back," *The News & Advance*, Lynchburg, VA, January 16, 2011.

Glanton, Dahleen, 2009, "Inside Nutty Nut Processor," *Chicago Tribune*, http://articles.chicagotribune.com/2009-02-04/news/0902030962_1_peanut-butter-peanut-corp-american-peanut-council (accessed 8 January 2013).

Gray, Janet B., Investigator (and investigative team members Darcy E. Brillhart, Sandra I. Gaul, Robert P. Neligan, Lesley K. Satterwhite and Theresa L. Stewart), 2009, "Inspectional Observations Form 483, Department of Health and Human Services, Food and Drug Administration," addressed to Sammy L. Lightsey, Plant Operations Manager, Peanut Corporation of America, Blakely, GA, February 5, 2009.

Greenhalgh, Michelle, 2010, "'Team D' is Model for Outbreak Investigations," Food Safety News Blog, June 3, 2010.

Gust, Bill, 2012, telephone interview with Mark Clark, March 2012.

Harris, Gardner, 2009a, "Peanut Products Sent Out Before Tests," *New York Times*, http://topics.nytimes.com/topics/reference/timestopics/organizations/p/peanut_corporation_of_america/index.html, February 11, 2009 (accessed 18 December 2012).

Harris, Gardner, 2009b, "Plant Recall Leads to Criminal Investigation," *New York Times*, http://topics.nytimes.com/topics/reference/timestopics/organizations/p/peanut_corporation_of_america/index.html, January 31, 2009 (accessed 18 December 2012).

Harris, Gardner, 2009c, "After Tests Peanut Plant in Texas is Closed," *New York Times*, http://topics.nytimes.com/topics/reference/timestopics/organizations/p/peanut_corporation_of_america/index.html, February 11, 2009 (accessed 18 December 2012).

Hauter, Wenonah, executive director, 2009, "FDA Issues Recall Press Release for Peanut Corporation of America, Food and Water Watch," http://www.foodandwaterwatch.org, January 14, 2009 (accessed 18 December 2012).

Hill, Catey, 2009, "Officials Want Criminal Probe of Peanut Corp. for Shipping Salmonella Tainted Peanut Butter," *New York Daily News*, http://www.nydailynews.com/news/money/officials-criminal-probe-peanut-corp-shipping-salmonella-tainted-peanut-butter-article-1.388767 (accessed 8 January 2013).

Jalonick, Mary Claire, 2009, "Possibly Tainted Peanut Butter Sent to Schools," Associated Press story published in the News and Advance, Lynchburg, VA, February 6, 2009.

Jenkins, David A., Frank B. Hu, Linda C. Tapsell, Andrea R. Josse and Cyril W. C. Kendall, 2007, "Possible Benefits of Nuts in Type 2 Diabetes," *Journal of Nutrition, University of Toronto*, February 28, 2007.

Lakefield Standard, 2009, Lakefield, MN, February 5, 2009.

Larkin, Catherine and Lorraine Woellert, "Peanut Corp. Knew of Salmonella Contamination in 2006, Bloomberg News Website, http://www.bloomberg.com/apps/news?pid=newsarchive&sid=al5WYfyNHRx0 (accessed 8 January 2013).

Layton, Lyndsey, 2009, "Obama Faults FDA on Food Safety", Washington Post, http://www.washingtonpost.com/wp-dyn/content/article/2009/02/02/AR2009020202967.html (accessed 8 January 2013).

Layton, Lyndsey and Nick Miroff, 2009, "Peanut Company at Center of Salmonella Scare Files for Bankruptcy Protection," *Washington Post*, February 14, 2009.

Layton, Lyndsey and Nick Miroff, 2009, "The Rise and Fall of a Peanut Empire," *Washington Post*, February 15, 2009.

Layton, Lyndsey, 2011, "Salmonella Cases Rise in United States, Federal Report Shows," *Washington Post*, June 7, 2011.

Lindsey, Sue, 2009, "Peanut Firm Linked to Salmonella Closes Suffolk Plant," Associated Press story published in *The News and Advance*, Lynchburg, VA, February 18, 2009.

Martin, Andrew, 2009, "Peanut Plant Says Audits Declared It in Top Shape," *New York Times*, February 5, 2009.

Martin, Andrew, 2009, "Peanut Corporation of America to Liquidate," *New York Times*, February 14, 2009.

Maugh II, Thomas H. and Mary Engel, 2009 "Peanut Company Lied on Salmonella Testing, FDA Finds," Los Angeles Times, articles.latimes.com/2009/feb/07/science/sci-peanut-fda7 (accessed 8 January 2013).

Mayo Clinic Staff, 2012, "Red Wine and Resveratrol: Good for Your Heart?" MayoClinic.com.

Milbank, Dana, 2009, "Mr. (Tainted) Peanut Butter Pleads the Fifth," *Washington Post*, www.washingtonpost.com, February 12, 2009.

Miller, Mary Helen, 2011, "Skippy Peanut Butter Recall for Possible Salmonella," *The Christian Science Monitor*, March 7, 2011.

Moss, Michael and Andrew Martin, 2009, "Food Problems Elude Private Inspectors," *New York Times*, March 5, 2009.

Moss, Michael and Robbie Brown, Andrew Martin and Margot Williams, 2009, "Peanut Case Shows Holes in Safety New," *New York Times*, February 9, 2009.

Parnell, Stewart, 2012, telephone interview with Mark Clark, March 2012.

Powell, Doug, 2010, "Nosestretcher alert: After 9 dead, 700 sick from PCA Salmonella, CEO Parnell's lawyer says 'I never intentionally harmed anyone' Victim says 'I don't want no peanut butter in my house'," http://barfblog.foodsafety.ksu.edu/barfblog, February 1, 2010 (accessed 18 December 2012).

Pritzker, Fred, 2009, "Death, Injury," *Minnesota Lawyer*, January 29, 2009.

Pritzker, Fred, 2011, *Food Poisoning Law* Blog: Collection of posts, dated from January 27, 2009 to February 13, 2011, http://foodpoisoning.pritzkerlaw.com/archives/cat-peanut-butter-lawsuit.html (accessed 18 December 2012).

Raban, Roni Caryn, 2009 "Peanut Plant Had History of Health Lapses," *New York Times*, January 27, 2009.

Reed, Ray, 2009, "Despite Hearing, Questions about Peanut Corp. Remain," *The News & Advance*, Lynchburg, VA, March 12, 2009.

Reinberg, Steven, 2009, "Obama Orders Review of FDA in Salmonella Outbreak," U.S. News and World Report, February 2, 2009.

Richardson, Renee, 2009, "Salmonella Outbreak Hits Brainerd," *Brainerd (MN) Dispatch*, January 14, 2009.

Rubenstein, Sarah, 2009, "Mold and Roaches Found in Peanut Butter Plant Linked to Salmonella," *Wall Street Journal* Health Blog, January 29, 2009.

Scott-Thomas, Caroline, 2009, "FDA Joins Criminal Investigation into Peanut Corporation Salmonella," *Food Navigator* Blog, http://www.foodnavigator-usa.com, February 2, 2009 (accessed 18 December 2012).

Severson, Kim, 2009, "List of Tainted Peanut Butter Items Points to Complexity of Food Production," *New York Times*, January 23, 2009.

Severson, Kim, 2009a, "Who's Sticking with Us?" *New York Times*, February 3, 2009.

Shaffer, David, 2009b, "A Heartbreaking Lawsuit: Surviving Cancer, Done In By Salmonella," *Minneapolis Star-Tribune*, January 26, 2009.

Shute, Nancy, 2009, "Peanut Company Knew Its Plant Was Contaminated with Salmonella," *US News and World Report*, January 28, 2009.

Smith, Andrew F., *Peanuts: The Illustrious History of the Goober Pea*, University of Illinois Press, Champaign, Illinois, 2002.

Subcommittee on Health, Committee on Energy and Commerce, US House of Representatives, 2011, "How Do We Fix Our Ailing Food Safety System?" March 11, 2009, Hearing Transcript, Serial No. 111-12, Government Printing Office.

Subcommittee on Oversight and Investigations, Committee on Energy and Commerce, US House of Representatives, 2009, "The Salmonella Outbreak: The Continued Failure to Protect the Food Supply," February 11, 2009, Hearing Transcript, Serial No. 111-2, Government Printing Office.

Sundlof, Stephen and Michael Rogers, US Food and Drug Administration, and Robert Tauxe, Centers for Disease Control and Prevention, 2009, "Anatomy of an Outbreak," YouTube, https://www.youtube.com/watch?v=kKDNEW8XHvs, uploaded February 6, 2009, by the FDA (accessed 8 January 2013).

Sustainable Business Forum, 2011, "Senator Leahy Asks DOJ for Update on Peanut Corporation of America Investigation," http://sustainablebusinessforum.com/billmarler/49984/senator-leahy-asks-doj-update-peanut-corp-america-investigation (accessed 8 January 2013).

Tavernese, Sabrina, 2013, "Charges Filed in Peanut Salmonella Case," *New York Times*, http://www.nytimes.com/2013/02/22/business/us-charges-former-owner-and-employees-in-peanut-salmonella-case.html?_r=1&, February 21, 2013 (accessed 24 February 2013).

Theimer, Sharon and Ricardo Alonzo-Zaldivar, 2009, "Gov't Launches Criminal Probe in Peanut Recall," Associated Press story published in the *Seattle Times*, Seattle, WA, January 31, 2009.

US Food and Drug Administration, 2009, Peanut Butter and Other Peanut Containing Products Recall List, www.accessdata.fda.gov/scripts/peanutbutterrecall/PeanutButterProducts2009.pdf, June 18, 2009 (accessed 18 December 2012).

US FDA News Release, 2011, "Unilever Announces Recall of Skippy Reduced Fat Peanut Butter Spread Due to Possible Health Risk – Limited Recall of 6 Best-if-used-by Dates," http://www.fda.gov/safety/recalls/ucm245897.htm, March 4, 2011.

US FDA News Release, 2013, "FDA Proposes New Food Safety Standards for Foodborne Illness Prevention and Produce Safety," http://www.fda.gov/NewsEvents/Newsroom/PressAnnouncements/ucm334156.htm, (accessed 4 January 2013).

US FDA, 2012, "The New Food Safety Modernization Act (FSMA)," description of law, blogs, links, http://www.fda.gov/Food/FoodSafety/FSMA/ucm257986.htm, page last updated December 28, 2012.

USA Today, 2009, "Officials Call for Criminal Probe into Salmonella Recall," http://usatoday30.usatoday.com/news/health/2009-01-27-peanut-salmonella_N.htm (accessed 8 January 2013).

Waters, Rob, 2009, "Peanut Corp. Shipped Product After Finding Salmonella (Update 1)," Bloomberg News, http://www.bloomberg.com/apps/news?pid=newsarchive&sid=aQ_T2VxTLyU&refer=home (accessed 8 January 2013).

15 Ethical aspects of nanotechnology in the area of food and food manufacturing

Herbert J. Buckenhüskes

15.1 INTRODUCTION

Nanotechnology, along with information technology and biotechnology, is one of the key technologies of our age. There are great expectations for nanotechnology in a number of fields, both in terms of general technical and technological development and in relation to food technology in particular. These three extremely different technologies have one thing in common, namely that their potential is euphorically overestimated by optimists, while pessimists see only the actual or supposed risks associated with them. Nanotechnology, a typical cross-disciplinary field, is driven above all by physical processes (top down), complex chemistry (bottom up) and our ever greater understanding of biological processes.

Since risks are associated with every human action, we need to investigate the background and interconnections with regard to possible advantages and disadvantages (i.e. risks) so that we can reach well-informed decisions. As far as nanotechnology is concerned, however, we face a fundamental problem: in the field of occupational safety or consumer protection, for example, a risk is principally calculated according to the formula risk = danger × exposure and is thus based on experience of comparable cases (extent of damage) and on probability considerations (likelihood of occurrence). The problem with nanotechnology is that, first, there are no real comparable cases and, second, the possible (i.e. supposed) risks appear to be simply endless. We therefore return to the point at which rational, ethical reflection is required. Such a requirement normally arises at times of upheaval when our cultural habits, our moral intuition and our social institutions can no longer cope with the problems and need to be examined, strengthened or redesigned (Vogt, 2003).

The fundamental and traditional objective of ethics is to discuss generally applicable standards and values, such as general principles of good practice. Should these then be applied to a specific case, practical powers of judgment and a trained conscience are required.

According to Jungmichel (2005) ethics, if it has not already, takes on a new social role in the nanotechnology discussion; whereas at the start of the 20th century ethics could be

Practical Ethics for Food Professionals: Ethics in Research, Education and the Workplace, First Edition.
Edited by J. Peter Clark and Christopher Ritson.
© 2013 John Wiley & Sons, Ltd. Published 2013 by John Wiley & Sons, Ltd.

found primarily in a single superstructure that controlled society, in the post-war period it took on a regulatory role with regard to the welfare state. In the 1970s and 1980s, ethics took on a critical role with regard to society and technology, such as in the environmental movement or in "liberation theology." Today it has a primarily advisory role when a decision between alternatives has to be made, for example in policy advice or ethics commissions. The new aspect is the role of ethics in providing constructive impetus, contributing expertise on designing value patterns and their contexts, mediating between diverging interests and enhancing the communication skills of the parties involved. Ethics is therefore becoming a pioneer of a technological development that incorporates soft factors into the structures of innovation processes and creates solutions to genuine needs.

15.2 DEFINITION

The problems of ethically rational reflection in nanotechnology begin as soon as you start to describe what the technology actually is. Although there have been many attempts to describe this subject, there is no official and generally accepted definition of nanotechnology. As an illustration, the definition is given here that is used by the German Federal Ministry of Education and Research in its Nano Initiative – Action Plan 2010 (Berghofer, 2010). According to this definition:

> Nanotechnology is the investigation, application, and production of structures, molecular materials, and systems with a dimension or production tolerance of less than 100 nanometres. The minute scale of the system components alone enables the realisation of new functionalities and properties for improving existing products and applications or developing new ones.

Other definitions specify the size range of nanotechnology more precisely, namely at approximately 1 nm to 100 nm. However, the German Federal Institute for Occupational Safety and Health states that the size limits are fluid because, at the lower end of the scale, complex molecules are also investigated and, at the upper end, agglomerates and aggregates of nanoparticles, nanofibers, nanotubes, etc. need to be considered that may be larger than 100 nm (Federal Institute for Occupational Safety and Health, 2007).

What is interesting about nanotechnology is that nanoscale materials exhibit different physical and chemical properties to larger-scale materials. Essentially, this may have two causes. On the one hand, nanomaterials have a significantly larger surface area compared to other substances of the same mass, which increases their chemical reactivity and may influence their forces and electrical properties. On the other, quantum effects start to dominate the substances' behavior in this dimension, thereby influencing their optical, electric and magnetic properties. It is also interesting, however, that, if molecules are supplied with energy, they can be activated to arrange themselves into defined structures whose properties can differ from macrostructures. For example, crystals and surface coatings have been developed from nanostructures.

15.3 NATURE AND NANO

The nano dimension is nothing new in either animate or inanimate nature. In fact, all living creatures contain a multitude of nanostructures. And since our food – with the exception of water and table salt – comes from living organisms (predominantly plants and animals), it is also largely made of nanostructures. For example, casein micelles in milk are approximately 20–500 nm in size, while whey proteins are around 3 nm. Moreover, in various manufacturing and preservation procedures used in the traditional production of food nanoparticles are involved or they can result in the creation of nanoparticles and nanostructures. In this context, natural nanostructures play a key role because they are either purposely destroyed or modified or because new and therefore synthetic nanostructures are formed that do not occur in nature. However, neither these nor naturally existing nanoparticles are covered by the term "nanotechnology". Rather, nanotechnology requires deliberate manipulation that purposely and intentionally produces or results in nanoparticles and nanostructures.

15.4 NANOETHICS

In view of the revolutionary potential assigned to nanotechnology, a debate has opened up about its ethical, social and legal implications that has been taken up by numerous commissions and working groups at various levels. The need for ethics in and for nanotechnology was recognized at a relatively early stage, and terms such as "nanoethics" were coined. In an initial analysis, however, Grunwald (2004) concluded that there were hardly any genuinely new ethical aspects in nanotechnology. He believes that most points of discussion relevant to nanotechnology are gradual shifts in accent and relevance of questions that principally already exist in fields such as technological ethics, bioethics, medical ethics and also in the theoretical philosophy of technology. The same applies to discussions about the tools and methods used, for example, in safety investigations, such as animal testing and research on humans. In contrast, specific aspects resulting from nanotechnology have not been identified so far. In his opinion, truly new questions are raised above all by the fact that the nanosciences and fields of nanotechnology break down the traditional barriers between physics, chemistry, biology and engineering. In practice, this means that previously separate ethical lines of reflection come together in nanotechnology, such as those in bioethics and technological ethics. That is why both Grunwald (2004) and Rippe (2007) believe that there is no justification for wanting to establish "nanoethics" as a new branch of applied ethics.

15.5 RISK DEBATE

The risk debate surrounding nanotechnology will be crucially important to its future. As with other major new technologies, the threat from nanotechnologies and nanomaterials perceived by the layperson is fundamentally different to the one seen by experts. Nanotechnology has not yet become an emotionally charged, vehemently disputed topic of public discussion. According to the Eurobarometer 2010, only about a quarter of Europeans have even considered this topic. People are generally positively disposed towards nanotechnology, yet its acceptance quickly declines, the more direct an impact that products have on the human body. Consumers are already very sceptical and even hostile towards ingredients and additives that have been produced or modified using nanotechnology.

The reason for this is that there are still many unanswered questions related to health considerations and environmental compatibility. Since biological and inorganic nanoparticles also occur in nature, our bodies do have defense mechanisms and barriers against nanoparticles. So it really does seem that Theophrastus Bombast von Hohenheim (known as Paracelsus) was even right about nanomaterials when he said "the dose makes the poison," although he would possibly have to add something about the size dimension.

The strengths and advantages of nanotechnology are also the cause of its potential risks. Although nanoparticles are not dangerous *per se*, the switch to the nanodimension changes the properties of many materials – sometimes drastically – even though their chemical composition stays the same (Berghofer, 2010):

- As the particle size shrinks, the specific surface area increases exponentially. It is above all the resulting reactivity of the nanoparticles that can give rise to a broad range of possible threats to people and the environment. For example, nanoparticles may lead to increased production of reactive oxygen species and cause oxidative stress, inflammation and thus damage to proteins, lipids, membranes and DNA.
- At the nanometer scale, the properties of substances sometimes change entirely.
- It is not possible to ascertain the properties of nanoparticles from the material's properties in non-nano form.
- Toxicological data from conventional materials and the properties of substances cannot simply be applied to nanostructures.
- Due to their size, nanoparticles can interact with living cells, creating the risk of nanoparticles being absorbed and accumulating in the body. It is largely unknown what this may mean for our health in specific cases.
- Artificial nanoparticles are not found in nature in the same form, raising the question as to whether they may trigger possible incorrect reactions in the immune system.

When it comes to regulating the risk of new technologies, the precautionary principle is mostly applied as a matter of course. However, how it is to be applied – in other words, whether a strong or a weaker approach should be taken – is often disputed (Rippe, 2007). The strong precautionary principle is applied when there is a lack of knowledge, which in practice results in the demand for the burden of proof to be reversed. This means that the manufacturer must prove that its product is safe. In contrast, the burden of proof is retained when the weak precautionary principle is applied (i.e. the aggrieved party must prove that it has been harmed due to a particular product). However, this version of the principle may only be applied if the manufacturer can prove that it has carried out a careful risk analysis. Which of the two versions of this principle is to be applied to regulate risk in the food industry will certainly depend on the form in which nanomaterials and nanoparticles are used. A key question in this context is how and where the particles exist or whether they have been absorbed into or attached to other materials. The latter is usually the case in packaging technology or the design of surfaces. As far as ingredients and additives are concerned, the answer to this question should be clear (i.e. because only the strong precautionary principle is applied here, as has been the case so far with the approval of new additives).

In 1995, Haller and Allenspach developed a three-level model that makes it easier to understand the factors influencing the risk debate, including in relation to the use of nanotechnology in the food industry. They postulate that communication and perception of risk generally take place at three different levels, with each individual level based on specific logic and each generating its own objectivity. The first level is the risk debate from a scientific and technical perspective, which involves an objective, knowledge-based risk analysis. At the second level, however, an emotional debate takes place that is strongly influenced by individuals' different perceptions of risk as well as their hopes and fears. Whether a risk is considered acceptable depends on attitudes at this level, in other words on people's individual judgment criteria. Finally, the third level – the social or ethical level – addresses group-specific interpretations of risk and ethical aspects. The different objectivities at the three levels are often contradictory and are therefore the cause of conflicts. According to Haller and Allenspach (1995), the intensity of these conflicts rises from the first to the third levels. At the final level, the social/ethical level, conflicts can only be resolved by enabling comprehensive dialog on the risks and involving scientists, politicians and the general public.

15.6 FURTHER ETHICAL ASPECTS

The general debate about the ethics of nanotechnology frequently encompasses other aspects that, however, are mostly irrelevant to the food industry. One of these is "technical improvement of human beings," which is perhaps relevant to the use of nanotechnology in medicine. Other aspects include distributive justice and the "control and privacy" issue, which is again applicable to medicine but also undoubtedly touches on electronics.

As far as the food industry is concerned, an aspect that must be seen as increasingly important is the relationship between technology and life, irrespective of any possible toxicological risk. Food has a highly symbolic value in almost all cultures. In fact, the culture surrounding eating is often an integral element of a country's or region's culture. In view of the trends towards increasingly processed, artificial convenience products and new types of food, greater importance needs to be attached to the question of eating culture and the debate about cultural values – the question "how do we want to live?". However, this is not an issue specific to nanotechnology, even though many developments in this field will accelerate the pace of debate around related issues (Grunwald, 2006).

15.7 CONCLUSION

Although there is no question about whether nanotechnology should be used in the food industry, its further development doubtlessly needs to be accompanied by critical, constructive and ethical analysis. According to Kordecki *et al.*, 2007, the main criteria will be:

- assessment of the consequences, including analysis throughout the lifecycle of the nanomaterial
- evaluation of the risks, as part of which nanomaterials need to be classified, labeled and monitored

- cost/benefit analysis
- consideration of alternatives
- fairness and the involvement of the public in decision-making processes.

Consumers' acceptance of nanotechnology in future will depend above all on the following factors (Berghofer, 2010):

- Manufacturers need to be open and transparent regarding their use of nanomaterials and nanotechnology methods (labeling as far as is possible and practical), *vis-à-vis* both authorities and consumers.
- Authorities should take a prospective approach, i.e. they should not approve the use of a nanomaterial until they have examined and rated its potential risks.
- Consumers must gain direct and immediate personal benefits from the use of nanotechnology in the area of food and nutrition and must be able to see these benefits.

It must therefore be reiterated that the call for transparency in the food industry must finally be implemented unconditionally. The right to information is one of the fundamental ethical demands that the food industry in the widest sense must meet. Although communication has not been one of the food industry's strengths so far, it is one of the most important demands made by consumers as far as health considerations are concerned. In this age of communication, not communicating is fatal in the medium term!

ACKNOWLEDGMENT

This chapter is based on a presentation given at the "Symposium on the ethical issues of high tech applications in food processing" organized by HighTech Europe, Berlin, Germany (2011).

REFERENCES

Berghofer, E (2010): Nanotechnologie im Bereich Lebensmittel und Ernährung. Vienna Chamber for Non-Salaried and Salaried Employees.
Eurobarometer 2010: Report of the European Commission's Directorate-General for Research – Europeans and Biotechnology in 2010 – Winds of change? Report by Gaskel, G *et al*. (2010). Online. Available from http://ec.europa.eu/public_opinion/archives/ebs/ebs_341_winds_en.pdf (accessed 7 January 2013).
Federal Ministry of Education and Research (2010) Nano Initiative – Action Plan 2010. Online. Available from http://www.bmbf.de/pubRD/nano_initiative_action_plan_2010.pdf (accessed 17 December 2012).
Federal Institute for Occupational Safety and Health (2007) Nanotechnology: Health and environmental risks of nanomaterials – research strategy. Online. Available from http://www.bmu.de/files/pdfs/allgemein/application/pdf/nano_forschungsstrategie.pdf (accessed 17 December 2012).
Grunwald, A (2004) Ethische Aspekte der Nanotechnologie. Eine Felderkundung. Technologiefolgenabschätzung. Theorie und Praxis 13 (2) 71–78
Grunwald, A (2006): Nichts Ungeheuerliches. In: Politische Ökologie 101, Oekom Verlag, Munich: 27–29.
Haller, M, Allenspach, M (1995) Kompetent – Inkompetent? Zur Objektivität des Urteils über Grösstrisiken. In: Thome, JP (ed.): Managementkompetenz. Die Gestaltungsansätze des NDU/Executive MBA der Hochschule St. Gallen. Versus Verlag, Zurich.

Jungmichel, N (2005) Moleküle, Märkte und Moral. Nanotechnologie und Ethik im Zusammenspiel. Presentation at the Annual Meeting of IWS e.V. (30 April 2005).

Kordecki, G, Knüppel, R, Meisinger, H (2004) Ethical aspects of nanotechnology. A Statement by the Working Group of Commissioners for Environmental Affairs of the Protestant Church in Germany (EKD). Akzente No. 14. Institute for Church and Society, Iserlohn.

Rippe, K.P. (2007) Braucht es eine NanoEthik? Lecture at the 2nd NanoConvention on 29 June 2007, Bern, Switzerland.

Vogt, M (2003) Ethische Aspekte der Nutzung nachwachsender Rohstoffe. In: Baden-Württemberg Ministry of Food and Agriculture (ed.): Nachwachsende Rohstoffe für Baden-Württemberg, Stuttgart, 1–11.

16 Food commodity speculation – an ethical perspective

Chris Sutton

16.1 INTRODUCTION

Over the past few years rising food commodity prices and accompanying price volatility have become a significant political and economic issue around the globe. The impact on the world's poorest is devastating, with the 2007/2008 food price crisis thought to have pushed over 40 million people into hunger (De Schutter, 2010, p. 2). After a brief respite following the 2008 crisis, prices increased rapidly again, with the index of international food prices compiled by the UN Food and Agriculture Organization (FAO) reaching an all time high in February 2011 (FAO, 2011a, p. 1). Although food commodity prices have subsequently fallen back somewhat, they remain relatively high and "extremely volatile" (FAO, 2011b, p. 1). This situation of high and rapidly fluctuating food commodity prices looks set to continue for years to come (FAO, 2011c, p. 12).

Although high and volatile food commodity prices affect everyone, they have a disproportionate effect on the poorest and most vulnerable. In developed countries the impact is moderated because food commodities tend to be a relatively small component of food retail prices, and because overall food purchases only constitute an average 10–15% of household spend (OECD, 2008, p. 8). Even so, the rising food prices are likely to have an impact on low income households. A report by the Joseph Rowntree Foundation in the UK found that recent increased hardship had been caused in part by the 2007/8 commodity price shocks in food and fuel (Hossain *et al.*, 2011, p. 11). In developing countries, where the poorest households can spend 60 or 70% of income on food, often in the form of basic commodities, the impact is much greater (FAO, 2011c, p. 14). Most at risk are those in food importing countries reliant on basic food commodities purchased on international markets. The OECD estimates that a 10% increase in the price of cereals adds \$4.5 billion to the food import bill of net importing developing countries (OECD, 2008, p. 8).

There is an ongoing and vigorous debate around the extent to which financial speculation in food commodity markets contributes to food price rises and price volatility. A joint report on food price volatility written for the G20 by a number of agencies including the FAO, International Monetary Fund (IMF), the United Nations Conference on Trade and Development (UNCTAD), the World Bank and the International Food

Policy Research Institute (IFPRI) illustrates the difficulty in forming a conclusive rec-ommendation on this issue. They conclude that most analysts recognize that increased financial sector involvement in food commodity markets "*probably* acted to amplify short term price swings and *could have* contributed to the formation of price bubbles in certain circumstances" (FAO, IFAD *et al.*, 2011, p. 12, *italics* added). This does not however constitute firm evidence that speculation is a determinant of price volatility and the recommendation of the report, recognizing the extent of disagreement that remains, is for more research to assist regulators in assessing whether regulatory responses are required (FAO, IFAD *et al.*, 2011, p. 22).

While the debate about speculation continues to rage, volatile food commodity prices continue to cause real suffering. Policy makers are divided on whether to take action, with Nicolas Sarkozy using France's presidency of the G20 in 2011 to promote greater regulation but with more sceptical governments such as the UK awaiting definitive proof that speculation causes harm before acting (Sarkozy, 2011a, b; Hoban, 2011). With reso-lution of the debate a long way off, the default position of taking no action while waiting for further evidence in a discourse that is already three years old seems inadequate.

Although centred on the impact of speculation, this debate is also a microcosm of a much larger and older discourse about social justice for the world's poor. That an estimated 925 million people are undernourished (FAO, 2010, p. 4) sits in sharp contrast to the financial activities undertaken by some of the wealthiest people in the developed world. This wider debate, although extremely important, is beyond the scope of this chapter.

Here, we argue that the debate can be moved forward in three areas. First, this can be done by investigating the burden of proof required before action is taken. In complex situations where definitive proof may never be provided, is the correct policy response always to stick to the status quo? A growing body of evidence suggests that speculation plays some role in food price volatility. Given the very real human suffering at stake, this chapter suggests that adopting a more precautionary approach and limiting the extent of speculation is the prudent action to take.

Second, the debate can be broadened from its narrow focus on whether or not specu-lation causes harm. What happens if we turn the question around the other way and ask whether speculation helps? Rather than considering whether speculation has a negative impact, can we assess whether speculation has a positive impact in terms of benefiting society as a whole? This chapter concludes that while some speculation can provide what might be termed "social value" by improving market liquidity and taking risk from other market participants, additional value is unlikely to be provided by the massive scale of speculation currently taking place in food commodity markets.

Finally, the debate can be moved forward by recognizing that speculation in food mar-kets is not an isolated occurrence but part of a wider trend taking place in the economic system, a process of financialization that has seen the financial services sector become increasingly dominant over other sectors of the economy. Financial interests are active across the food system, as investors in and owners of various food related organizations. Most relevant to the speculation debate is the increasing interest of investment funds in acquiring agricultural land for investment – so-called "land grabbing." The implications of this broader issue for the food system are beyond the scope of this chapter. Suffice it to say that the global food system is unlikely to meet the needs of people, especially the most vulnerable people, if the driver of powerful interests within it is limited only to investment return.

In summary, given that speculation has the capability to cause harm, and has questionable value to society as whole, the chapter recommends that policy makers support regulatory initiatives to impose limits to speculative activity in food commodities markets. If speculation is a significant cause of volatility this will save lives. If speculation is only a marginal cause, little of value will have been lost, and policy makers can focus on other causes of food price volatility.

The concepts of a precautionary approach and social value, outlined above in the context of speculation, may also be valuable to policy makers in other areas, ensuring that the vulnerable are protected and that wider social benefits are considered alongside the potential for wealth creation. Ultimately the creation of a fairer and more sustainable food system is dependent on the creation of economic and financial structures that also consider people and planet alongside profits.

This chapter begins with a brief overview of how financial markets are used in agriculture before looking at the debate about the impact of speculation and alternative approaches policy makers can consider when making decisions about speculation and food commodity markets. The chapter concludes by considering regulatory responses to speculation in food markets.

16.2 FINANCIAL DERIVATIVES AND AGRICULTURE

At the center of the debate on financial speculation and food prices are complex derivative products, known as such because their value is derived from some other underlying asset. Markets for a particular type of derivative known as a "future" enable the producers, purchasers and others involved in the supply chain of certain agricultural commodities to "hedge" against the risk that commodity prices will move unfavorably.

To offer a simplified example, a farmer sows wheat in the spring but expects to sell her harvested crop in October. At the time she plants her crop, she does not know what the price will be when she comes to sell and therefore cannot know what level of profit she will make or even if she will make a loss. If she has a buyer arranged, she could enter into a forward contract with him, agreeing up-front the price at which she will sell the wheat when the crop is ready in October. This removes the risk that she will lose money due to the price of wheat having fallen when she comes to sell, and passes it on to the buyer. Of course, he may be concerned that the price of wheat is going to rise during the year, and is therefore happy to hedge against his exposure to this risk by securing his October purchase price in the Spring.

An alternative way of managing this risk is to enter into a futures contract traded on a central exchange. The farmer entering the futures market will use standardized contracts which include specifications covering commodity grade, quantity, delivery location and delivery date. These standardized contracts provide a more "liquid" market by appealing to a larger number of buyers and sellers than would be interested in trading directly with the farmer. The main futures markets for commodities are found in the United States, with Chicago prominent in several commodities including wheat and maize. Within the EU, the main markets are London and Paris (European Commission, 2009, p. 4).

To return to our example, in order to hedge her risk using futures markets, the farmer will sell a number of standardized futures contracts, choosing those that most closely match the wheat she is planning to sell with an appropriate future delivery date. This transaction effectively locks in the price that she will receive for her wheat. As the delivery month for the futures contract approaches, the farmer will close out her "hedge" position by purchasing futures contracts equivalent to the ones she originally sold, and will proceed to sell her wheat harvest on the cash market. If the price of wheat on the cash market has fallen, this should also be reflected in falling prices for similar futures contracts. The lower revenue the farmer receives from selling the wheat is compensated for by a broadly equivalent gain from the trades she has made in the futures market. Selling and buying futures contracts has effectively enabled the farmer to manage the risk of adverse price movements. As in this example, futures contracts very rarely result in actual delivery of the commodity being traded (Kolb and Overdahl, 2007, p. 24).

16.3 SPECULATION OR FINANCIALIZATION?

The *Oxford English Dictionary* defines speculation as "investment in stocks, property, etc. in the hope of gain but with the risk of loss" (Oxford University Press, undated). As a definition to describe the activities that form the focus of this debate this is overly broad and could include pretty much any investment activity in capital markets. A definition by John Bogle, financial markets expert and founder of a US mutual fund, is more applicable, distinguishing speculation as different from, and in fact the opposite of, investment. For Bogle, investment is about long-term ownership of businesses and the creation of "intrinsic value" over time. This occurs as businesses produce goods and services that add value to society and increase wealth. By contrast, he defines speculation as short-term trading of *financial instruments* rather than *businesses*, held on the expectation of profit from increased *prices* rather than increased *intrinsic value* (Bogle, 2009, pp. 49–50, italics added).

An element of speculation has always existed in commodity futures markets and can play a useful role. Speculators looking to profit from price movements take on the risk of other market participants, provide market information to help set more accurate prices, and provide liquidity that enables markets to operate more efficiently (Angel and McCabe, 2010, p. 278). As financial speculators usually aim to buy when prices are low and sell when prices are high, they can even be seen as reducing the extremes of commodity prices (De Schutter, 2010, p. 4).

The debate about food commodities markets, rather than being simply about what might be termed "traditional" speculation, refers to a broader set of activities undertaken by non-commercial market participants. This ranges from very short-term speculation by high-velocity traders, and active trading by hedge funds and other financial players, to longer and more passive engagement by institutional investors looking for exposure to commodities via complex financial products such as commodity index funds.

All these investment based activities are focused on achieving a financial return based on changes in commodity futures prices, as opposed to participating in the markets to hedge risk inherent in food production or with regard to the production or distribution of the underlying food commodity. Using Bogle's terms the focus is on price, not value. In keeping with other literature on the topic, this paper will refer to these activities as "speculation". However a more apt term to use might be 'financialization' as a description of "the increasing influence of financial motives, financial markets and financial actors in the operation of commodity markets" (UNCTAD, 2011, p. 13).

16.4 COMPLEXITY

The current debate is struggling to find a consensus about the extent to which speculation contributes to price volatility or price rises. Given the complex mechanisms at work in both food and financial market mechanisms, definitive proof is very difficult to produce.

The impact of speculation needs to be seen in the context of longer-term structural changes in demand and supply conditions, which mean food prices are likely to remain high over the coming years. Population growth, economic growth and biofuels stimulate demand for food exerting upwards pressure on prices. While increased prices may encourage much needed investment in agriculture, increased oil prices will increase the costs of agricultural production, while constraints on the availability of water and productive land and the impacts of climate change all serve to constrain global supply. (FAO, 2011c, p. 14). Isolating the impact of financial actors on short-term prices against the backdrop of these longer-term trends and the myriad of day to day supply and demand conditions that impact food prices at international, national and local levels is a very challenging task.

Within financial markets further complexity makes it difficult to identify the impact of financial actors on futures markets, and the impact of futures markets on day to day cash prices for food commodities. In order to prove that speculation does contribute to price volatility it must be proven that speculative activity influences prices in financial futures markets for food commodities and that prices set in financial futures markets, for the delivery of food commodities at some date in the future, alter the real or spot prices for food commodities traded today. This task is made harder as currently there is little transparency around financial contracts traded over the counter (OTC), bilateral agreements between financial institutions and investors that constitute much of the influx of funds into food commodity markets.

The complexity involved in assessing the impact of speculation means that correlations alone do not provide reliable proof that speculation is a cause. While the correlation of rapidly increasing commodity index fund investment and rapidly rising food commodity prices provides a disturbing picture, it has been difficult to prove beyond any doubt a causal relation between the two. Complexity also means that accounts of the impact of speculation are as dependent on theoretical explanations of how financial and commodity markets work as they are on empirical observations. As economic theory is not immutable fact, the debate about speculation is as much about our beliefs about how the world works as it is about what is really happening on the ground.

16.5 BURDEN OF PROOF

Even though consensus in this debate has not been reached, a broad body of research at the very least casts doubt on the view that the massive inflow of funds into food commodity markets is a totally benign influence.

There is no doubt that food commodity markets have been subject to a large influx of funds by financially motivated participants such as swaps dealers, commodity index funds and money managers (FAO, IFAD *et al.*, 2011, p. 22). Alongside deregulation, which encouraged the growth of OTC derivatives products, commodities have become attractive to investors looking to diversify their portfolios and hedge against inflation (UNCTAD, 2011, p. 13). Olivier De Schutter, the UN Special Rapporteur on the right to food, argues that in the run-up to the 2007–2008 food price crisis, the failure of returns elsewhere in the financial system as a result of the sub-prime lending crisis increased further the demand for commodities from institutional investors (De Schutter, 2010, pp. 5–6).

Those who see financial speculation as a cause of the 2007–2008 food crisis have been especially critical of commodity index funds. These financial products are designed to enable investors to have exposure to a wide range of commodity futures markets including energy, metals and agricultural commodities without having to invest in individual commodities futures contracts. Commodities have become especially attractive to large institutional investors such as pensions and hedge funds as they are believed to move in the opposite way to shares and bonds and provide a hedge against inflation (Masters, 2008). While the purchase of a commodity index fund in itself does not directly affect the commodities futures markets, the financial institutions that sell them are able to use futures markets to hedge the risk they are exposed to as a result of selling them.

Commodity index fund investors treat commodities futures markets as a passive investment in the same way they might invest in stocks or bonds, with the aim of profiting from increases in price over a period of time. As a result, financial institutions accumulate large "long only" (buy only) positions in commodity markets, which Olivier De Schutter argues, generates upward momentum in food commodity futures prices that ultimately translates into increased food commodity spot prices (2010, p. 6). Exemptions from position limits created by the Commodity Futures Trading Commission to protect the market from manipulation mean that these institutions can assume very large positions that can be more than 10 times the size of positions held by other market participants (Masters, 2008; Sanders *et al.*, 2008, p. 8).

Those arguing that speculation is having a significant negative impact come from a broad variety of backgrounds and include economists, hedge fund managers, prominent financiers and businessmen alongside campaigning organizations (for example see Worthy, 2011; Christian Aid, 2011; Masters, 2008, Ghosh, 2010; Soros, 2008; Branson *et al.*, 2008). In a joint letter organized by the World Development Movement (WDM) before a G20 meeting in 2011, over 450 economists including academics from Oxford, Cambridge, and the London School of Economics argued that "Excessive financial speculation is contributing to increasing volatility and record high food prices" and "that prices have moved too much to be based on fundamental supply and demand factors" (WDM, 2011).

At the least it seems that food commodity markets are not behaving as they should. US wheat farmers and elevators are increasingly unable to use futures markets to hedge production and distribution activities (US Senate, 2009, pp. 44–49). A report from UNCTAD observed that commodity markets are becoming increasingly linked to information flows in financial markets, indicating that factors other than supply and demand are driving price movements. Commodities with little in common are starting to move together in response to announcements about economic indicators (UNCTAD, 2011, p. viii).

The view that speculation is having a significant impact is not unanimous. As well as economists arguing that there is little evidence that speculation causes harm, some commentators point to the impact of more fundamental factors including stock levels, oil prices, bio-fuels, export restrictions and macro-economics factors (for example see Bobenrieth and Wright, 2009; Piesse and Thirtle, 2009, p. 124; Headey and Fan, 2010, p. xv). Nevertheless the number of respected commentators arguing that speculation is an issue is significant and calls into question the level of evidence policy makers require before acting. What burden of proof is required? The onus is currently on critics of speculation to prove beyond doubt that harm is caused. Given the potential risk of human suffering, should more energy be expended to confirm that harm is not caused?

16.6 ECONOMIC ASSUMPTIONS

Economic approaches are central in this debate, with academics on both sides of the debate using economic modelling to illustrate how speculative activity influences markets. By making assumptions about how markets work and how people behave, economic models provide a simplified representation of the real world, with the aim of enabling the key mechanisms at work to be identified. Conventional financial theory assumes that markets are characterized by a large number of informed buyers and sellers, and that prices reflect all available information. According to this theory, speculation is unlikely to be a significant influence on commodity prices. If a price moves up from the fundamental value which reflects this information, informed participants would see an opportunity to make risk-free profit by exploiting the mispricing, with their actions stabilizing prices back at their fundamental values (Rapsomanikis, 2009, p. 19; Gilbert, 2009, p. 19).

In reality, predicting the outcome of activities in futures markets is more complicated. Information may not be readily available, and rational behavior may not entail moving prices back down to a theoretical equilibrium. In futures markets, prices could theoretically be pushed above their fundamental values (set by information about supply and demand conditions), because the information on which trades are based only becomes apparent after a period of time. While futures commodity price increases may cause immediate action in terms of planting crops, there is a lag before this information results in increased grain inventories, and a further delay before these are reflected in the inventory information that would inform the futures market (Lagi et al., 2011, pp. 5–6). This delay in information could result in so-called herding behavior, where market participants judge their own information to be incomplete and follow the behavior of other traders, acting on the belief that others in the market have better information than them (UNCTAD, 2011, p. 22).

Even where other market actors know that prices are above their fundamental level, they may be limited in the extent to which they engage in arbitrage. This could be a totally rational stance, as there would be a significant risk of loss should prices continue to move upwards against that arbitrage position. As economist Christopher Gilbert points out, "In practice, the informed investors are likely to sit on the sidelines until sense returns to the market since there is no easier way to lose money than to be right but to be right too early" (2010, p. 4). Traders may even engage in "positive feedback trading," purchasing contracts in the expectation that others will follow, pushing up the price and enabling them to sell at a profit (UNCTAD, 2011, p. 22; De Long *et al.*, 1990, p. 394).

The quantification of herding or positive feedback trading behavior is difficult, though it has been attempted using econometric modeling (for example see Gilbert, 2010; Tokic, 2011). Observation of market data tells us what trades were made, but doesn't necessarily help in assessing the motivations of traders or the information they actually used to make their decisions. Although less scientific in approach, interviews with market participants are therefore also important in understanding what is driving market activity. Commodity market participants interviewed for an UNCTAD report generally felt that financial investors had become more important and could move prices in the short term, thereby increasing volatility (2011, p. 48).

Although economic models provide a useful way of understanding a complex environment, they are only theoretical, and may not capture the real world characterized by "human beings and their interests, ideologies and normative convictions" (Archer and Fritsch, 2010, p. 120). Difficulty in quantifying the impact of financial investment, then, does not constitute proof that there is no impact.

16.7 TOWARDS A PRECAUTIONARY APPROACH

The complexity of this debate means that in all likelihood additional research will not lead to a consensus. Therefore policy approaches that await conclusive proof prior to action may not provide an adequate response to this issue. The current policy discourse is premised on assessing the likelihood that financial speculation causes harm. Yet without a consensus around the evidence, such an assessment is seemingly impossible, and policy inertia is the result. In the absence of conclusive evidence, and given the seriousness of the consequences if speculation does cause harm, the adoption of a more precautionary approach may be advisable.

The precautionary principle is a contested but widely used concept that provides an alternative basis for making decisions about financial speculation. There is no universally agreed definition of the precautionary principle and the wording used to describe it varies. However, drawing together examples of the term's use in various international treaties, the World Commission on the Ethics of Scientific Knowledge and Technology, an advisory body to UNESCO, devised a working definition. It states that "[w]hen human activities may lead to morally unacceptable harm that is scientifically plausible but uncertain, actions shall be taken to avoid or diminish that harm." (2005, p. 14). Risk governance expert Andrew Stirling (2007, p. 310) argues that whilst a risk-based approach is suitable where there is strong confidence in the assessed outcomes and probabilities, it is not applicable to situations characterized by uncertainty, ambiguity or ignorance. It is in these circumstances that the precautionary principle is valuable in providing guidance by "giv[ing] the benefit of the doubt to the protection of human health and the environment, rather than to competing organizational or economic interests" (Stirling, 2007, p. 312).

Although controversial, the precautionary principle is widely used by policy makers to protect people and the environment. It is recognized in the field of environmental policy following its inclusion in Principle 15 of the Rio Declaration, which states that a precautionary approach will be applied by states in order to protect the environment (UNCED, 1992). It is also widely used in European settings, with the precautionary principle forming part of the EU's legislative approach to food-related issues such as GM crops, food safety and the Common Fisheries Policy (European Council, 1990, 2002, 2008). It is deployed in Chapter 15 of this book in the context of nanotechnology in food production. The European Commission published a communication in 2000 clarifying how the precautionary principle should be used in EU policy making. Invoking the precautionary principle is seen as an appropriate approach in situations where a hazard is identified but where scientific evaluation does not allow risk to be evaluated with "sufficient certainty," either because of insufficiency of data, or where the nature of the data is imprecise or inconclusive (European Commission, 2000). These conditions appear similar to those apparent in assessing the impact of speculation on food commodity markets.

Adoption of the precautionary principle might apply to financial speculation in food commodity markets in two ways. First, it could be used to reassess the burden of proof required before action is taken, given the high stakes involved in terms of human suffering. In the absence of conclusive evidence, is there sufficient evidence to act? Second, it could be used to question who has responsibility for proving that harm is caused. Is the relevant responsibility here to determine beyond doubt that financial activity is causing harm, as is currently assumed, or to determine beyond doubt that *no* harm is caused? In other words, should the onus be on those benefiting from financial investments in food commodity markets to prove that they are not causing harm, rather than on the critics of speculation? A third way of thinking about this is to turn the question around completely. Rather than asking if speculation causes harm, we might ask whether it creates good.

16.8 SPECULATION AND SOCIAL VALUE

The debate about the value of speculation is an old one. The economist Amartya Sen argues that Adam Smith, writing in 1776, had in mind speculators seeking excessive risk when he described the activities of "projectors and prodigals" who, given access to capital, "were most likely to waste and destroy it" (Sen, 2010; Smith, 1986, p. 457). In a critique of financial speculation written in 1902, John A. Ryan, a Catholic theologian and economist, argued that whilst the miller adds utility by turning wheat into flour, and an investor adds utility by providing capital for use in productive business, a speculator "add[s] nothing to the utility of any property" (Ryan, 1902, pp. 335–6).

Speculation can have an important and socially valuable role in helping markets function efficiently. By participating in commodity markets, speculators take on the risk of producers, enabling them to produce more food than they otherwise would (Angel and McCabe, 2010, pp. 280–281). They are also seen to provide benefits in terms of aiding price discovery (the interaction between buyers and sellers which determines a markets price), liquidity (the ease with which contracts can be bought and sold) and market deepening (the extent to which the market can absorb a large volume of transactions without this affecting price) (Angel and McCabe, 2010, p. 281; FSA and

HM Treasury, 2009, p. 35). In other words, the presence of speculators makes it easier for those looking to hedge to find someone to trade with, reduces the transaction cost of trading, and can make the market more stable. Traditional speculators can also ease market volatility, because they tend to buy when prices are low and sell when prices are high (De Schutter, 2010, p. 4).

One of the main arguments against the imposition of limits to speculative activity is that it may reduce these benefits and harm the operation of commodity markets. In the UK, the Treasury and Financial Services Authority (FSA) opposes limiting participation in commodity markets, arguing that this is potentially detrimental to "efficient markets and price formation . . . " (FSA and HM Treasury, 2009, p. 35). Yet, while these benefits may apply to traditional forms of speculation, it is more questionable whether they apply to the types of financial activity taking place in food commodity markets on a significant scale. In a recent paper, Lord Turner, Chairman of the UK FSA, questions some of the assumptions held by his own organization in the run-up to the financial crisis: that financial innovation and market liquidity were always good, and regulation, except in cases of specific market failure, to be avoided (Turner, 2010, p. 15). Although the subject of his paper is reforming the wider banking system, his insights seem very relevant to the debate around speculation in commodity markets. Similar perspectives, particularly with regard to the benefits of market liquidity, often seem to underpin policy responses (for example see Hoban, 2011).

While recognizing the benefits of increased liquidity, Turner argues that the benefit it provides is subject to diminishing marginal returns. In other words, as markets become more liquid, the value added by further liquidity decreases. Additionally in certain markets, the increased number of speculators and position takers required to provide this increased liquidity can have a negative effect itself, leading to momentum type effects. Uncertainty, a lack of information, and complex principal/agent relationships can lead to participants taking rational decisions that contribute to instability in the market as a whole (Turner, 2010, pp. 39–40). Therefore while speculation can contribute to liquidity, a social good, it is far from clear that this is always beneficial.

While the social value provided by speculation is questionable, the contribution to societal wellbeing of financial organizations participating in the complex financial markets is also being challenged. In classical economic theory financial intermediaries play a neutral role, connecting buyers with sellers. Yet this does not represent the complex set of relationships that exist in financial markets or describe the activities of a financial sector which has grown over recent decades, such that in the run-up to the financial crisis it provided around 25% of UK corporation tax receipts (Darling, 2011, p. 7). Paul Woolley of the London School of Economics argues that financial intermediaries actually play a dominant role in setting market prices. When investors participate in financial markets they essentially delegate their involvement to the intermediaries. This asymmetry of information between the financial intermediary and end investor leads to mispricing in the market, but also enables the financial intermediary to extract rents or excessive profits at the expense of the end investor. Rather than providing social good in the form of efficient markets, the asymmetry of information leads to social bad in terms of mispricing and rent seeking (Woolley, 2011, p. 125, p. 131). In relation to commodities, Woolley argues that these should be rejected by investors as ultimately they offer a long-term return of less than 0% after financial fees, are subject to herding behavior, and with regard to commodity indices "can be gamed by the investment banks that maintain them" (Woolley, 2011, p. 139).

Policy-makers considering whether speculation causes harm might also consider the extent to which the current high levels of speculative activity in food commodity markets are likely to provide value to society at large or even the wider economy. If they do not, should this not have as much bearing on the decision to regulate as the profitability of the banking sector?

16.9 FINANCIALIZATION

Speculation in food markets is not an isolated occurrence but should be seen in the context of the wider economic system. Demand for exposure to commodities increased as returns from elsewhere in the financial system dried up, first in property markets and then in stock markets (De Schutter, 2010, pp. 5–6; Lagi *et al.*, 2011, p. 7). Rather than seeking excessive returns, the motivations of institutional investors during difficult economic times may have been one of risk minimization and the seeking of any available return.

The need to earn a return is linked to a fundamental need for growth in our economic system. In his book *Prosperity without Growth*, sustainable development expert Tim Jackson talks of a dilemma between economic growth, which is unsustainable, and 'de-growth', which is unstable. Failure to pursue a growth policy currently leads to recession, and consequently losses in livelihoods and wellbeing. Yet the downsides of growth in our current economic system include environmental destruction and the exacerbation of social disparities (Jackson, 2009, pp. 62–64). The attractiveness of commodities can be seen as reflecting a wider characteristic of our current economic system: that it depends on continual economic growth, seemingly regardless of how that growth is achieved.

The need for growth may provide one of the reasons why the financial services industry has been able to grow to the extent that it can contribute 40% of corporate profits in the UK and the US (Woolley, 2010, p. 121). Financialization has seen rapid growth of the financial services sector in relation to the real economy in terms of its share of national income, corporate profits and market capitalization (Turner, 2010, p. 14).

In the context of the need for growth and the dominance of the financial system, speculation is not an isolated phenomenon. Financial firms are directly engaged in the food system; Warren Buffet's Berkshire Hathaway was until recently the largest investor in Kraft, whilst 3G, a Brazilian private equity firm, purchased Burger King outright (*New York Times*, 2011; Arnold *et al.*, 2010). Even the distinction between food corporations and financial organizations is becoming blurred; the food conglomerate Cargill runs its own asset management company, Black River Asset Management, and may register as a swap dealer in the US derivatives markets, making it subject to similar rules to investment banks (Cargill, undated; Meyer, 2011a).

While financialization is apparent across the food system, perhaps the closest related phenomenon to speculation in commodity markets is the increasing interest in agricultural land as an investment – so-called "land grabbing." The rise in commodity prices in 2008, and the decline in investment returns elsewhere, also had the effect of increasing investment interest in agricultural land, particularly in sub-Saharan Africa. Financial companies are attracted by the likely appreciation of land values, the use of land as a hedge against inflation, and the potential for long-term returns (Deininger *et al.*, 2011, p. xxv, p. xxxii, p. 2). Emergent Asset Management, a UK based company, reportedly owns or leases 100 000 hectares in Africa and targets annual returns of 25% (Schaffler *et al.*, 2011). Interest is not limited to Africa however, with Galtere, a US-based fund manager, hoping to attract $1 billion of investment in a fund focused on agricultural projects in Brazil, Uruguay and Australia (Reuters, 2010).

16.10 POLICY APPROACHES TO SPECULATION

Addressing the impact of wider structural economic influences and policy approaches to address them is well beyond the scope of this chapter, as is any detailed investigation into the impact of financial investors in land. However, it is possible to consider policy responses to speculation in food commodity markets.

Critics of financial speculation advocate a strong regulatory approach to place limits on speculative activity. In response to the 2008 financial crisis, the Dodd Frank Act was passed in the United States. The aim of this legislation was to improve transparency and reduce risk in financial markets by standardizing hitherto opaque OTC transactions and requiring that trades be cleared using central bodies (Gensler, 2010). As this description includes products like commodity index funds, this will result in improved transparency of food commodity futures markets, providing better data about the impact of speculation. More controversially, the act also mandated the introduction of new position limits, which are aimed at curbing excessive speculation by restricting the size of positions that any single actor can hold in certain financial markets. Although the measures were reportedly watered down in response to financial sector lobbying, the Commodity Futures Trading Commission who are responsible for implementing the reforms, finally agreed towards the end of 2011 to impose limits on 28 commodities futures contracts of which 19 were food related (Meyer, 2011b; CFTC, 2011).

Within Europe, European Commissioner Michel Barnier has proposed similar reforms to those passed by the United States Congress and changes to the way financial markets work are proposed under various legislative proposals including the European Market Infrastructure Regulation (EMIR) and the review of the Markets in Financial Instruments Directive (MiFID) (Barnier, 2010). While these proposals are likely to increase transparency in financial markets, consensus about more direct interventions to limit speculation such as the implementation of position limits is likely to be more difficult to achieve. The UK government currently does not support a mandatory approach, arguing that the case that position limits are effective at controlling prices has not been made (FSA and HM Treasury 2009, pp. 31–32). It remains to be seen how the United States or European legislation will be implemented in practice and what effect it will have on food price volatility.

The concepts of a precautionary approach and social value provide alternatives to conventional economic arguments common in the debate around speculation, Given that speculation has the capability to cause harm, and has questionable value to society as whole, the implication for policy makers is that regulatory initiatives to impose limits to speculative activity in food commodities markets should be supported. If speculation is a significant cause of volatility this will save lives. If speculation is only a marginal cause, little of value will have been lost, and policy makers can focus on other causes of food price volatility.

These ethical perspectives may also provide useful tools for thinking about wider issues in the food system, giving the benefit of the doubt to people rather than profit, and considering social values as well as investment return. Ultimately the creation of a fairer and more sustainable food system is dependent on the creation of an economic and financial structure which, in distributing investment funds, recognizes these values alongside income generation.

REFERENCES

Angel, J. and McCabe, D. (2010), 'The Ethics of Speculation', *Journal of Business Ethics*, Vol. 90, pp. 277–286.

Archer, C. and Fritsch, S. (2010), 'Global fair trade: Humanizing globalization and reintroducing the normative to international political economy', *Review of International Political Economy*, Vol. 17, No. 1, pp. 103–128.

Arnold, M., Lucas, L., Bevins, V. (2010), 'Burger King approves 3G Capital's bid', *Financial Times*, 2nd September 2010 [online]. Available at: http://www.ft.com/cms/s/0/a7a4e112-b6b9–11df-b3dd-00144feabdc0.html#axzz1cNBFLi00 (accessed 5 March 2012).

Barnier, M. (2010), *'Remarks at "Securities Industry and Financial Markets Association' (SIFMA) Securities Industry and Financial Markets Association, New York'* [online] October 28 2010. Available at: http://europa.eu/rapid/pressReleasesAction.do?reference=SPEECH/10/612&format=HTML&aged=0&language=EN&guiLanguage=en (accessed 19 March 2012).

Bobenrieth, E. and Wright, B. (2009), 'The Food Price Crisis of 2007/2008: Evidence and Implications', paper presented at Joint Meeting of the Intergovernmental Group on Oilseeds, Oils and Fats (30th Session), the Intergovernmental Group on Grains (32nd Session) and the Intergovernmental Group on Rice (43rd Session) Santiago, Chile, 04/11/2009 - 06/11/2009 [online]. Available at: http://www.fao.org/es/esc/common/ecg/584/en/Panel_Discussion_paper_2_English_only.pdf (accessed 5 March 2012).

Bogle, J. (2009) *Enough* (John Wiley & Sons, Inc.: New York).

Branson, R., Masters, M., Frenk, D. (2010), 'Swaps, spots and bubbles' in Letters, *The Economist*, 29th July 2010 [online]. Available at: http://www.economist.com/node/16690679 (accessed 5 March 2012).

Cargill (undated), 'Financial and Risk Management', page from website [online]. Available at: http://www.cargill.com/products/financial-risk/index.jsp (accessed 5 March 2012).

Commodity Futures Trading Commission (2011) 'Position Limits for Futures and Swaps' *Federal Register*, Vol. 76, No. 23 pp. 71626–71703 [online]. Available at: http://www.cftc.gov/ucm/groups/public/@lrfederalregister/documents/file/2011–28809–1a.pdf (accessed 19 March 2012).

Christian Aid (2011), *Hungry for justice: fighting starvation in an age of plenty* [online]. Available at: http://www.christianaid.org.uk/resources/policy/christian-aid-week-report-2011.aspx (accessed 5 March 2012).

Darling, A. (2011), *Back from the Brink* (Atlantic Books: London).

Deininger, K., Byerlee, D., Lindsay, J., Norton, A., Selod, H., Stickler, M. (2011), *Rising global interest in farmland – Can it yield equitable benefits?*, (The International Bank for Reconstruction and Development/The World Bank: Washington).

De Long, J., Shleifer, A., Summers, L, and Waldman, R. (1990), 'Positive Feedback Strategies and Destabilising Rational Speculation', *The Journal of Finance*, Vol. 45, No. 2, pp. 379–395.

De Schutter, O. (2010), 'Food Commodities Speculation and Food Price Crises. Regulation to reduce the risks of price volatility', Briefing Note 02 - September 2010 [online]. United Nations Special Rapporteur on the Right to Food. Available at: http://www.srfood.org/images/stories/pdf/otherdocuments/20102309_briefing_note_02_en_ok.pdf (accessed 5 March 2012).

European Commission (2000), *Communication from the Commission on the precautionary principle /* COM/2000/0001 final * [online]. Available at: http://eur-lex.europa.eu/LexUriServ/LexUriServ.do?uri=CELEX:52000DC0001:EN:NOT (accessed 5 March 2012).

European Commission (2009) Commission Staff Working Document. Agricultural commodity derivatives markets: the way ahead 28 October 2009. Brussels. European Commission SEC(2009) 1447.

European Council (1990), *Council Directive 90/220/EEC of 23 April 1990 on the deliberate release into the environment of genetically modified organisms* [online]. Available at: http://eur-lex.europa.eu/smartapi/cgi/sga_doc?smartapi!celexapi!prod!CELEXnumdoc&lg=EN&numdoc=31990L0220&model=guichett (accessed 5 March 2012).

European Council (2002), Regulation (EC) No 178/2002 of the European Parliament and of the Council of 28 January 2002 laying down the general principles and requirements of food law, establishing the European Food Safety Authority and laying down procedures in matters of food safety [online]. Available at: http://www.food.gov.uk/multimedia/pdfs/1782002ecregulation.pdf (accessed 5 March 2012).

European Council (2008), Directive 2008/56/EC of the European Parliament and of the Council of 17 June 2008 establishing a framework for community action in the field of marine environmental policy (Marine Strategy Framework Directive) [online]. Available at: http://eur-lex.europa.eu/LexUriServ/LexUriServ.do?uri=CELEX:32008L0056:en:NOT (accessed 5 March 2012).

Food and Agriculture Organisation of the United Nations (FAO) (2010), *The State of Food Insecurity in the World* (Food and Agriculture Organization of the United Nations: Rome). Available at: http://www.fao.org/docrep/013/i1683e/i1683e.pdf (accessed 5 March 2012).

Food and Agriculture Organization of the United Nations (FAO) (2011a), *Food Outlook – Global Market Analysis*, June 2011 [online]. Available at: http://www.fao.org/docrep/014/al978e/al978e00.pdf (accessed 5 March 2012).

Food and Agriculture Organization of the United Nations (FAO) (2011b), *Food Outlook – Global Market Analysis*, November 2011 [online]. Available at: http://www.fao.org/docrep/014/al978e/al978e00.pdf (accessed 5 March 2012).

Food and Agriculture Organisation of the United Nations (FAO) (2011c), *The State of Food Insecurity in the World 2011* (Food and Agriculture Organization of the United Nations: Rome). Available at: http://www.fao.org/docrep/014/i2330e/i2330e.pdf (accessed 5 March 2012).

FAO, IFAD, IMF, OECD, UNCTAD, WFP, the World Bank, the WTO, IFPRI, and the UN HLTF (2011), *Price Volatility in Food and Agriculture Markets: Policy Responses* [online]. Available at: http://www.oecd.org/document/20/0,3746,en_2649_37401_48152724_1_1_1_37401,00.html (accessed 5 March 2012).

Financial Services Authority and HM Treasury (2009), *Reforming OTC Derivative Markets A UK perspective* (Financial Services Authority & HM Treasury: London). Available at: http://www.fsa.gov.uk/pubs/other/reform_otc_derivatives.pdf (accessed 5 March 2012).

Gensler, G. (2010) 'Remarks before the Annual Dinner of the National Economists Club, Washington DC' [online] November 9 2010 Available at: http://www.cftc.gov/PressRoom/SpeechesTestimony/opagensler-58.html (accessed 19 March 2012).

Ghosh, J. (2010), 'The unnatural coupling: food and global finance', *Journal of Agrarian Change*, Vol 10, No. 1, pp. 72–86.

Gilbert, C. (2009), 'How to understand high food prices', Discussion Paper No. 23 Dipartimento di Economia, Università degli Studi di Trento. Available at: http://www.unitn.it/files/23_08_gilbert.pdf (accessed 5 March 2012).

Gilbert, C. (2010) 'Speculative Influences on Commodity Futures on Commodities Futures Prices 2006–2008', *United Nations Conference on Trade and Development Discussion Paper No. 197*, March 2010 [online]. Available at: http://www.unctad.org/en/docs/osgdp20101_en.pdf (accessed 5 March 2012).

Headey, D. and Fan, S. (2010), 'Reflections on the Global Food Crisis', Research Monograph 165 (International Food Policy Research Institute: Washington DC).

Hoban, M. (2011), House of Commons written answers 13 June 2011, Hansard HC vol 529, part 168. Available at: http://www.publications.parliament.uk/pa/cm201011/cmhansrd/cm110613/text/110613w0003.htm#11061330000238 (accessed 5 March 2012).

Hossain, N., Byrne, B., Campbell, A., Harrison, E., McKinley, B. and Shah, P. (2011), *The impact of the global economic downturn on communities and poverty in the UK*, Joseph Rowntree Foundation [online]. Available at: http://www.jrf.org.uk/publications/impact-global-economic-downturn-communities-and-poverty-uk (accessed 5 March 2012).

Jackson, T. (2009) *Prosperity without grow* (Earthscan: London).

Kolb, R. and Overdahl, J. (2007) *Futures, Options and Swaps* 5 edition (Blackwell Publishing: Oxford).

Lagi, M., Bar-Yam, Y., Bertrand, K., Bar-Yam, Y., (2011), *The Food Crises: A Quantitative Model of Food Prices Including Speculators and Ethanol Conversion*, New England Complex Systems Institute [online]. Available at: http://necsi.edu/research/social/foodprices.html (accessed 5 March 2012).

Masters, M. (2008), *Testimony before the Committee on Homeland Security and Governmental Affairs. United States Senate*, May 20 2008. Available at: http://hsgac.senate.gov/public/_files/052008Masters.pdf (accessed 5 March 2012).

Meyer, G., (2011a), 'Cargill faces jump in trading costs', *Financial Times*, 1 March 2011 [online]. Available at: http://www.ft.com/cms/s/0/350004de-443a-11e0–931d-00144feab49a.html#axzz1cNBFLi00 (accessed 5 March 2012).

Meyer, G. (2011b), 'CFTC approves new caps on speculators', *Financial Times*, 18 October 2011 [online]. Available at: http://www.ft.com/cms/s/0/a1532cc8-f98a-11e0-bf8f-00144feab49a.html (accessed 19 March 2012).

New York Times (2011), 'What does Buffet Think', DealBook, *New York Times*, August 4 2011 [online]. Available at: http://dealbook.nytimes.com/2011/08/04/what-does-buffett-think/ (accessed 5 March 2012).

Organisation for Economic Cooperation and Development (2008), *Rising Food Prices – Causes and Consequences* [online]. Available at: http://www.oecd.org/dataoecd/54/42/40847088.pdf (accessed 5 March 2012).

Oxford University Press (undated), Oxford Dictionaries, definition for 'speculation' [online]. http://oxford dictionaries.com/definition/speculation (accessed 5 March 2012).

Piesse, J. and Thirtle, C. (2009), 'Three bubbles and a panic: An explanatory review of recent food commodity price events', *Food Policy*, Vol. 34, pp. 119–129.

Rapsomanikis, G. (2009), *The 2007–2008 food price swing: Impact and policies in Eastern and Southern Africa*, FAO Commodities and Trade Technical Paper 12 (Food and Agriculture Organization of the United Nations: Rome).

Reuters (2010), 'Galtere says raising $1 bln agribusiness fund', *Reuters*, UK edition, 1st September 2010 [online]. Available at: http://uk.reuters.com/article/2010/09/01/commodities-fund-galtere-idUKN0113842720100901 (accessed 5 March 2012).

Ryan, J. (1902), 'The Ethics of Speculation', *International Journal of Ethics*, Vol. 12, No. 3 (Apr., 1902), pp. 335–347.

Sanders, D, Irwin, S and Merrin, R. (2008), *The Adequacy of Speculation in Agricultural Futures Markets: Too Much of a Good Thing?*, Marketing and Outlook Research Report 2008-02, Department of Agricultural and Consumer Economics, University of Illinois at Urbana-Champaign, June 2008 [online]. http://www.farmdoc.illinois.edu/marketing/morr/morr_08–02/morr_08–02.pdf (accessed 5 March 2012).

Sarkozy, N. (2011a), Address by the President of the French Republic to the European Commission Conference on Commodities and Raw Materials, Tuesday 14 June 2011, Brussels. Transcript available online at: http://ec.europa.eu/bepa/pdf/conferences/president-sarkozy-speech-14-june-2011_en.pdf (accessed 5 March 2012).

Sarkozy, N. (2011b), Speech to G20 Ministers of Agriculture, Wednesday 22nd June 2011, Élysée Palace, Paris. Transcript available online at: http://www.g20-g8.com/g8-g20/g20/english/for-the-press/speeches/nicolas-sarkozy-adresses-a-speech-to-the-g20.1402.html (accessed 5 March 2012).

Sen, A. (2010), 'The Economist Manifesto' in *The New Statesman*, 23rd April 2010 [online]. Available at: http://www.newstatesman.com/ideas/2010/04/smith-market-essay-sentiments (accessed 5 March 2012).

Schaffler, R., Ramirez, R., Macfarlane, S., Cropley, E. (2011), 'Special Report - In global land rush, a search for fair returns', *Reuters*, U.S edition, 31st January 2011 [online]. Available at: http://www.reuters.com/article/2011/01/31/idINIndia-54527420110131 (accessed 5 March 2012).

Smith, A. (1986), *The Wealth of Nations*, Books I–III (Penguin Classics: London).

Soros, G. (2008), interviewed for *Stern* magazine, 3rd July 2008 [online]. Available at http://www.stern.de/wirtschaft/news/maerkte/george-soros-we-are-in-the-midst-of-the-worst-financial-crisis-in-30-years-625954.html (accessed 5 March 2012).

Stirling, A. (2007), 'Risk, precaution and science: towards a more constructive policy debate – talking point on the precautionary principle', *European Molecular Biology Organization Reports*, Vol. 8, No. 4, pp. 309–315.

Tokic, D. (2011), 'Rational destabilizing speculation, positive feedback trading, and the oil bubble of 2008', *Energy Policy*, Vol. 39, No. 4, pp. 2051–2061.

Turner, A. (2010), 'What do banks do? Why do credit booms and busts occur and what can public policy do about it?', in *The Future of Finance – The LSE Report*, London School of Economics and Political Science [online] Available at: http://www.futureoffinance.org.uk/ (accessed 5 March 2012).

United Nations Conference on Environment and Development (1992), *Report of the United Nations Conference on Environment and Development*, Annex 1, Rio Declaration on Environment and

Development (Rio de Janeiro, 3–14 June 1992) [online]. Available at: http://www.un.org/documents/ga/conf151/aconf15126–1annex1.htm (accessed 5 March 2012).

United Nations Conference on Trade and Development (2011), *Price Formation in Financialized Commodity Markets: The Role of Information* (UNCTAD: New York and Geneva, June 2011) [online]. Available at: http://www.unctad.org/en/docs/gds20111_en.pdf (accessed 5 March 2012).

United States Senate (2009), *Excessive speculation in the Wheat Market*, Majority and Minority Staff Report, June 24th, 2009, Washington DC Permanent Subcommittee on Investigations, United States Senate.

Woolley, P. (2010) 'Why are financial markets so inefficient and exploitative and a suggested remedy' in *The Future of Finance – The LSE Report*, London School of Economics and Political Science [online]. Available at: http://www.futureoffinance.org.uk/ (accessed 5 March 2012).

World Commission on the ethics of Scientific Knowledge and Technology (2005) *The Precautionary Principle*, United Nations Educational, Scientific and Cultural Organization: Paris [online] Available at http://unesdoc.unesco.org/images/0013/001395/139578e.pdf (accessed 5 March, 2012).

World Development Movement (2011), 'G20 Economists Letter' [online]. Available at: http://www.wdm.org.uk/sites/default/files/G20%20Economists%20letter%20LATEST%20VERSION.pdf (accessed 5 March 2012).

Worthy, M. (2011), *Broken Markets: How financial market regulation can help prevent another global food crisis*, World Development Movement [online]. Available at: http://www.wdm.org.uk/sites/default/files/Broken-markets.pdf (accessed 5 March 2012).

IV Conclusion

17 Reflections on food ethics

Christopher Ritson

17.1 INTRODUCTION

The word "ethical" is now in frequent use and appears in a wide range of contexts. Most people probably believe they have some understanding of what "behaving ethical" implies – that it is something to do with "doing good." It is therefore of some surprise to anyone who chooses to investigate the subject of ethics more formally to discover that definitions of it range from what might seem rather vague and over-simplistic – "Doing the right thing," to versions identifying ethical principles derived from moral philosophy, as outlined in Part I of this volume.

Yet, in Chapter 1, Seebauer points out that "people can do good based upon gut instinct alone."

Referring to the Chinese concept of *Renging*, Chau (Chapter 4) comments:

> It is the social obligation that one feels is attached to a situation, particularly as a return of favor. It is like a charitable doing that just seems right, regardless of logic or reasoning behind it.

Commenting on ethical practices in the workplace, J. Peter Clark notes that "Mostly, we make good choices intuitively." (Chapter 11)

This raises the issue of whether it is necessary in a book on "practical ethics" to devote so much attention to the fundamental principles underlying ethical behavior. There are, I believe, four reasons for this.

The first Seebauer provides in the sentence following his reference to "gut instinct".

> Many people find more satisfaction in understanding why they do what they do.

Chau develops this in a different way, showing how understanding the philosophical principles underlying Chinese ethical behavior can help food marketing companies to

Practical Ethics for Food Professionals: Ethics in Research, Education and the Workplace, First Edition.
Edited by J. Peter Clark and Christopher Ritson.
© 2013 John Wiley & Sons, Ltd. Published 2013 by John Wiley & Sons, Ltd.

meet subtle variations in the requirements of people in the food they purchase. He then goes on to show how this understanding could lead to food companies behaving in what would be considered, from a Western perspective, unethical behavior, in the context of childhood obesity.

The second reason, again referred to by Seebauer, is that "good ethical behavior usually leads to good consequences for ourselves, our organizations and the larger world." But for this to apply there needs to be some consensus about what constitutes good ethical behavior and the processes which lead this to have a positive impact on, in particular, the organization that adopts it. Several examples of this are given later in this chapter.

The third, and perhaps most important reason for studying ethical principles in detail is articulated by Seebauer as [understanding ethical principles] "offers the power to do what is good more easily"; and Mepham as "Can the discipline of food ethics serve a useful role in promoting more ethical practices in the various sectors of the food industry? . . . my answer to the question is affirmative."

People and organizations will not always know what is the "right" thing to do, without the aid of some guiding principles. Moreover these principles need to reflect some consensus in society over what constitutes ethical behavior and positive outcomes. Many of what we would regard as evil events have been undertaken by individuals believing that what they were doing what was "right", guided by their own set of principles far removed from justice, autonomy and wellbeing of others.

The fourth reason why the inclusion of a substantial section on fundamental principles is merited concerns, paradoxically, the words "practical ethics" in the title of this book. The Introduction refers to the book being intended as an aid to those responsible for important decisions about marketing, resources, sustainability, the environment, and people in the food industry. All the chapters in Part I give examples of how ethical principles can be applied in practice; but Chapters 2 and 3 present cases to show how to develop systematic ways of applying ethics to practical decision making. Both chapters begin by articulating similar ethical principles and demonstrate that a major problem facing the ethical decision maker is likely to arise as a consequence of conflict between the application of alternative principles and/or the impact on different interests in society.

Both chapters provided cases of how this problem might be resolved. Nairn takes us through four contemporary methods in bioethical decision-making concerned with a common occurrence relating to an end of life care dilemma faced in medical ethics. Mepham introduces the Ethical Matrix and applies it to a contemporary issue in milk production of relevance to the agricultural industries on both sides of the Atlantic. Both cases can be described as "practical ethics".

17.2 ISSUES IN FOOD ETHICS

The four chapters in Part I are all measured, almost restrained, in the way they introduce the subject of ethics and show how an understanding of ethical principles can aid ethical decision-making in the food industry. The first sentences in the first chapter of Part II immediately highlight a feature of the chapters on "Issues in Food Industry Ethics".

America is in a leadership crisis. Many of our prominent leaders in all walks of life view their positions of leadership as a right to power, money and privilege. They perceive their status as a mandate to loot, pillage and abuse those who they are chosen to lead.

This reflects a deep feeling – almost a passion – for the importance of ethical behavior relating to the issue covered. Here are some more examples:

This . . . really upset me because they tortured cattle in a piece of equipment I had designed. Employees were told to poke electric prods up a steer's anus but to never do it when the USDA inspector was around (Chapter 7).

. . . people are being persuaded to spend money we don't have, on things we don't need, to create impression that won't last, on people we don't care about (Chapter 8, commenting on the impact of food marketing on sustainability).

People who make food have always been exploited. The desire for cheap food drives down wages in food production. Whether it is bananas or coffee workers, they always bear the brunt. The rich want cheap food at any price (Chapter 10).

The degree to which it is expressed in emotive language varies across the chapters in Part II, but it is evident that all the authors have a deep interest and personal involvement in the topic they cover. The chapters contain a blend, to different degrees, in being descriptive (what is happening) and prescriptive (what should happen). Sometimes the descriptive element describes good practice – existing ethical behavior; more often it describes bad practice that the author argues requires correction. Only one chapter explicitly applies a systematic method to ethical decision-making, of the kind outlined above; but the prescriptive sections of the chapters do so implicitly.

Chapter 5 on "Ethics in Business" provides an overview of ethical behavior in business decisions. The author repeatedly refers to "doing the right thing" and interestingly concludes with statements which reflect the comments at the beginning of this chapter on "gut instinct" and "principle guided" ethics.

If an individual makes a firm personal commitment to operate in the areas of right and wrong, black or white, the choices are straightforward. He or she will always opt for "the right thing". It becomes second nature. These decisions are made without thought or consideration.

However many individuals don't possess the internal fortitude and courage to take such a firm stand. There are too many pressures that are too difficult to deal with. If an employee feels this way they must continually measure their choices and decisions against an ethical benchmark.

Chapter 6 on ethics in publishing food science research provides a review of potential pitfalls for authors relating to ethics, and concludes with a set of principles specific to this topic to guide authors to ensure that their publications are ethical.

Underpinning Chapter 7 is only one ethical principle – that farm animals should be treated humanely. ("First of all, it is the right thing to do.") There is a catalog of malpractice in this respect and practical advice on how to achieve farm animal welfare. There is an example here of the comment by Seebauer quoted at the beginning of this chapter that "good ethical behavior usually leads to good consequences for ourselves, our organizations and the larger world" in that:

> There are many situations where maintaining high standards of animal welfare will improve the quality of meat.

The author also refers to the potential impact on consumption of bad publicity and the growing movement from consumers only willing to purchase meat produced humanely. However, how will consumers know? High standards of animal welfare may improve meat quality but are likely to involve higher production costs. Recent examples in Europe concern laying hens in cages, and the use of crates in veal production. Economists have suggested that, by delineating three kinds of food product attributes, it is possible to identify circumstances in which ethical standards in agricultural production may not always bring their own reward. These attributes have been labeled "search", "experience" and "credence" (Ritson and Mai, 1996).

A search attribute describes something that a consumer can know at the time of purchase, for example the color, size, and shape of an item of fruit. The consumer is aware of an experience attribute only after consumption. This will be predominantly things like flavour, texture, cooking quality, but might extend to "experiences" such as food poisoning if it can be linked reliably to a particular meal. Consumers learn to link search attributes to experience ones – that green bananas are not ripe and black ones overripe. Credence attributes, though, are things that you simply have to believe in.

Many of the ethically related food production issues covered in Part II of this book represent attributes which consumers increasingly value in the food they purchase. But they are also typically credence attributes. Animal welfare is one. Chapter 8 on "sustainable food production and consumption" cites several more – location of production, use of herbicides and pesticides, nutritional composition of foods. One chapter is devoted entirely to labor conditions in agricultural and food production (Chapter 10).

The classic solution to the problem of credence attributes valued by consumers is food labeling (as well as other point of sale information, and increasingly information websites.) The food company which has adopted ethical values in production can label their produce accordingly, and can provide information on product composition that health conscious consumers will seek.

There are two problems with this neat solution. First, there is no incentive for the "non-ethical" company to label negative credence attributes (e.g. "contains pesticide residues" "animals kept in cages"), though sophisticated consumers may draw their own conclusions from a lack of positive credence attribute labeling. Worse, there *is* an incentive for the less scrupulous company fraudulently to label the presence of positive credence attributes; or at least make misleading claims (e.g. "90% fat free" instead of "contains 10% fat"). In these circumstances the ethical, but higher cost, producer is disadvantaged.

But even if all producers wish to behave ethically by being, for example, honest and open about product composition and nutritional benefits of the foods they sell, they may come up with such a wide variety of ways of communicating this that consumers are unable to make informed judgements in their purchase decisions. The chapter by Sue Davis on "Responsible Health and Nutrition Claims" (Chapter 9) both makes the case for regulation of health and nutrition claims in food marketing, and describes in detail the development of European regulation of this, and the necessary coordination of methods of nutrition labeling.

In one sense it may seem odd discussing the role of Government regulation in a book on food ethics, as the essence of the subject is individuals and organizations voluntarily adopting standards of behavior which they believe to be "right". The issue of health claims and food labeling highlights the fact that society may sometimes need to take steps to protect the interest of the food producer who behaves ethically, and regulate to provide for uniformity in their attempts to do so.

There are other aspects of ethical behavior that may require regulation. In confronting the question of "Can we decide whether any particular practice is "exploitative" [of labor] or whether it is 'ethical' ", Clutterbuck (Chapter 10) puts it succinctly:

> Who should decide – government through regulations, companies through their own standards or civil society though producers and consumers.

He then goes on to show how an approach based on the Ethical Matrix can help us to identify cases of worker exploitation in food production and distribution.

17.3 CASE STUDIES

Part III of the book is titled "Examples and Case Studies", though this is more a matter of emphasis than a major shift in focus. This change in emphasis is perhaps best illustrated by comparing the first chapters in Parts II and III. In Chapter 5, Bednarz introduced the *issue* of ethics in business. In Chapter 11 J. Peter Clark turns to the ethical challenges that individual employees may face and provides practical advice on how to face these challenges. He concludes that, although:

> Mostly, we make good choices intuitively. [and] have acquired an instinctive understanding of the virtues and codes of conduct. However, some choices are difficult and require courage to make. In such cases, it helps to have core beliefs and values.

This returns us to the link between ethical behavior and the ethical principles derived from moral philosophy. In Chapter 12 Louis Clark explores how the kinds of ethical challenges in the workplace outlined by J. Peter Clark, and the "core values" that help us to meet these challenges, which Louis Clark refers to as "a universal ethical sense in humans", can be explored in the work of philosophers, ranging from Aristotle to the present day. The editors have located this chapter here because, just as chapters introducing ethical principles provide a better appreciation of understanding issues in food ethics, so the detail of the ethical challenges faced by individuals in the workplace illuminates – puts flesh on, as it were – the insights provided by the works of the great philosophers.

Louis Clark concludes his chapter by commenting on a "contemporary challenge: supporting human rights violations through consumer behavior". This leads in to the subject of the next chapter, which reports on how one of the principles embodied in the Ethical Matrix of Mepham in Chapter 3 – "Fairness" – is currently being applied.

The case study on The Fair Trade Movement (Chapter 13) is also an example of positive outcomes relating to problems raised in three of the "Issues" chapters, those concerned with worker exploitation, sustainable production, and food labeling. Virtually all of the characteristics of fair trade products which distinguish them are credence attributes that consumers of fair trade products value. The success of the fair trade movement has been dependent on recognition by consumers of the FAIRTRADE mark, and their confidence in the certification of the standards required for products carrying the mark guarantee.

The first of the three principles of the ethical matrix is "wellbeing" and, crucially, this of course includes the wellbeing of those who consume the food product, as well as those who produce it. Food production processes that induce health hazards in food consumption have the potential to reduce the wellbeing of consumers – and the extreme version of this is death from food poisoning. The Case Study on "The Peanut Corporation of America" (Chapter 14) reports in detail on one example of this. Here, the important contribution is to understand "what went wrong". Did the problem arise because of a failure of the food producer to "do the right thing" in terms of applying commonly accepted standards of hygiene in food production, failure to comply with food safety legislation, failure in the system of enforcement of food safety legislation, or inadequate legislation in the first place? Everyone accepts that ethical behavior in the food industry implies supplying safe food. As with health claims, there remains a debate as to what extent this requires Government regulation to enforce ethical behavior.

The application of modern, advanced technologies in food production, such as nano-technology (Chapter 15) has added a new dimension to what we consider to be the scope of the subject of food ethics. The inclusion of a chapter on "Ethical aspects of nanotechnology in the area of food and food manufacturing" has its origin in an earlier development, the application of biotechnology to food production. In 1991, the emerging technology of genetic modification (GM) and its application to food products led the British Advisory Committee on Novel Foods and Processes (ACNFP) to establish a role for an "ethical" specialist, (a role I currently occupy) to complement the contribution of the range of scientific disciplines on the Committee required to assess the safety of novel foods coming onto the market.

In 1992, the ACNFP established an *ad hoc* "Committee on the Ethics of Genetic Modification in Food Use" under the Chairmanship of the Committee's first ethical specialist, Reverend Dr John Polkinghorne. Following consultation with a wide range of bodies, four main ethical areas of public concern were identified, namely:

1. The transfer of human genes to food animals.
2. The transfer of genes from animals, the flesh of which is forbidden for use by certain religious groups, to animals which they usually do eat (e.g. pig genes inserted in sheep).
3. The transfer of animal genes into crops, which would be of particular concern to vegetarians.
4. The use of organisms containing human genes as animal feed.

Subsequent research (e.g. Kuznesof and Ritson, 1996) has, however, revealed that public concern over the application of modern technologies in food production is much wider than this. In particular the willingness of a consumer to "accept" a GM food product is strongly influenced by who is perceived to benefit from it. Second, the ethical issues articulated by the Polkinghorne Committee tend to merge in consumer minds with broader aspects of food quality and safety. More than anything else, the development of hostility to the new technology in the UK was associated with fear of the unknown, a belief that it was "unnatural", and involved "scientists" imposing on people a new technology, which would only benefit food companies. Once a consumer benefit could be identified, then the GM product becomes more acceptable and, paradoxically, safer.

This last point raises another ethical issue – how to respond to consumer concerns about food safety, which scientific evidence suggests are groundless. Studies comparing "expert" and "lay" perspectives on food safety risks usually produce an inverse relationship between the two, with consumers fearing most food additives and chemical residues, followed by new technologies; with scientists concentrating on microbiological contamination (e.g. the peanut case study) and impact of poor diet on health (Ritson and Kuznesof, 2006).

Buckenhüskes begins Chapter 15 with

Nanotechnology, along with information technology and biotechnology, is one of the key technologies of our age.

The parallels with biotechnology in food production, and the subject of bioethics (Mepham, 2008) are considerable. He goes on:

The need for ethics in and for nanotechnology were recognized at an early stage, and terms such as "nanoethics" were coined". However: "Nanotechnology has not yet become an emotionally charged, vehemently disputed topic of public discussion.

After reviewing the potential for application of nanotechnology in the food industry, the chapter suggests that communication and perception of risk take place at three levels. First:

From a scientific and technical perspective which involves an objective, knowledge based analysis.

Second:

an emotional debate takes place that is strongly influenced by individuals' different perceptions of risk as well as their hopes and fears.

Third:

the social or ethical level addresses group-specific interpretations of risk and ethical aspects.

As with the discussion above of consumer and scientist perspectives on GM foods, there is contradiction between the levels which is a cause of conflict.

Buckenhüskes concludes with a comment, which again echoes the experience of biotechnology in agriculture:

Consumers' acceptance of nanotechnology in future will depend above all on:

1. Manufacturers need to be open and transparent regarding their use of nanomaterials.
2. Authorities . . . should not approve the use of a nanomaterial until they have examined and rated its potential risks.
3. Consumers must gain direct and immediate personal benefit from the use of nano-technology in the area of food and nutrition and must be able to see these benefits.

The second of these invokes the "precautionary principle" in the regulation of new food technologies in food production, and it is the precautionary principle that guides the recommendation in the final chapter in this section of the book, on food commodity speculation. The chapter begins by noting that:

There is an ongoing and vigorous debate around the extent to which financial speculation in food commodity markets contributes to food price rises and price volatility.

There is, however, no ambiguity over the adverse impact on the World's poor of the recent experience of rapidly rising prices for basic food commodities. Whether viewed from the principle of "wellbeing" or of "fairness", if commodity speculation does contribute to food price volatility, then this cannot be described as "ethical behavior".

Sutton (Chapter 16) argues that taking an ethical perspective allows the debate to shift from attempting to assess the extent to which price volatility and commodity speculation are related, to one in support of regulation, of the kind introduced by the United States in 2011 to curb excessive speculation, but still lacking in Europe.

His penultimate paragraph merits repeating here as an example of how application of the Ethical Matrix (albeit, in this case only implicit) – comparing the impact on different interests from different ethical perspectives, can lead to a convincing policy recommendation:

The concepts of a precautionary approach and social value provide alternatives to conventional economic arguments in the debate around speculation. Given that speculation has the capability to cause harm, and has questionable value to society as a whole, the implication for the policy maker is that regulatory initiatives to impose limits to speculative activity in food commodities markets should be supported. If speculation is a significant cause of volatility this will save lives. If speculation is only a marginal cause, little of value will have been lost, and policy makers can focus on other causes of price volatility.

17.4 CONCLUDING COMMENTS: FOOD MARKETING AND CONSUMER CHOICE

A number of references have been made to the relevance of the principle of wellbeing, and the principle of fairness/justice, to the case studies covered in Part III of the book. The principle of autonomy has been less visible. This chapter concludes with some comments on its relevance to food marketing, diet and health, an issue which recurred throughout this book, but particularly in Chapters 2, 4, 8, and 9.

On both sides of the Atlantic, the dominant health issue has become the incidence of obesity. Although the causes of this are complex, there is no doubt that a major factor is that people eat too much, and in particular, too much of calorie dense, high fat and sugary foods. The question is to what extent people freely choose "bad" diets? Navigating through the ethical issues raised by this is far from easy. At what point does the principle of wellbeing clash with that of autonomy?

The first point to make is that poor diet has an adverse impact on the wellbeing of everyone, via the increased national costs of health care, not just the wellbeing of the obese, and policies to improve diets are often justified on these grounds, probably because of a reluctance to become involved in a debate over consumer choice.

Second, there is a consensus that it is unethical for food companies to market unhealthy foods aggressively to children; but more generally in food marketing it is difficult to identify at what point manufacturing or selling a food product – making it available to consumers who wish to purchase it – steps over the line into becoming encouraging consumers, by for example high pressure advertising or sponsorship of sports events, to adopt unhealthy diets, as argued in Chapter 8.

In contrast, public policies encouraging consumers to adopt healthy diets – such as the British 5-a-day fruit and vegetable campaign – are usually welcomed. Even here, though not everyone agrees. The British Minister of Education is reported (Mail Online, 22 February 2012) as describing the British Food Standards Agency as having gone from:

> a body that was responsible for governing the safety of food to one that became yet another meddlesome and nanny organization that was telling us what we should eat and in what proportion.

In comparing our reaction to the activities of fast food restaurants and manufacturers of sugary drinks, with that of publically financed campaigns promoting consumption of fruit and vegetables, it is therefore wellbeing that takes precedence. Both are attempting to change what consumers choose to eat, but one is believed to have a positive effect on wellbeing, and the other a negative effect. In any case, there can be debate over whether either inhibits consumer autonomy – the extent to which advertising and other methods of food product promotion merely provides information to consumers, so that they can make more informed choices, or persuades them to take consumption decisions which are no longer autonomous.

There are of course more explicit ways of limiting consumer food choice. A recent example is the announcement by New York Mayor Michael Bloomberg that he planned to prohibit the sale of sugary drinks in containers above a certain size, in restaurants, movie theatres and street carts. The Center for Consumer Freedom echoed that of the British Minister of Education, with an advertisement in the *New York Times* depicting the Mayor as a Nanny under the caption:

> New Yorkers need a Mayor, not a Nanny.... What next? Limits on the width of a pizza slice, size of a hamburger or amount of cream cheese on your bagel.

The American Beverage Association called the plan "ridiculously unreasonable, unsound and incongruous" (Mail Online, 2 June 2012).

Another method of attempting to improve diets is by what has become known as a "fat tax" – a term which can conveniently be interpreted as either referring to the kind of product taxed, or the problem being addressed by the policy. Proposals to tax "unhealthy" foods also often produce a hostile reaction and there are two odd things about this.

In principle, a price policy aimed at improving diets could operate either by taxing unhealthy foods, or by subsidizing healthy ones. To many people, subsidizing healthy foods is much more acceptable – it is seen as not impacting on consumer choice. Yet taxpayers lose with subsidies and gain with taxes, and economists would argue that it is legitimate to raise the price of those food products which contribute to the social cost of health care caused by bad diets. Economists have estimated that taxes on soft drinks and confectionary throughout the US generate about $1 billion per year (Mazzocchi *et al.*, 2009). This money could be used to finance health campaigns or to subsidize fruit and vegetables.

Second, other Government policies influence food prices. For example, for many years the Common Agricultural Policy (CAP) of the European Union, as a consequence of its measures aimed at supporting prices for farmers, is estimated to have increased food prices in the UK on average by the equivalent of a 15% tax (Ritson, 1997). More fundamentally, however, the price raising effect of the Policy varied from product to product, pushing up most the consumer prices of dairy products and sugar - equivalent to a tax of 100% or more, with little or no price raising effect on fruit and vegetables. Quite by accident, the CAP was a kind of "fat tax", pitched at a level well in excess of any contemplated overtly as a way of improving diets. It seems difficult to argue that it is ethical to push up food prices for low income consumers in order to aid farming, but not to improve the wellbeing of consumers themselves.

Where does this leave the "ethical" food business? In its "Ethics: A toolkit for food businesses", the Food Ethics Council (2011) argues that taking ethics into account implies limiting the choice of products available to consumers, something which has become known (as explained in Chapter 8) as "choice editing":

> This can mean taking the worst products off the shelves.

This of course covers obvious issues such as products manufactured from endangered species or production involving worker exploitation, but also "unhealthy" foods, in particular high fat, high salt or high sugar versions of standard products. If a food business in New York chose to limit the size of container in which it served sugary drinks, then this would also be an example of choice editing. The Food Ethics Council challenges the view that this inhibits consumer choice, referring to research in the UK which "has found consumers are comfortable with this." There are some decisions that people want others to take for them and perhaps this includes achieving a better diet. According to the toolkit:

> It is getting harder [for food businesses] to play the choice card.

REFERENCES

Food Ethics Council (2011) "Ethics: A toolkit for food businesses", http://www.foodethicscouncil.org/node/392 (accessed 19 December 2012).

Kuznesof, S and Ritson, C (1996) "Consumer acceptability of genetically modified foods with special reference to farmed salmon" British Food Journal 98(4/5).

Mail Online (22 February 2012) www.dailymail.co.uk/news/article-2104406.

Mail Online (2 June 2012) www.dailymail.co.uk/news/article-2153881 (accessed 19 December 2012).

Mazzocchi, M, Traill, W B, and Shogren, J F (2009) "Fat Economics: Nutrition, Health and Economic Policy", Oxford University Press, Oxford.

Mepham, B (2008) "Bioethics. An introduction for the biosciences," second edition. OUP, Oxford.

Ritson, C (1997) "The Cap and the Consumer" in Ritson C, and Harvey, D R, The Common Agricultural Policy. CAB International, Wallingford.

Ritson, C and Li Wei Mai (1998) "The Economics of Food Safety" Nutrition and Food Science Number 5 September/October.

Ritson, C and Kuznesof, S (2006) "Food Consumption, Risk Perception and Alternative Production Technologies" in Eilenberg, J and Hokkanen, HMT (eds) "An Ecological and Societal Approach to Biological Control", Springer, New York.

Index

Practical Ethics for Food Professionals: Ethics in Research, Education and the Workplace, First Edition.
Edited by J. Peter Clark and Christopher Ritson.
© 2013 John Wiley & Sons, Ltd. Published 2013 by John Wiley & Sons, Ltd.